After Saddam
Prewar Planning and the Occupation of Iraq

Nora Bensahel, Olga Oliker, Keith Crane, Richard R. Brennan, Jr.,

Heather S. Gregg, Thomas Sullivan, Andrew Rathmell

Prepared for the United States Army

Approved for public release; distribution unlimited

RAND ARROYO CENTER

The research described in this report was sponsored by the United States Army under Contract No. DASW01-01-C-0003.

Library of Congress Cataloging-in-Publication Data

After Saddam : prewar planning and the occupation of Iraq / Nora Bensahel ... [et al.].
    p. cm.
  Includes bibliographical references.
  ISBN 978-0-8330-4458-7 (pbk. : alk. paper)
    1. Iraq War, 2003– 2. Military planning—United States. 3. Postwar
reconstruction—Iraq. 4. Coalition Provisional Authority. 5. Insurgency—Iraq.
6. National security—Iraq. I. Bensahel, Nora, 1971–

DS79.76A345 2008
956.7044'3—dc22

                                                                2008025846

The RAND Corporation is a nonprofit research organization providing objective analysis and effective solutions that address the challenges facing the public and private sectors around the world. RAND's publications do not necessarily reflect the opinions of its research clients and sponsors.

RAND® is a registered trademark.

Published 2008 by the RAND Corporation
1776 Main Street, P.O. Box 2138, Santa Monica, CA 90407-2138
1200 South Hayes Street, Arlington, VA 22202-5050
4570 Fifth Avenue, Suite 600, Pittsburgh, PA 15213-2665
RAND URL: http://www.rand.org/
To order RAND documents or to obtain additional information, contact
Distribution Services: Telephone: (310) 451-7002;
Fax: (310) 451-6915; Email: order@rand.org

Preface

Soon after Operation IRAQI FREEDOM (OIF) began in March 2003, RAND Arroyo Center began a research project at the request of the U.S. Army. This project set out to prepare an authoritative account of the planning and execution of combat and stability operations in Iraq and to identify key issues that could affect Army plans and goals, operational concepts, doctrine, and other Title 10 responsibilities.

The resulting body of work will interest those involved in organizing, training, and equipping military forces to plan for, deploy to, participate in, and support joint and coalition operations. Although focused primarily on Army forces and activities, the analysis also describes other aspects of joint and combined operations. RAND analysts collected the information in these volumes from many sources, including unit after-action reports, compilations of lessons learned, official databases, media reports, other contemporary records, and interviews with key participants in OIF.

The results of this project are documented in multiple volumes, some not available to the general public, as described below:

- *Decisive War, Elusive Peace: Operation IRAQI FREEDOM,* MG-641-A, Richard E. Darilek, Walter L. Perry, Laurinda L. Rohn, and Jerry M. Sollinger, editors. This volume is an overview of the research findings.
- *After Saddam: Prewar Planning and the Occupation of Iraq,* MG-642-A, Nora Bensahel, Olga Oliker, Keith Crane, Richard R. Brennan, Jr., Heather S. Gregg, Thomas Sullivan, and Andrew Rathmell. This volume is a treatment of the prewar planning for the postwar situation and of postwar military and reconstruction activities.
- *Operation IRAQI FREEDOM: Executive Summary,* MG-643-A, Walter L. Perry, Laurinda L. Rohn, and Jerry M. Sollinger. This volume, not available to the general public, presents an executive summary of the research findings.
- *Operation IRAQI FREEDOM: Volume I, The Genesis,* MG-643/1-A, Jefferson P. Marquis, Walter L. Perry, David E. Mosher, Stephen T. Hosmer, Andrea Mejia, Richard E. Darilek, Jerry M. Sollinger, Vipin Narang, Charles W. Yost, John Halliday, and John R. Bondanella. This volume, not available to the gen-

eral public, describes the political and military activities leading up to the operation.

- *Operation IRAQI FREEDOM: Volume II, Defeating Saddam,* MG-643/2-A, Bruce R. Pirnie, John Gordon IV, Richard R. Brennan, Jr., Forrest E. Morgan, Alexander C. Hou, and Charles W. Yost. This volume, not available to the general public, covers major combat operations in Iraq.
- *Operation IRAQI FREEDOM: Volume III, Managing the War,* MG-643/3-A, Walter L. Perry, Edward O'Connell, Miranda Priebe, Forrest E. Morgan, Lowell H. Schwartz, and Alexander C. Hou. This volume, not available to the general public, describes the command and control (C2) of the forces and supporting operations.
- *Operation IRAQI FREEDOM: Volume IV, Prewar Planning and the Occupation of Iraq,* MG-643/4-A, Nora Bensahel, Olga Oliker, Keith Crane, Richard R. Brennan, Jr., Heather S. Gregg, Thomas Sullivan, and Andrew Rathmell. This volume, not available to the general public, describes the prewar planning for the postwar situation and postwar military and reconstruction activities.
- *Operation IRAQI FREEDOM: Volume V, Sustaining the Force,* MG-643/5-A, Eric Peltz, David Kassing, Jerry M. Sollinger, Marc Robbins, Kenneth J. Girardini, Peter Schirmer, Robert Howe, and Brian Nichiporuk. This volume, not available to the general public, covers the mobilization and sustainment of the forces.

This report provides an unclassified treatment of the post–major combat military and stabilization activities. It begins by examining prewar planning for postwar Iraq, in order to establish what U.S. policymakers expected the postwar situation to look like and what their plans were for stabilization. The report then examines the role of U.S. military forces after major combat officially ended on May 1, 2003. Finally, the report examines civilian efforts at reconstruction, focusing on the activities of the Coalition Provisional Authority (CPA) and its efforts to rebuild structures of governance, security forces, economic policy, and essential services prior to June 28, 2004, the day that CPA dissolved and transferred authority to the Iraqi Interim Government. The research for this volume was completed in September 2004 and the final draft was submitted in October 2004.

The purpose of this analysis is to find out where problems occurred and to suggest possibilities to improve planning and operations in the future. The results of such analysis can seem therefore to be overly focused on the negative. This should not be taken to mean that no good was done. In fact, dedicated U.S. and coalition personnel, both military and civilian, engaged in many positive and constructive activities, individually and collectively. That this analysis does not highlight all those activities should not in any way detract from their value. Our focus, however, remains on finding ways to improve.

This research was co-sponsored by the Deputy Chief of Staff, G-3, U.S. Army, and the Deputy Chief of Staff, G-8, U.S. Army. It was conducted in RAND Arroyo Center's Strategy, Doctrine, and Resources Program. RAND Arroyo Center, part of the RAND Corporation, is a federally funded research and development center sponsored by the United States Army.

The Project Unique Identification Code (PUIC) for the project that produced this document is DAMOAX003.

For more information on RAND Arroyo Center, contact the Director of Operations (telephone 310-393-0411, extension 6419; FAX 310-451-6952; email Marcy_Agmon@rand.org), or visit Arroyo's web site at http://www.rand.org/ard/.

Contents

Figures

Tables

Summary

Major combat operations in Iraq lasted approximately three weeks, but stabilization efforts in that country are, as of this writing, ongoing. The U.S. Army and the U.S. Marine Corps are increasingly taxed by the demands of the continuing insurgency, with more than 100,000 troops expected to remain in Iraq for the foreseeable future.

The evidence suggests that the United States had neither the people nor the plans in place to handle the situation that arose after the fall of Saddam Hussein. Looters took to the streets, damaging much of Iraq's infrastructure that had remained intact throughout major combat. Iraqi police and military units were nowhere to be found, having largely dispersed during combat. U.S. military forces in Baghdad and elsewhere in the country were not prepared to respond rapidly to the initial looting and subsequent large-scale public unrest. These conditions enabled the insurgency to take root, and the Army and Marine Corps have been battling the insurgents ever since.

Why was the United States so unprepared for the challenges of postwar Iraq? As part of a larger study of Operation IRAQI FREEDOM (OIF), RAND Arroyo Center examined prewar planning for postwar Iraq and the subsequent occupation to seek an answer to this question and to draw lessons and recommendations from the Iraq experience.

It is not the case that no one planned for post-Saddam Iraq. On the contrary, many agencies and organizations within the U.S. government identified a range of possible postwar challenges in 2002 and early 2003, before major combat commenced, and suggested strategies for addressing them. Some of these ideas seem quite prescient in retrospect. Yet few if any made it into the serious planning process for OIF.

They were held at bay, in the most general sense, by two mutually reinforcing sets of assumptions that dominated planning for OIF at the highest levels. Although many agencies and individuals sought to plan for post-Saddam Iraq, senior policymakers throughout the government held to a set of fairly optimistic assumptions about the conditions that would emerge after major combat and what would be required thereafter. These assumptions tended to override counterarguments elsewhere in the government. Meanwhile, senior military commanders assumed that civilian authorities would be responsible for the postwar period. Hence they focused the vast majority of their attention on preparations for and the execution of major combat operations. That both sets of assumptions proved to be invalid argues for the development of a new and broader approach to planning military operations, and perhaps a louder military voice in shaping postwar operations.

Military Planning for Phase IV

The notion of a "Phase IV" in OIF came out of the war planning process that commenced in the fall of 2001, shortly before the fall of the Taliban in Afghanistan. On November 27, 2001, the Secretary of Defense directed U.S. Central Command (CENTCOM) to develop a plan to remove Saddam Hussein from power. The plan that emerged for OIF, later called OPLAN 1003V, outlined four phases: establishing international support and preparing for deployment; shaping the battlespace; major combat operations; and post-combat operations. The final version of OPLAN 1003V provided guidance and responsibilities for Phase IV operations, giving CENTCOM's land component, the Combined Forces Land Component Command (CFLCC), primary responsibility for post-combat operations.

Both CENTCOM and CFLCC developed supporting OPLANs in early 2003 that focused on Phase IV operations. Elements of each of these plans appear fairly prescient in retrospect. Yet they were always a low priority at CENTCOM, which focused the vast bulk of its time, attention, and resources on major combat operations. Although CENTCOM's commander, General Tommy Franks, refers to Phase IV frequently in his memoirs, for example, he never identifies the specific mission that U.S. forces should have had during that time. To the contrary: He expresses the strong sentiment that his civilian superiors should focus on postwar operations while he focused on the war itself.[1] He goes on to argue that civic action sets the preconditions for security rather than the other way around.[2] And he justifies his decision to retire right after combat ended because the mission was changing and a new commander should be there throughout Phase IV.[3]

---

[1] Franks states, "While we at CENTCOM were executing the war plan, Washington should focus on policy-level issues . . . I knew the President and Don Rumsfeld would back me up, so I felt free to pass the message along to the bureaucracy beneath them: *You pay attention to the day* after and *I'll pay attention to the day* of." Tommy Franks, *American Solider*, New York: Regan Books, 2004, p. 441. Emphasis in the original.

[2] Franks writes, "As I had said throughout our planning sessions, civic action and security were linked—*inextricably* linked. There was a commonly held belief that civil action would not be possible in Iraq without security. I would continue to argue that there could be no security without civic action." Franks, p. 526. Emphasis in the original.

[3] Franks, p. 530.

In short, General Franks saw major combat operations during Phase III as fundamentally distinct from Phase IV stability and reconstruction requirements, and as the military's primary task. That mindset reinforced an understandable tendency at CENTCOM to focus planning on major combat as an end in itself rather than as a component part of a broader effort to create a stable, reasonably democratic Iraq. The result, arguably, was a military operation that made the latter, larger goal more difficult to achieve.

Civilian Planning for Phase IV

General Franks was correct in seeing the need for greater civilian involvement in the stabilization of Iraq, since civilian agencies possess many of the capabilities needed for post-conflict operations. In fact, several U.S. government organizations, particularly the Office of the Secretary of Defense (OSD), the State Department, the U.S. Agency for International Development (USAID), and the National Security Council (NSC) conducted separate studies of postwar possibilities. The problem, therefore, was not that no one in the U.S. government thought about the challenges of post-Saddam Iraq. Rather, it was the failure to coordinate and integrate these various thoughts into a coherent, actionable plan.

At the center of the interagency planning process lay the NSC, which, starting in the summer of 2002, oversaw several interagency working groups that brought together representatives from the Department of Defense (DoD), the Department of State, the Central Intelligence Agency (CIA), and other organizations. Most of these working groups focused on the conduct of the war, but the working group on Iraq Relief and Reconstruction (IR+R) focused on postwar plans. This group produced fairly detailed humanitarian relief plans, but its reconstruction plans remained vague, reflecting a sense that reconstruction would not be necessary and stabilization would be handled by the Iraqis themselves.

If the NSC staff failed to consider alternative scenarios that might pose differing requirements, neither did it provide strategic guidance on various aspects of U.S. policy during the postwar period. Repeated requests for policy guidance from CENTCOM, Task Force IV, the Office of Reconstruction and Humanitarian Assistance (ORHA), and others went unanswered, leaving each agency to make its own assumptions about key aspects of the postwar period. Key questions, such as whether the U.S. postwar authority would be military or civilian in nature, went unanswered throughout the planning process. When the NSC issued strategic guidance in late March 2003 (as will be discussed in Chapter Three), the war was already under way. As a result, the various planning processes that occurred across the U.S. government were neither coordinated nor guided by a set of consistent goals and objectives.

Above all, the NSC seems not to have mediated persistent disagreement between the Defense Department and the State Department that existed throughout the planning process. Secretary of State Colin Powell influenced a few key diplomatic decisions—notably the decision to take the case for war with Iraq to the United Nations in September 2002—but the Defense Department controlled most planning decisions. State's main postwar planning effort, the Future of Iraq project, may not have been a workable plan for post-Saddam Iraq, but it raised many of the right questions about that phase of OIF. Yet the Defense Department largely ignored this project, to the point of preventing Tom Warrick, the study's leader, from working for ORHA in the weeks just before the war began.

The Defense Department created a new office to handle the increased workload associated with potential military operations in Iraq. It was called the Office of Special Plans (OSP), so as not to draw attention to the preparations for a possible war while President Bush simultaneously sought international support at the United Nations. OSP developed policy guidance on a wide range of issues, including the question of postwar governance, the future of the Iraqi army, and the de-Ba'athification process. Because the DoD exercised a great deal of control over planning for OIF, and ultimately took full control of the operation in January 2003, OSP exerted substantial influence over U.S. planning for Iraq.

Two particular sets of assumptions guided U.S. prewar planning for the postwar period. First, administration officials assumed that the military campaign would have a decisive end, and would produce a stable security situation. They intended to shrink the U.S. military presence down to two divisions—between 30,000 and 40,000 troops—by the fall of 2003. Deputy Secretary of Defense Paul Wolfowitz succinctly expressed this assumption during congressional testimony on February 27, 2003, when he stated, "It's hard to conceive that it would take more forces to provide stability in post-Saddam Iraq than it would take to conduct the war itself and to secure the surrender of Saddam's security forces and his army."[4] Second, they assumed that the Iraqi population would welcome U.S. forces. Three days before the war, Vice President Richard Cheney clearly articulated this view by stating, "My belief is we will, in fact, be greeted as liberators."[5] Iraqi exiles supported this belief by emphasizing that the Iraqis would greet U.S. forces with "sweets and flowers."[6]

The one post-Saddam challenge for which the U.S. government actually planned was that of a possible humanitarian emergency brought on by the possibly massive flow of refugees, combined with shortages of food, water, and medicine. An interagency planning team started meeting in the fall of 2002 and worked with in-

---

[4] Paul Wolfowitz, testimony to the House Budget Committee, February 27, 2003.

[5] Vice President Richard Cheney, remarks to *Meet the Press*, March 16, 2003.

[6] Kanan Makiya, as quoted in Joel Brinkley and Eric Schmitt, "Iraqi Leaders Say U.S. Was Warned of Disorder After Hussein, but Little Was Done," *New York Times*, November 30, 2003.

ternational organizations (IOs) and nongovernmental organizations (NGOs) to generate detailed humanitarian relief plans across a range of possible scenarios. As it turned out, because of the speed of military operations, which left supply networks largely intact, the war in Iraq did not generate significant humanitarian requirements.

Task Force IV

Significantly, observers at CENTCOM's Internal Look exercise, held in December 2002, noted that the warplans for Iraq did not include detailed planning for the postwar period. Later that month, the Chairman of the Joint Chiefs of Staff ordered Joint Forces Command to create a new organization, based on the Standing Joint Force Headquarters (SJFHQ) concept, that would plan for Phase IV and form the nucleus of a postwar military headquarters in Iraq. This new organization, called Task Force IV (TFIV), was placed under CENTCOM's operational control and started assembling in Tampa in January 2003.

Although the Joint Staff had identified an extremely important problem with the existing warplans, Task Force IV proved to be an unworkable solution to it. Having been created very late in the planning process, and coming from outside CENTCOM, it had little influence. The fact that the task force's director was a one-star general, outranked by key players in CENTCOM's planning process, only compounded the problem. By March 2003, it was clear that Task Force IV would not become the nucleus of a postwar military headquarters, and it was officially disbanded by the end of the month.

The Office of Reconstruction and Humanitarian Assistance (ORHA)

On January 20, 2003, the National Security Council issued NSPD-24, which gave the Department of Defense primary responsibility for postwar Iraq and tasked DoD to form a new office to take charge of planning. Retired Army Lieutenant General Jay Garner was named to lead this new office, which became known as ORHA. Many of ORHA's early staff members were military personnel because U.S. agencies proved reluctant to provide staff for ORHA, though its composition grew more balanced over time.

ORHA personnel soon discovered that the many administrative issues involved in setting up their organization left little time to deal with substantive issues and long-term planning. ORHA did plan for possible humanitarian relief operations, drawing on interagency relief plans prepared elsewhere. It also developed the concept of Ministerial Advisory Teams to ensure that Iraqi ministries continued to function between the fall of Saddam Hussein and the establishment of a new permanent government. These concepts were discussed at a meeting held at the National Defense University on February 21 and 22, 2003, which included representatives from every U.S. government agency that would have a role in reconstruction. The meeting revealed several serious shortcomings in preparations for dealing with postwar Iraq: U.S. agencies were reluctant to provide personnel for the ministerial teams, and the question of who would provide postwar security in Iraq remained unaddressed. Both of these issues would later pose significant problems for both ORHA and its successor, the Coalition Provisional Authority (CPA).

ORHA deployed to Kuwait in mid-March 2003, although many staff members would have preferred to remain in Washington longer to continue developing working relationships with their counterparts throughout the U.S. government. Once in Kuwait, ORHA learned that, for security reasons, the CFLCC commander did not want ORHA collocated with his forces at Camp Doha. ORHA thus set up its headquarters at the Kuwait Hilton, approximately 45 minutes away from Camp Doha and lacking rudimentary communications and infrastructure.

Significantly, ORHA's personnel were not privy to the warplans until shortly before the war started. ORHA had planned to enter Basra and start reconstruction efforts as soon as coalition military forces secured that city, but during the second week of March, Garner learned that the warplans called for most military forces to go straight to Baghdad instead of remaining in rear areas to provide security, thus rendering many of ORHA's plans obsolete. CFLCC directed ORHA to remain in Kuwait while major combat operations were conducted throughout Iraq. Not only did this render its plans ineffective, but once Baghdad fell and the looting started, it exposed ORHA to charges that it was doing nothing to stop destruction around the country.

ORHA began entering Baghdad on April 21, after Garner personally asked for and received permission to do so from General Franks. ORHA quickly discovered that conditions in Iraq were markedly different from those originally anticipated. The expected humanitarian crisis never materialized, while extensive looting damaged much of the infrastructure that the military campaign had deliberately left intact. Furthermore, the unsettled security situation significantly hindered ORHA's reconstruction efforts.

ORHA's planning problems quickly became irrelevant, however, as on April 24, three days after Garner arrived in Baghdad, the Secretary of Defense informed him that President Bush intended to appoint L. Paul Bremer as his permanent envoy to Iraq. U.S. officials announced Bremer's appointment on May 6, and he arrived in Baghdad on May 12 with a mandate to create a new Coalition Provisional Authority. Unlike ORHA, CPA would possess all the powers of an occupation authority. ORHA's staff shrank as CPA's grew, with few ORHA personnel choosing to stay on and work for CPA. Garner left Iraq on June 1, almost two weeks after ORHA had been superseded by CPA.

The Coalition Provisional Authority

In May 2003, the Coalition Provisional Authority took over from ORHA, and L. Paul Bremer became the administrator of Iraq. From then until Bremer handed power over to the Iraqis on June 28, 2004, the United States and the United Kingdom were the legal occupiers of Iraq. They had two simultaneous and sometimes competing missions: to run the country and to build up Iraqi institutions that would enable self-rule. The November 2003 decision to accelerate the handover of power by July 1, 2004, exacerbated the tension between the two missions.

Although CPA was the governing body of occupied Iraq, it was not the only coalition structure in country, and it did not have authority over all other structures. Combined Joint Task Force 7 (CJTF-7), the military command, which reported to CENTCOM, functioned separately, as did various intelligence agencies (including the CIA), and the Iraq Survey Group, which continued its hunt for weapons of mass destruction (WMD). In the absence of a detailed plan, these groups had to work out relations on the spot. While personal relations were often good, failures of coordination and information sharing sometimes created significant tensions, most commonly between the civilian and the military arms of the occupation.

This problem was further exacerbated by the structural weaknesses of the CPA. It remained limited throughout its existence by the fact that the United States had never planned to be an occupying authority, and that it was quickly assembled on an ad hoc basis. It was staffed at half its authorized level, and many on its staff lacked government experience and only served short rotations. The lack of personnel, combined with the deteriorating security situation, also meant that CPA had a negligible presence outside Baghdad, leaving military forces throughout the rest of the country to fill the gap left by the lack of civilian authority and reconstruction capacity. The Army and Marine Corps thus carried the major share of the stability and reconstruction missions outside Baghdad.

Building governance structures. CPA worked hard to build governance structures under the tremendous strain of a deteriorating security situation that did not welcome exiles. At the same time, the CPA staff's lack of access to other Iraqis resulted in continued reliance on exiles in the building of a new Iraq. The CPA appointed the Iraqi Governing Council (IGC)—a multi-ethnic, multi-sectarian, and exile/Kurdish-dominated body—on July 13, 2003, after considerable debate and discussion about what form the new Iraqi government should take. While it was never popular with Iraqis, this 25-person body became over time an increasingly independent actor and CPA's primary Iraqi interlocutor.

The IGC and CPA jointly issued the November 15 agreement, promising that the CPA would transfer authority to an interim Iraqi government by July 1, 2004, and requiring that a "basic law" or interim constitution be drafted by February 28, 2004. The process of drafting the basic law, or Transitional Administrative Law (TAL), as it came to be called, took place largely in the first two months of 2004. A variety of issues surfaced during the TAL discussions, which kept the IGC from reaching full agreement on the TAL by the deadline. Ambassador Bremer and his staff pushed the drafters to continue work on the document into the early hours of March 1, 2004, when agreement was finally reached.

On June 1, 2004, the Iraqi Interim Government (IIG) was formed, with Iyad Allawi, formerly chair of the IGC security committee, as prime minister and Sheikh Ghazi al-Yawer as president. Two deputy presidents, as prescribed in the TAL, and a new cabinet were also selected. This new government then worked with CPA, the United Nations, and coalition capitals to facilitate the transfer of authority, which took place on June 28.

Creating security forces and institutions. One of the greatest challenges faced by CPA and CJTF-7 was the creation of new Iraqi security forces. Prewar planning assumptions—that the old Iraqi military could undergo a process of disarmament, demobilization, and reintegration (DDR) while helping ensure security during the interim period, and that police forces would remain largely intact and ensure law and order—proved deeply flawed. CPA was soon rebuilding both a military and a police force.

CPA Order Number 2, issued on May 23, 2003, formally dissolved Iraq's armed forces and its defense ministry, along with a number of other Saddam-era security-related structures. The Iraqi police service, historically a powerless and corrupt structure, was suddenly expected to be the front line for internal security—in a deteriorating security situation. CPA's advisors to the interior ministry, which had responsibility for police, were short-staffed and constantly torn between the effort to build effective structures and the need to get police on the streets and patrolling. This tension was exacerbated by a failure on the part of coalition capitals to recognize the crucial nature of the police mission and allocate sufficient resources to it.

Military training was better structured, since there was less immediate need for Iraqi military forces and more prewar planning existed (CENTCOM had always planned on a new military for Iraq, but had expected to be able to rely more on the structures of the old one). Military personnel could readily be hired, trained, and then deployed. In addition to the Iraqi armed forces, coalition troops also developed the Iraqi Civil Defense Corps (ICDC), which served as an Iraqi auxiliary of various sorts to coalition troops.

Several problems plagued the building of Iraqi defense ministry forces. Crucial was the question of mission—whether such forces should be built for defense against external threats, or to help in the current conflict. Early efforts to use units domestically led to refusals to fight and desertions. Militias also posed ongoing challenges for the CPA, since they had to be either disbanded or somehow brought under the control of the new Iraqi government. Although discussed for many months, efforts in this area did not begin in earnest until February 2004. Critical to these efforts was the adoption of the TAL, which would make all armed forces and militias not under federal control illegal in the new Iraq, except as provided by law. However, the implementation of this process had barely begun at the time the IIG took power, and its future remained in doubt.

Economic policy and reconstruction. Economic policy was another area for which there was little planning prior to the war. Under the CPA, coalition advisors sought to create an economic structure that would foster entrepreneurship and foreign investment. They faced opposition in some of these efforts from the IGC, which tended to prefer the status quo.

The CPA was successful in reviving the Central Bank of Iraq, implementing a new currency and exchanging it for the old. It declared a tax holiday and lifted tariffs and import restrictions for 2003, and it issued a law on foreign direct investment. CPA also defined a budget for the second half of 2003 and for 2004: the first in dollars, the second in dinars. More problematic were efforts to liberalize prices, particularly for gasoline and fuel, to reform the food rationing system fully, and to restructure state-owned companies so that they could function in a modern economy. Plans to downsize and close such structures encountered stiff opposition from the Iraqi Governing Council. CPA's failures to reform Iraq in these areas led to both continued economic waste and potentially slowed reconstruction.

CPA also had the task of restoring essential services. It hoped to improve provision of services to about what it had been under Saddam Hussein, but soon found that the best it could do was to focus on the basic provision of water, oil, and electricity. Iraqi infrastructure was damaged both by the 1991 Persian Gulf War and by years of sanctions and neglect afterward. OIF and particularly the looting that ensued did additional damage to the capacity to produce electricity, oil, and water. This was a surprise to coalition forces, who expected to provide food and water to refugees and to protect the oil sector. They did not, however, expect to carry out large-scale reconstruction.

Reconstruction was mostly pursued through contracting mechanisms. Because there was some expectation of work in this area, USAID awarded a number of contracts early on. Kellogg, Brown and Root (a Halliburton subsidiary), Bechtel, and other contractors were awarded large contracts to work on the oil fields, electricity, government buildings, ports, airports, and so forth. Other contracts were let throughout 2003. These were funded through a variety of mechanisms, including U.S. appropriations; Iraqi oil export earnings, deposited in the Development Fund for Iraq; accrued assets, including seized assets of Saddam or the Ba'ath party; funds from the UN Oil for Food account; and promises of assistance from other donors.

The success of reconstruction during the occupation period was mixed. At the time CPA handed over power to the IIG, electric power generation was near prewar levels, while oil production was below its preconflict peak and hampered by sabotage. Water provision, however, had improved, and mobile telephone service helped compensate for lagging fixed-line provision.

The Army and Postwar Planning

Looking back, we can see that the failure to plan for and adequately resource stability operations had serious repercussions that affected the United States throughout the occupation period and continue to affect U.S. military forces in Iraq. Because U.S. forces were not directed to establish law and order—and may not have had enough forces for this mission anyway—they stood aside while looters ravaged Iraq's infrastructure and destroyed the facilities that the military campaign had taken great pains to ensure remained intact. Because Iraq's own police and military evaporated shortly after Saddam fell, ordinary Iraqis lived in a basically lawless society for months, during which, among other things, insurgents, terrorists, and criminal gangs assembled with impunity. And because U.S. forces have had to focus on providing security for their own personnel (both military and civilian) as much as for Iraqis, the buildup of coalition forces did not bring the degree of safety and security it might have brought had order been imposed from the start.

The situation has only gotten worse since the insurgency began. U.S. forces have had to assume that ordinary citizens may be potential belligerents, often leaving Iraqi civilians in the crossfire. A consistent majority of the Iraqi population identified security and safety as the most urgent issue facing Iraq throughout the occupation period.[7] The failure to stabilize and secure Iraq has therefore had the inadvertent effect of strengthening the insurgency, as Iraqis witness many of the negative effects of the U.S. military presence without seeing positive progress on the issues that matter to them most. The insurgency has also been aided by the failure of U.S. military forces to emphasize the mission of sealing the country's borders—a mission that still ranks relatively low on the list of important coalition missions—enabling critical foreign support to flow into Iraq.

---

[7] This trend continued after the June 28, 2004 transfer of authority. Results do vary somewhat by city. Between January and August 2004, the percentage of the population identifying safety and security as the most urgent issue averaged 63 percent in Baquba; 60 percent in Mosul; 53 percent in Baghdad; 47 percent in Najaf; and 30 percent in Basra. When asked "How safe do you feel in your neighborhood?" the number of respondents who answered "not very safe" or "not safe at all" averaged 63 percent in Basra; 58 percent in Baquba; 57 percent in Baghdad and Najaf; 46 percent in Mosul; and 33 percent in Karbala. See "Opinion Analysis," U.S. Department of State Office of Research, M-106-04, September 16, 2004, Appendix 6A.

This is not to say that stability and order would rule in Iraq today had U.S. planners only spent more time planning for post-Saddam operations. Counterfactuals like this lie beyond proof. Still, a strong inference can be drawn that, had security been imposed across Iraq from the moment Saddam fell, the insurgency that so afflicts Iraq today would not have had the political "space" in which to take root. And the Iraqi people themselves, however resentful they might have been of occupation forces, could have at least thanked those forces for enforcing law and order and thus a degree of public safety. In terms of its status with ordinary Iraqis, after all, U.S. forces were in the worst possible situation: there in numbers sufficient to be resented as occupiers but insufficient to impose order. It seems highly likely that the situation in Iraq today would be more manageable had U.S. planners spent more time thinking through post-Saddam scenarios and planning for both combat and post-combat with the worst of those scenarios in mind.

Instead, U.S. government planning was based on a set of optimistic assumptions that was never seriously challenged: that the military campaign would have a decisive end and would produce a stable security environment; that U.S. forces would be greeted as liberators; that Iraq's government ministries would remain intact and continue to administer the country; and that local forces, particularly the police and the regular army, would be capable of providing law and order. Those assumptions channeled the interagency planning process, such as it was, into a focus on humanitarian relief, on the assumption that reconstruction and stabilization would not be required. And they made it very difficult—because they made it seem unnecessary—to assign responsibility and resources for providing security in the immediate aftermath of major combat operations, perhaps the single most important failure of the prewar planning process.

In a very real sense, key officials predicted the future with sufficient confidence to rule out alternative plans. In fact, of course, the future is always unpredictable, which is why planners routinely explore alternative scenarios in search of the "worst cases" that can pose the greatest challenges to their plans. Their plans then reflect actions that either cover or hedge against those possibilities. This is in some sense the basis for the standard military planning and decision process.

Yet in this case, few military voices besides that of Army Chief of Staff General Eric K. Shinseki called attention to the possibility of a major, long-term security challenge in post-Saddam Iraq. One reason other military voices remained muted was that the military operated within the prevailing assumptions set by senior civilian officials, which did not identify security as a problem. Also, as General Franks makes clear in his memoirs, the senior Army planner for OIF was reluctant to take responsibility for security and stabilization missions in the aftermath of major combat. This was not seen as the military's role or mission.

Yet it is precisely through General Franks that the military could have voiced its concerns, since it is the combatant commander, far more than the "institutional" services, who plays a strong role in the interagency planning process. Yet the institutional services are not irrelevant, since it is from them that the combatant commanders are drawn. And those commanders reflect a view of war and stabilization that can only be taught in service schools and other institutions. What Franks lacked was a complete view of what his forces were about to undertake. A more holistic view would be informed by three key assumptions:

- First, it should be clear from U.S. interventions not just in Iraq, but in Afghanistan, Kosovo, and Bosnia, that wars do not end when major conflict ends. Wars emerge from an unsatisfactory set of political circumstances, and they end with the successful creation of new and more favorable political circumstances—in this case, circumstances more favorable to U.S. interests. Creating those new circumstances may not involve continuing conflict, and even if conflict is present, it may not be as intense as the counterinsurgency operations confronting U.S. forces in Iraq today. But given the likely security vacuum following major conflict, planners cannot avoid considering a variety of forms of conflict.

- Second, these post-conflict missions will almost unavoidably fall to forces present on the ground at the time. To some extent the security missions that follow major conflict are legitimate tasks for ground forces that, by virtue of their possession of the instruments of violence, can impose security in such situations. But the absence of security makes it unlikely that the civilian organizations that would normally handle reconstruction tasks will be available quickly to take on those roles. In the immediate aftermath of major conflict, and perhaps for a good deal longer, "civilian" as well as "military" missions will fall to forces on the ground.

- Finally, it should be clear that the way the actual conflict unfolds exerts enormous influence over the situation that emerges and evolves after the major conflict ends. To provide security in the aftermath of Saddam's fall, the invading force needed more troops. A larger force might also have been able to force Saddam's military to surrender rather than simply melt away, weapons in hand. These observations testify to the dangerous artificiality of the distinction between Phase IV, on the one hand, and the phases that preceded it. They are not distinct phases; planning for each in sequence can produce unhappy outcomes.

These lessons have significance for the U.S. Army's Title 10 role of organizing, training, and equipping forces for use by combatant commanders in major conflicts. The Army must put real meaning into the phrase "full spectrum force." It must be able to fight and dominate an adversary in major conflict. But as we can see in Iraq, Army forces must also be prepared to provide security to a civilian populace, reconstruct infrastructure as necessary, escort children safely to school, perhaps even help clear raw sewage from the streets. They will usually do so in a cultural environment foreign to them, yet those missions will require them to have at least enough cultural awareness to avoid undermining the mission.

But the more crucial significance of these basic lessons comes at the level of military and strategic planning. Clearly these lessons produce a very different view of the military planning process than the one for OIF. Military planners must start with a view of the desired outcome of the war—not the outcome of major conflict, but the creation of the desired political circumstances that signal the real end of the war. They must do so both because their forces, and especially forces on the ground, will be intimately involved in creating those circumstances, and because the way in which military action unfolds will heavily shape the way the rest of the war unfolds.

One way to capture this lesson is to say that military planners must "start with Phase IV." But a more accurate solution is to dispense with phases, which inevitably produce sequenced plans that risk missing crucial connections from phase to phase. Planners must start with strategic guidance from the civilian leadership on where they want to be, strategically, when the war ends. They can then work backward to points of major conflict, shaping plans for those in ways that contribute to the larger and longer-term strategic goal.

Starting planning this way will ensure that "Phase IV" will not be ignored or underplayed in the planning process. But as planning for OIF makes clear, it is essential that planners entertain a full array of possible scenarios for getting to that strategic end point. Even the most reasonable assumptions must be challenged, and hedging actions must be an integral part of the plan. Recognizing that military forces—largely U.S. Army forces—will play a role in these activities should give the combatant commander good reason to force this conversation into the planning process.

Acknowledgments

This report could not have been written without the extensive cooperation of many individuals who were involved in the planning and execution of postwar reconstruction activities in Iraq.[1] In particular, we wish to thank Lieutenant General (ret.) Ron Adams; Colonel John Agoglia, USA; Jon Alterman; Colonel Thomas Baltazar, USA; Colonel Kevin Benson, USA; Brigadier General David Blackledge, USA; Scott Castle; Julie Chappell; James Clad; Joseph Collins; A. Heather Coyne; Roger Corneretto; Major Ray Eiriz, USA; Mike Eisenstadt; Greg Gardner; Lieutenant General (ret.) Jay Garner, USA; Brigadier General (P) Steven Hawkins, USA; Colonel Tom Hayden, USA; Major Chris Herndon, USA; Lieutenant Colonel Chris Holshek, USA; Colonel Paul Hughes, USA; Bernard Kerik; Lieutenant Colonel Chris Kinnan, USAF; Lewis Lucke; William Luti; Roman Martinez; Dayton Maxwell; Michael McNerney; Frank Miller; Meghan O'Sullivan; Lieutenant General David Petraeus, USA; Lieutenant Colonel Bob Polk, USA; Colonel Tony Puckett, USA; Ambassador Robin Raphel; Gordon Rudd; Lieutenant Colonel Steve Seroka, USAF; Abe Shulsky; Walt Slocombe; Commander David Tarantino, USN; Bob Teasdale; Gerry Thompson; George Ward; Tom Warrick; Tom Wheelock; Ross Wherry; and others who chose to remain nameless. Any errors remain the authors' own.

Joseph Collins, James Dobbins, Jim Dewar, and Tom McNaugher provided thorough reviews of earlier drafts, which greatly improved the quality of the final report. We also thank several RAND colleagues who provided comments and insights throughout the research and writing of this report, including Richard Darilek, Lynn Davis, Audra Grant, Andy Hoehn, Terry Kelly, Karl Mueller, Walt Perry, Lauri Rohn, David Shlapak, Jerry Sollinger, and Peter Wilson. Sarah Harting provided excellent administrative assistance.

---

[1] Please note that all military ranks here reflect what they were at the time of the original research.

List of Acronyms and Abbreviations

AOUSC	Administrative Office of the U.S. Courts
APA	Arab Police Academy
BBG	Broadcasting Board of Governors
C5	Director of Policy, Plans, and Strategy (Combined Forces Land Component Command)
CBI	Central Bank of Iraq
CENTCOM	U.S. Central Command
CERF	Commander's Emergency Response Funds
CERP	Commander's Emergency Response Program
CFLCC	Combined Forces Land Component Command
CG	Commander's Guidance
CIA	Central Intelligence Agency
CJCS	Chairman of the Joint Chiefs of Staff
CJTF-7	Combined Joint Task Force 7
CJTF-Iraq	Combined Joint Task Force—Iraq
CMAD	Collection, Management, and Analysis Directorate
CMATT	Coalition Military Assistance Training Team
CPA	Coalition Provisional Authority
CPC	Constitution Preparation Committee
CRS	Congressional Research Service
CTEG	Counter Terrorism Evaluation Group
DART	Disaster Assistance Response Team
DDR	Disarmament, Demobilization, and Reintegration
DFI	Development Fund for Iraq
DFID	Department for International Development (UK)

DLA	Defense Logistics Agency
DoD	U.S. Department of Defense
DOE	U.S. Department of Energy
DOJ	U.S. Department of Justice
DOT	U.S. Department of Transportation
DPS	Diplomatic Protection Services
EB	Bureau of Economic and Business Affairs (U.S. State Department)
ESG	Executive Steering Group
FAA	Federal Aviation Administration (U.S.)
FCC	Federal Communications Commission (U.S.)
FCO	Foreign and Commonwealth Office (UK)
FEMA	Federal Emergency Management Agency (U.S.)
FEST	Forward Engineering Support Teams
FIF	Free Iraqi Forces
FMIS	Financial Management Information System
FORSCOM	Forces Command (U.S. Army)
FPS	Facilities Protection Services
FY	Fiscal Year
GAO	General Accounting Office (renamed the Government Accountability Office in July 2004)
GDP	Gross Domestic Product
HDRs	Humanitarian Daily Rations
HHS	U.S. Department of Health and Human Services
HOC	Humanitarian Operations Center
HPT	Humanitarian Planning Team
IATA	International Air Transport Association
ICDC	Iraqi Civil Defense Corps
ICG	International Crisis Group
ICP	Iraqi Communist Party
ICRC	International Committee of the Red Cross
ID	Infantry Division
IDPs	Internally Displaced Persons
IEDs	Improvised Explosive Devices

IGC	Iraqi Governing Council
IIG	Iraqi Interim Government
ILO	International Labor Organization
IMF	International Monetary Fund
INA	Iraqi National Accord
INC	Iraqi National Congress
INIS	Iraqi National Intelligence Service
INL	Bureau of International Narcotics and Law Enforcement (U.S. State Department)
INR	Bureau of Intelligence and Research (U.S. State Department)
IOs	International Organizations
IPMC	Iraq Political-Military Cell
IPS	Iraqi Police Service
IPU	International Postal Union
IR+R	Iraq Relief and Reconstruction
IRDC	Iraqi Reconstruction and Development Council
ISA	International Security Affairs
IST	Iraqi Special Tribunal
IT	Iraqi Technocrat
J5	Director for Strategic Plans and Policy (Joint Staff)
JFCOM	U.S. Joint Forces Command
JNEPI	Joint NGO Emergency Preparedness Initiative
KBR	Kellogg, Brown and Root
KDP	Kurdistan Democratic Party
LOGCAP	Logistics Civil Augmentation Program
MCNS	Ministerial Committee for National Security
MEF	Marine Expeditionary Force
MEI	Middle East Institute
MNF-I	Multinational Force-Iraq
MoD	Ministry of Defense (Iraq)
MoI	Ministry of the Interior (Iraq)
MP	Military Police
MW	Megawatts
NCO	Noncommissioned Officer

NED	National Endowment for Democracy
NESA	Near East and South Asia
NGO	Nongovernmental Organization
NSC	National Security Council
OCHA	Office for the Coordination of Humanitarian Affairs (UN)
OFF	Oil for Food Program
OIC	Organization of Islamic Conference
OIF	Operation IRAQI FREEDOM
OPCON	Operational Control
OPEC	Organization of the Petroleum Exporting Countries
OPLAN	Operations Plan
ORHA	Office of Reconstruction and Humanitarian Assistance
OSD	Office of the Secretary of Defense
OSHA	Occupational Health and Safety Administration (U.S. Labor Department)
OSP	Office of Special Plans
OUSD(P)	Office of the Under Secretary of Defense for Policy
OVP	Office of the Vice President
PDS	Public Distribution System
PMO	Project Management Office
POLAD	Political Advisor
POWs	Prisoners of War
PRB	Program Review Board
PRM	Bureau of Population, Refugees, and Migration (U.S. State Department)
PSYOP	Psychological Operations
PUK	Patriotic Union of Kurdistan
RIE	Restore Iraqi Electricity
RIO	Restore Iraqi Oil
RRRP	Rapid Regional Response Program
RTI	Research Triangle Institute
SAMS	School of Advanced Military Studies (U.S. Army)
SCIRI	Supreme Council for the Islamic Revolution in Iraq
SJFHQ	Standing Joint Force Headquarters

SOEs	State Owned Enterprises
SOF	Special Operations Forces
TAL	Transitional Administrative Law
TBD	To Be Determined
TFIV	Task Force IV
TIP	Transitional Integration Program
UK	United Kingdom
UN	United Nations
UNDP	United Nations Development Program
UNESCO	United Nations Educational, Scientific, and Cultural Organization
UNFAO	United Nations Food and Agriculture Organization
UNHCR	United Nations High Commissioner for Refugees
UNICEF	United Nations Children's Fund
UNMOVIC	United Nations Monitoring, Verification, and Inspection Comission
UNSCR	United Nations Security Council Resolution
USA	U.S. Army
USACE	U.S. Army Corps of Engineers
USAF	U.S. Air Force
USAID	U.S. Agency for International Development
USDA	U.S. Department of Agriculture
USPS	U.S. Postal Service
WFP	World Food Program
WHO	World Health Organization
WMD	Weapons of Mass Destruction

CHAPTER ONE

Introduction

After more than 15 months of planning, Operation IRAQI FREEDOM (OIF) commenced in March 2003. Major combat operations in Iraq lasted approximately three weeks, but stabilization efforts in that country are, as of this writing, ongoing. The U.S. Army and the U.S. Marine Corps are increasingly taxed by the demands of the continuing insurgency, with more than 100,000 troops expected to remain in Iraq for the foreseeable future. How did Iraq get to this point? Why was the United States so unprepared for the challenges of postwar Iraq?

The evidence suggests that the United States had neither the people nor the plans in place to handle the situation that arose after the fall of Saddam Hussein. Looters took to the streets, damaging much of Iraq's infrastructure that had remained intact throughout major combat. Iraqi police and military units were nowhere to be found, having largely dispersed during combat. U.S. military forces in Baghdad and elsewhere in the country were not prepared to respond rapidly to the initial looting and subsequent large-scale public unrest. These conditions enabled the insurgency to take root, and the Army and Marine Corps have been battling the insurgents ever since.

It is *not* the case that no one planned for postwar Iraq. On the contrary, many agencies and organizations within the U.S. government *did* identify a range of possible postwar challenges in 2002 and early 2003, before major combat commenced, and suggested strategies for addressing them. Some of these ideas seem quite prescient in retrospect. Why, then, were they not incorporated into the planning process? As part of a larger study of OIF, RAND Arroyo Center examined prewar planning for postwar Iraq and the subsequent occupation, and drew lessons and recommendations from the Iraq experience.

U.S. civilian planning was driven by a particular set of assumptions, held by senior policymakers throughout the government, about the conditions that would emerge after major combat and what would be required thereafter. These assumptions—which included U.S. forces being greeted as liberators, the emergence of a stable security situation, and the continued functioning of the Iraqi government ministries—remained largely unchallenged. No contingency plans were developed in case these assumptions proved to be incorrect.

Furthermore, senior military commanders assumed that civilian authorities would be responsible for the postwar period. They focused the vast majority of their attention on preparations for and the execution of major combat operations, and assumed that their responsibilities largely ended there. They assumed, incorrectly as it turned out, that the war would have a clearly defined end and they would quickly transfer responsibility for Iraq to civilians. This overlooked the lack of a standing civilian organization capable of taking such responsibility, and the possible requirement that military forces provide basic law and order during any transition period. Furthermore, civilians did not participate in the highly classified war planning process, which would have made coordinated planning for the transition period extremely challenging even if such a standing authority had existed.

This report examines the range of U.S. government planning efforts for postwar Iraq, as well as the challenges that emerged for both military and civilian authorities during the occupation period. Chapters Two through Six examine prewar planning efforts for postwar Iraq. Chapter Two examines military planning, including the plans developed by U.S. Central Command (CENTCOM) and the Combined Forces Land Component Command (CFLCC). Chapter Three examines civilian planning, starting with an overview of the interagency process and examining the specific roles of the Office of the Secretary of Defense (OSD), the State Department, the U.S. Agency for International Development (USAID), and the National Security Council (NSC) staff. It also summarizes some of the reports and recommendations issued by think tanks and academic institutions that were published before the war. Chapter Four describes Task Force IV (TFIV), an organization created by the Joint Chiefs of Staff to fill some of the gaps in CENTCOM's postwar planning efforts. Chapter Five tells the story of the Office of Reconstruction and Humanitarian Assistance (ORHA), the organization created within the U.S. Department of Defense (DoD) after it had been designated the official lead agency for postwar Iraq. Chapter Six focuses specifically on planning for humanitarian assistance. Such assistance is often considered to be part of reconstruction, but it warrants a separate discussion because humanitarian assistance planning proved to be far more coordinated and effective in this case than reconstruction planning.

Chapter Seven provides an overview of combat operations between the middle of April 2003 and August 2004. It examines the security situation in Iraq in the immediate aftermath of major combat operations, looks at the military organization in theater for Phase IV stability and support operations, describes the major types of attacks conducted by and against coalition forces, and analyzes events reported as significant. This analysis provides a snapshot of continuing combat operations in Iraq through the occupation period.

Chapters Eight through Twelve examine the Coalition Provisional Authority (CPA), the organization established in May 2003 to oversee the reconstruction of Iraq. CPA possessed a much more robust mandate than ORHA, one that confirmed its role as an occupying authority. CPA's orders had the force of law throughout Iraq during its 14-month existence. Its goal was to create a democratic and free Iraq by the end of the occupation period. Chapter Eight starts by discussing the origins, goals, structure, and functions of CPA. The next four chapters examine each of CPA's four core "foundations": Chapter Nine focuses on building Iraq's new security forces; Chapter Ten addresses governance and political reconstruction; Chapter Eleven assesses economic policy; and Chapter Twelve examines the restoration of essential services in Iraq.

Chapter Thirteen concludes the report by analyzing why the United States failed to prepare adequately for the challenges and remained unprepared for the conditions that emerged in postwar Iraq. The chapter identifies some of the unchallenged assumptions and political constraints that limited the planning process, indicates some of their consequences, and concludes with recommendations for U.S. policy and for the U.S. Army.

The purpose of this analysis is to find out where problems occurred and to suggest possibilities to improve planning and operations in the future. The results of such analysis can seem therefore to be overly focused on the negative. This should not be taken to mean that no good was done. In fact, dedicated U.S. and coalition personnel, both military and civilian, engaged in many positive and constructive activities, individually and collectively. That this analysis does not highlight all those activities should not in any way detract from their value. Our focus, however, remains on finding ways to improve.

This volume draws on a wide range of sources, including government documents, press reports, numerous interviews with U.S. civilian and military officials, and, for Chapters Eight through Twelve, the personal experiences of several authors who worked for CPA. Most of those interviewed chose to remain anonymous, but their affiliations are noted so that their statements can be put into context. We have tried to corroborate their statements wherever possible, and we have noted when information was supported by multiple interviews or through published accounts.

CHAPTER TWO

Military Planning Efforts

The rapid and decisive defeat of Iraq's military forces, and the subsequent advance to Baghdad, Tikrit, Kirkuk, and Mosul, clearly demonstrated the dominance of the U.S. military on the battlefield. The success of its campaign plan during major combat operations ensured that coalition forces simultaneously attacked Iraqi forces throughout the depth and breadth of Iraq, including major operations in the north with the Kurds, in the western desert, along the eastern border with Iran, and throughout the central Tigris/Euphrates river valley from Umm Qasr to parts of the Sunni Triangle north of Baghdad. Initially, most Iraqis viewed coalition forces as liberators. Before long, however, an organized resistance began to undermine the early military success and erode Iraqi public support.

From May 2003 and continuing beyond June 2004, an insurgency mounted within Iraq. This new enemy, consisting of loose coalitions of former Ba'athists, Iraqi Islamists, and foreign fighters, has waged a relentless war against coalition operations and the new Iraqi government by attacking infrastructure, government officials, civilian targets, and coalition military. Moreover, by June 2004 an overwhelming majority of the Iraqi public had come to view U.S. military forces as foreign occupiers rather than liberators.[1] The question becomes, then, to what extent did planning shortfalls contribute to this situation? In retrospect, what could have been done to prevent or mitigate the difficulties the United States and its coalition partners began experiencing in Iraq? What does this mean for Army forces, and for Army planning?

This chapter examines the military planning for postwar operations—often referred to as Phase IV—that took place at U.S. Central Command (CENTCOM) and at Combined Forces Land Component Command (CFLCC). The chapter concludes with conclusions and recommendations concerning the planning process in general and the specifics of Phase IV planning for OIF in particular.

---

[1] Interview with NSC official, July 2004.

CENTCOM Operational Planning

On November 27, 2001, shortly before the fall of the Taliban in Afghanistan, the Secretary of Defense directed CENTCOM to develop a plan that would forcibly remove Saddam Hussein from power.[2] As part of its deliberate planning process, CENTCOM had developed Operations Plan (OPLAN) 1003 in 1998 in the event that the United States found itself in another war with Iraq. On December 7, 2001, the CENTCOM Commander, General Tommy Franks, presented Secretary of Defense Donald Rumsfeld with the first iteration of his Commander's Concept of Operations.[3] According to Franks, although the existing plan provided an operational construct for removing Saddam Hussein and eliminating the threat of weapons of mass destruction (WMD) use by Iraq, it did not address the current disposition of U.S. forces, advances in precision-guided munitions made since the end of the Persian Gulf War, or the lessons being learned from U.S. operations in Afghanistan. By the time Franks's new OPLAN was finished it grew to 89 pages, with thousands of pages of specialized appendices.

Early in the iterative process of developing his Commander's Concept, General Franks limited participation to a small number of senior officers. By December 8, 2001, he had developed a working matrix that highlighted what he called "Lines and Slices." The slices can best be understood as target sets he wanted to affect, while the lines were the means he would use against particular targets. Together they represent what General Franks considered to be the primary focus areas of the operation. This matrix demonstrates that even at this early stage, Franks was envisioning a plan that focused not only on defeating the Iraqi military and removing Saddam Hussein from power, but also on doing it in a way that achieved his overall strategic goals, particularly as they related to the Iraqi population. According to Franks, the challenges associated with the "day after the war" were being considered early in the planning process.[4]

On December 28, 2001, General Franks briefed his Commander's Concept to President George W. Bush and identified four major phases of operations. Each phase had specific end-state objectives that had to be achieved before moving to the next phase. Phase I included establishing international support and creating an "air bridge" that would be used to transport forces and capabilities into the theater. Phase

---

[2] Tommy Franks, *American Solider*, New York: Regan Books, 2004, p. 315.

[3] The Commander's Concept of Operations communicates the basic principles that should guide detailed planning efforts. As Franks describes, it contains "the philosophical underpinnings of what might eventually become a plan." See Franks, p. 329.

[4] Franks, pp. 340–341. The "Lines and Slices" matrix developed by General Franks in December 2001—which is reproduced in his book on p. 340—remained the foundation for campaign planning for OIF. As late as April 2003, CFLCC referenced a more refined variant of this template when it published a draft of OPLAN ECLIPSE II.

II was designed to "shape the battlespace" before ground operations were initiated. Phase III identified two primary goals: "regime forces defeated or capitulated" and "regime leaders dead, apprehended, or marginalized." Finally, Franks briefed his overarching concept for Phase IV, "post hostility operations," arguing that this phase would be the longest—"years, not months" in duration. In fact, the briefing chart used to discuss Phase IV split the arrow representing a timeline into segments and identified its duration as "unknown." The end-state objectives for Phase IV were identified as "the establishment of a representative form of government, a country capable of defending its territorial borders and maintaining its internal security, without any weapons of mass destruction."[5] While many details had yet to be resolved, the President approved the overarching concept of a four-phased campaign.[6]

On January 7, 2002, General Franks assembled his small group of planners engaged in compartmentalized planning for OPLAN 1003V and charged them with developing options in the event that President Bush decided to initiate an attack in response to action taken by Iraq. For the next three months, a handful of officers on the CENTCOM staff continued to plan for all four phases of OPLAN 1003V. As the Commander's Concept matured, it eventually involved a five-pronged attack with ground forces simultaneously advancing into Iraq from Kuwait and Turkey, special operations forces (SOF) moving into the western desert areas to prevent SCUD missiles from being employed, a comprehensive information and psychological operations (PSYOP) "front" being launched to erode the resolve of the Iraqi military, and an operational fires attack targeting Baghdad and the Republican Guard forces defending the city. At this point in the planning process, Phases I through III were expected to take up to 135 days.[7] A central part of the Phase IV plan was the continued deployment of additional forces until a sufficient number of forces were in theater to accomplish the mission. As General Franks stated to President Bush in February 2002, "[a]s stability operations proceed, force levels would continue to grow—perhaps to as many as two hundred and fifty thousand troops, or until we are sure we've met our end-state objectives."[8] This aspect of the plan is critically important to highlight, because military leaders in the operational headquarters and on the Army staff assumed that forces would continue to flow into theater even after the end of Phase III.

---

[5] Franks, p. 351.

[6] The briefing prepared for the President did not include a discussion of phasing. However, after General Franks concluded his briefing, Secretary of State Colin Powell asked a question that resulted in the need to refer to back-up slides detailing CENTCOM's phasing concept. It was not until Franks's February 3, 2002, briefing to the President that these slides on phasing made their way into the formal briefing.

[7] Franks, p. 366.

[8] Franks, pp. 361–363, 366, and 376–377.

It was not until March 21, 2002, that General Franks met with his component commanders for the first time to discuss the "shape and scope" of a potential military operation against Iraq. While the compartmentalized nature of the planning process continued, small planning staffs for each of the components were established to begin working in parallel on their supporting OPLANs.

On August 5, 2002, Franks again briefed President Bush and the National Security Council (NSC) on OPLAN 1003V. What had begun as a concept nine months earlier had now become a full campaign plan. The expected length of Phase III had been shortened to 90 days, but the time required to meet Phase IV objectives remained indeterminate. Franks raised what he called the potential for "catastrophic success," which would occur if the Iraqi military resistance fractured early in the campaign, if a military coup toppled Saddam Hussein, or if Shi'ite and Kurdish rebellions occurred in Iraq.[9] He expressed particular concern about the potential for lawlessness and violence in the immediate aftermath of military operations. There was wide agreement within the NSC that, if this occurred, coalition forces would continue with the campaign plan until order was restored and the Iraqis were able to govern themselves. At the conclusion of the briefing, Franks stated that he envisioned having a maximum of 250,000 troops in Iraq at the end of Phase III.[10] These forces would be necessary to help create a new Iraqi military and establish a constabulary force.[11] Franks further stated that "well-designed and well-funded reconstruction projects that put large numbers of Iraqis to work and quickly meet community needs—and expectations—will be the keys to our success in Phase IV." He made it clear that it was important to enable Iraqis to gain control of their own governance as soon as possible. Significantly, Franks told President Bush that the U.S. exit strategy had to be linked to effective Iraqi governance rather than to any artificial timeline, a conclusion not challenged by any member of the NSC.[12]

---

[9] Franks, pp. 380–393.

[10] Franks first mentioned that Phase IV might require as many as 250,000 troops in a February 2002 briefing to Secretary Rumsfeld. Franks, p. 366.

[11] It is important to note that the total number of troops in Iraq never reached 250,000. Moreover, as soon as U.S. forces marched into Baghdad, General Franks made the decision to stop the deployment of the 1st Cavalry Division, even though the end-state objectives of his plan had not yet been met. It is also unclear how it was determined that 250,000 troops would be sufficient for the tasks likely to be required in a postwar Iraq. As will be discussed later, U.S. military experience in postwar situations in Bosnia and Kosovo suggest that postwar operations require a ratio of 20 soldiers for every 1,000 inhabitants. With an Iraqi population of 25 million, historical experience suggests that a minimum of 500,000 soldiers would be required. That number is in addition to the numbers of police, gendarmerie, and carabinieri units that would be of vital assistance in maintaining public order. Obviously, the number of troops and police required for any post-conflict operation will be affected by a number of different factors, including population, terrain, assigned missions, level of violence and organized resistance, and the degree to which neighboring countries and the local populace are supporting the resistance. For a more detailed analysis of this subject, see James Dobbins et al., *America's Role in Nation-Building: From Germany to Iraq*, Santa Monica, CA: RAND Corporation, MR-1753-RC, 2003, pp. 149–153.

[12] Franks, pp. 392–393.

Franks and his planning staff had numerous discussions about what a postwar Iraq might look like and what type of interim political government might best be established. While members of his staff argued that rebuilding the infrastructure could not take place without security, Franks believed that security would not be achieved without reconstruction and civic action. There was, however, wide agreement that the two issues were inextricably linked.[13]

CENTCOM assumed it would help the interim government establish a paramilitary security force that would be drawn from some of the better units of the defeated Iraqi army. It was envisioned that these units would work in conjunction with coalition military forces to help restore order and prevent armed conflicts among ethnic, religious, or tribal factions. Franks concluded that "this model had been used effectively in Afghanistan," and CENTCOM planners believed it was the best solution they had to the challenge of providing security in the immediate aftermath of major combat operations.[14] Questions remained, however, about the effectiveness of the Afghanistan model, especially outside Kabul, and whether that model would work in a country that is largely urbanized, has clear fault lines among three major ethnic and religious groups, and had no clear successor government.

CENTCOM planners believed that additional forces would be required for Phase IV operations. However, there was substantial agreement among the planning staffs in theater that surrendering Iraq divisions and corps could be quickly employed to help maintain security, and that they would facilitate a rapid transition back to Iraqi control. The final version of OPLAN 1003V assumed that these surrendering Iraqi forces would be available to the coalition, and that additional U.S. forces would continue arriving in theater as stability operations commenced after Phase III, or, as the OPLAN stated, "until we are sure we've met our end-state objectives."[15]

After OPLAN 1003V was published in February 2003, the CENTCOM future planning staff began developing a detailed plan for Phase IV operations called OPLAN IRAQI RECONSTRUCTION. It was developed in collaboration with the planning staffs of the main ground force units in Iraq, the 1st Marine Expeditionary Force (I MEF) and the U.S. Army's V Corps. This plan identified seven focus areas for the postwar period: maintain the rule of law, provide security, support civil administration, provide necessary assistance to civilian governance, maintain and enlarge the coalition, provide emergency humanitarian assistance as required, and assist in the assessment, restoration, and repair of critical life support infrastructure.[16]

---

[13] Franks, pp. 422 and 424.

[14] Franks, p. 419.

[15] Franks, p. 366.

[16] Interview with CENTCOM official, May 2005. See also Kenneth R. Timmerman, "Details of the Postwar Master Plan," *Insight on the News,* November 24, 2003.

OPLAN IRAQI RECONSTRUCTION suffered from three significant limitations. First, it was completed at the end of April 2003, after the fall of Baghdad and after subsequent looting destroyed much of the infrastructure left intact by the military campaign and expanded the range of reconstruction requirements. Second, it lacked any additional resources; the forces and capabilities that had conducted major combat operations were the ones that would be available for postwar stability and reconstruction operations. Finally, the plan envisaged military forces supporting civilian reconstruction efforts, not playing the lead role. Yet, as will be discussed in Chapter Five, civilian reconstruction efforts suffered from unchallenged assumptions, a very short planning time, and a lack of coordination with the military.

Combined Forces Land Component Command (CFLCC) Phase IV Planning

As specified in OPLAN 1003V, the Combined Forces Land Component Command (CFLCC) was given primary responsibility for Phase IVa—stability operations. Lieutenant General Paul Mikolashek, then the commander of the U.S. Third Army, which had been designated the CFLCC, was informed of his responsibilities for Phase IV when the Third Army was directed to begin planning in January 2002, just days after General Franks met with President Bush at Crawford, Texas. Colonel Kevin Benson, who had been selected to be the Director of Policy, Plans and Strategy (C5), was given access to the developing warplans.[17] He immediately concluded that the CFLCC needed to put more emphasis on Phase IV planning and, consequently, spent the next two months making contact with people—inside both government and academia—who were involved in projects that examined issues associated with postwar Iraq.[18]

The C5 also led a group of strategic planners who developed CFLCC's supporting plan for OPLAN 1003V—called OPLAN COBRA II. From the very beginning, COBRA II was an all-inclusive OPLAN that addressed all aspects of the ground campaign, beginning in Phase I and continuing through Phase IV redeployment. During the first few months, however, access to OPLAN 1003V was limited to a handful of senior officers at CFLCC. Colonel Benson was the only officer below the rank of brigadier general who was "read-on" to OPLAN 1003V. Consequently, the CFLCC C5 planning team was constrained in its efforts to build a comprehensive

---

[17] Within Joint and Combined Headquarters, the J5 and C5 respectively provide political-military oversight for all aspects of the operation, including interacting with the host nation, nongovernmental organizations, civil affairs, and other U.S. government and coalition civilian organizations. For a more detailed description of the duties and responsibilities of the J5 and C5, see Joint Publication 5-00.2, *Joint Task Force Planning Guidance and Procedures*, January 13, 1999.

[18] Interview with CFLCC official, August 2004.

supporting plan.[19] In October 2002, the classification of OPLAN 1003V was downgraded to Secret and a larger number of the CFLCC planning staff gained access to it.

As CFLCC started to ramp up for the possibility of war, the size and composition of its staff, including the planning staff, grew. By January 2003, nine graduates from the Army's School of Advanced Military Studies (SAMS) had been assigned to CFLCC because of their unique skills.[20] In fact, the Army transferred many of these officers from other duty assignments well in advance of their normal rotation dates specifically because of their schooling.[21]

As we shall discuss in Chapter Four, a newly created Task Force IV (TFIV) headquarters, commanded by Brigadier General Steven Hawkins, forward deployed in late January 2003 from Tampa to Kuwait and was placed under the operational control of the CFLCC. TFIV had been established by order of the Chairman of the Joint Chiefs of Staff (CJCS), who intended it to be the nucleus around which Phase IV operations would be planned and conducted. However, by the time TFIV arrived in theater, CFLCC had already done considerable planning for post-conflict operations, and there was considerable confusion over exactly what this new headquarters would do.[22]

CFLCC's planning staff envisioned that TFIV would only assume responsibilities for Phase IV operations after the CFLCC redeployed to the United States. However, the commander of TFIV believed he was given full responsibility for Phase IV planning. As a result, the CFLCC C5 and TFIV conducted parallel planning for the same mission. Little direct coordination occurred between the two planning staffs, though the TFIV commander met regularly with the C5 and the deputy commander of CFLCC to share information. It is important to note that as late as mid-February 2003, the exact role of TFIV was undefined. On February 15, 2003, Lieutenant General David McKiernan (who had taken command of U.S. Third Army in September 2002) was presented with two options for employing TFIV: embedding the task force within a three-star headquarters or building a three-star headquarters around TFIV. The ultimate decision was to do neither: TFIV was disbanded and the Phase IVb mission was assigned to V Corps.[23]

Once COBRA II was completed, the C5 began a series of wargaming efforts to test the plan's assumptions and to identify any potential shortcomings that could be

---

[19] This is a common occurrence in the development of OPLANs.

[20] Interview with CFLCC official, August 2004.

[21] Lieutenant Colonel Steven W. Peterson, "Central but Inadequate: The Application of Theory in Operation Iraqi Freedom," a paper presented in partial completion of the course of study at the National Defense University, undated, p. 3.

[22] Interview with CFLCC official, August 2004.

[23] At mission assumption, this military headquarters became known as Combined Joint Task Force 7 (CJTF-7).

rectified before the initiation of hostilities. To facilitate this wargaming effort, Lieutenant Colonel Steven Peterson, the chief of intelligence planning within C5, directed an aggressive "Red Team" effort.[24] By the middle of February 2003, thanks in part to this effort, some in the C5 staff concluded that, as planned, "the campaign would produce conditions at odds with meeting the strategic objectives" established by CENTCOM.[25] Members of the C5 staff further concluded that the "joint campaign was specifically designed to break all control mechanisms of the regime and that there would be a period following regime collapse in which [CFLCC forces] would face the greatest danger to [U.S.] strategic objectives." The assessment went on to describe the "risk of an influx of terrorists to Iraq, the rise of criminal activity, probable actions of former regime members, and the loss of [any weapons of mass destruction] that was believed to exist."[26] This assessment did not foresee all challenges that would confront CFLCC during the transition from Phase III to Phase IV, but it did identify a number of actions that needed to be addressed in the OPLAN, including "planning to control the borders, analyzing what key areas and infrastructure should be immediately protected, and allocating adequate resources to quickly re-establish postwar control throughout Iraq."[27]

Although a complete consensus within the C5 staff concerning the magnitude of the problem could not be reached, it was clear that if the ground war went as fast as some expected, and if the regime collapsed suddenly, then coalition forces most likely would not be in place where needed for Phase IV stability operations.[28] When this assessment was communicated to the commander of CFLCC, C5 staff members reported that Lieutenant General McKiernan chose not to change the ground combat plan that had been developed with CENTCOM and subordinate commands because the Iraqi military remained the greatest immediate threat to coalition forces. After they briefed CFLCC commander Lieutenant General McKiernan in February, no one on the staff was willing to bring the issue back to the surface later and make the argument that combat forces needed for the warfighting effort should be repositioned in order to better prepare the forces to respond to situations that *might* occur after the battle had been won.[29] Although members of the planning staff recognized that forces would have to conduct major repositioning at the end of major combat to be in place for their assigned Phase IV missions, their first priority remained the destruction of Iraqi forces, and they believed that the existing plan offered the best way

---

[24] Interview with CFLCC official, August 2004.

[25] Peterson, p. 11.

[26] Peterson, p. 10.

[27] Peterson, p. 10.

[28] Interview with CFLCC official, January 2005.

[29] Peterson, pp. 10–11.

to accomplish this objective. In essence, sufficient forces were not available to conduct both missions simultaneously. In the end, Lieutenant General McKiernan and his staff decided to accept risk during the early stages of Phase IV rather than during major combat operations.[30]

In fact, the CFLCC planners had correctly anticipated that coalition forces would not be in position to address the immediate security challenges brought about by the collapse of the Iraqi government. As anticipated and planned for in ECLIPSE II, a major repositioning of forces was required during the latter part of April and early May to ensure even a minimum level of security throughout the country.[31]

Even though the C5 staff was unwilling to press CFLCC's commander to fundamentally reshape the conduct of Phase III operations, their analysis of Phase IV did convince Lieutenant General McKiernan to create a sequel OPLAN in the event that the end-state conditions envisioned for Phase III did not materialize. With CFLCC's approval, the C5 began writing OPLAN ECLIPSE II as a sequel to the existing OPLAN. The plan had been through 15 revisions by the middle of March 2003, and the final coordinating draft was released on April 12.[32] As with all other OPLANs, the subordinate and supporting commands all produced their own annexes to this sequel. Each draft plan was shared with V Corps as it was being written, in order to inform the corps' planning efforts for Phase IV. This type of distributed and parallel planning for Phase IV ensured that V Corps and I MEF—along with their subordinate divisions and brigades—were near-equal partners in the planning process.[33]

While it is clear that CFLCC was gaining a realistic appraisal of the potential security challenges that would confront coalition forces in the postwar period, the planning staff never formally challenged some of the basic assumptions outlined in COBRA II. For example, ECLIPSE II assumed that Iraqi civil authorities would "continue to run local and regional essential services." Moreover, although this was not listed as a planning assumption, CFLCC envisioned that civil order would be controlled "through published proclamations and the existing legal system where possible." ECLIPSE II also overestimated the degree to which the remnants of the Iraqi government would provide essential services and security in the immediate aftermath of major combat operations. Consequently, military resources were allocated based upon a fundamentally flawed view of both the friendly and enemy situations that would exist during the transition from Phase III to Phase IV. It would, however, be inaccurate to conclude that military planners accepted the rhetoric coming out of

---

[30] Interview with CFLCC official, January 2005.

[31] See Chapter Seven for more on the repositioning of these forces.

[32] Coalition Forces Land Component Command (CFLCC) OPLAN ECLIPSE II BASE PLAN, Coordinating Draft, April 12, 2003.

[33] Interview with V Corps and CJTF-7 official, January 2005.

certain portions of the policy community in Washington that military forces would be welcomed with open arms. Indeed, they had numerous discussions about the possibility of an insurgency. However, they did not believe that the potential insurgency against U.S. forces and the future Iraqi government would affect any of the courses of action being developed.[34]

Planning at V Corps and Subordinate Commands

It is important to note that planning for Phase IV operations occurred at all levels of military command down to division and sometimes brigade level. Modern military planning occurs simultaneously at various levels of command through a distributed and collaborative planning process. This type of distributed planning is more difficult once units have forward deployed into a theater of operations. For OIF, however, division battle staffs were brought into the planning process during the summer of 2002. As noted earlier, these plans were developed without any clear strategic guidance for integrating political, economic, and military efforts for stabilization and reconstruction. Military commanders at all levels were unprepared for the magnitude of the policy and legal issues that they confronted during the initial stages of Phase IV operations.

V Corps also faced a unique challenge associated with transforming itself into a combined joint task force for Phase IVb. Until the beginning of combat operations in March, no decision had been made about which headquarters would assume responsibility for Phase IV operations once CFLCC redeployed at the conclusion of Phase IVa. Thus, while V Corps was fighting the war during Phase III and conducting stability operations during Phase IVa, it also had to reorganize its staff to become a CJTF headquarters. The lateness of this decision suggests the degree to which senior civilian and military leaders within DoD underestimated the challenges that would confront coalition military forces after the defeat of Iraqi forces.

Observations

The coalition achieved a quick and decisive military victory that resulted not only in the destruction or collapse of the Iraqi military, but also in the disintegration of the command and control capacity of the Iraqi government. Despite the success it produced on the battlefield, the planning process failed to include sufficient flexibility to enable the CFLCC to respond rapidly to Phase IV security requirements.

---

[34] Interview with CFLCC official, February 2005.

Although collaborative, iterative, and continuous planning took place at all levels of command, no record exists showing that CENTCOM or CFLCC participated in a similar process with civilian governmental agencies, international organizations, or nongovernmental organizations; military planners believed such collaboration would not be necessary for stability, reconstruction, and transition activities to succeed. Additionally, and not inconsequentially, CENTCOM and OSD did not systematically assess the forces and capabilities that would be required for Phase IV operations.[35] Finally, the decision by General Franks and Secretary Rumsfeld to stop the deployment of the 1st Cavalry Division and other reinforcing forces when coalition forces entered Baghdad further exacerbated a shortfall in the number of troops required to simultaneously complete Phase III and begin Phase IV.

As noted earlier, the CENTCOM plan specified that the force level would continue to grow until end-state objectives had been met. The decision to stop the flow of forces into theater prior to that point was, in effect, a change to the plan that had been agreed to by subordinate commanders. The planning staffs within CENTCOM, CLFCC, and V Corps were never asked to provide input for this change. According to one officer, the V Corps staff would have preferred for the force flow to continue as originally envisaged.[36]

The potential for unforeseen challenges during Phase IV was anticipated prior to the initiation of hostilities. General Franks first expressed his concern about the possibility for "catastrophic success" in a meeting with President Bush and the NSC on August 5, 2002. According to Franks, he and Secretary Rumsfeld used this phrase to refer to the possibility that large-scale combat operations could be over much sooner than anyone imagined and, consequently, that functional plans and policies needed to be created in Washington to be prepared for the "occupation and reconstruction."[37] In essence, with the exception of immediate security concerns and actions necessary for emergency restoration of critical infrastructure, the majority of activities required for Phase IV were perceived by the Department of Defense to be the responsibility of civilian agencies and departments. The Department of Defense, of course, was responsible for providing resources to support these efforts—to the extent they were available and did not conflict with ongoing military operations.

---

[35] According to one official, OSD did not conduct an estimate of the number of troops required for Phase IV operations because that was the responsibility of CENTCOM planners and the Joint Staff. However, senior officials within OSD questioned why more troops might be required for Phase IV operations than were needed for major combat operations. For example, Deputy Secretary of Defense Paul Wolfowitz testified to Congress that General Eric Shinseki's estimate of "several hundred thousand soldiers" to conduct post-conflict operations was "wildly off the mark." Wolfowitz went on to state that he found it "hard to conceive that it would take more forces to provide stability in a post-Saddam Iraq than it would take to conduct the war itself and to secure the surrender of Saddam's security forces and his army—hard to imagine." See testimony of Deputy Secretary of Defense Paul D. Wolfowitz to the House Budget Committee, February 27, 2003.

[36] Interview with V Corps and CJTF-7 official, January 2005.

[37] Franks, pp. 392 and 442.

The CFLCC planning staff had come to similar but more ominous conclusions. From their wargaming efforts in January and February 2003, members of the C5 planning staff had come to believe that not only was there a possibility that warfighting would be over much faster than anyone had anticipated, but a possibility also existed that the operation would cause disintegration of the entire Iraqi regime. They further concluded that this governmental collapse could result in some form of civil unrest, lawlessness, or a rise in acts of terrorism. Finally, the C5 planning staff concluded that the rapid advance of maneuvering units would place ground forces in positions that might not enable them to adjust rapidly to the immediate postwar situation. A troop-to-task analysis was conducted to determine how many brigade equivalents would be necessary to maintain security, but the analysis was predicated on the existence of an effective local police force. No one anticipated, and therefore no one planned for, the requirements for maintaining security in the absence of local police. Most importantly, the troop-to-task analysis was constrained by the number of battalion- and brigade-sized units available in theater to conduct Phase IV operations. According to one official, an unconstrained analysis to determine requirements was not undertaken.[38] In effect, planners took the forces that they had available and spread them throughout Iraq, even though the number of forces were insufficient for simultaneous combat and stabilization operations during the transition from Phase III to Phase IV.

If the possibility of "catastrophic success" was shared by both CENTOM and CFLCC, why were they not prepared for the postwar looting and violence that took place in Baghdad and other urban areas in Iraq in the immediate aftermath of major combat operations? While critics have argued that this was the result of a lack of planning, it is clear that detailed Phase IV planning was conducted at CENTCOM, CFLCC, V Corps, and subordinate commands. Both OPLAN 1003V and ECLIPSE II established end-state conditions and directed specific actions to ensure a successful transition to post-combat operations. In fact, having been designated the supported command within CENTCOM for Phase IV operations, CFLCC developed OPLAN ECLIPSE II specifically to address Phase IV operations as a sequel to major combat operations. Thus, it was not a lack of planning for Phase IV that led to the coalition's military forces being unprepared for the immediate postwar challenges. Instead, problems arose from the *ineffectiveness* of the planning process in identifying the likely requirements for the transition from Phase III to Phase IV, and the failure to challenge set assumptions about what postwar Iraq would look like. Thus, the real question becomes: Why was the planning process that resulted in the quick and decisive defeat of the Iraqi military so ineffective in preparing for postwar operations?

---

[38] Interview with CFLCC official, January 2005.

The full answer to this question requires an analysis of civilian policy and planning within and among the numerous U.S. governmental agencies and departments with responsibilities relating to reconstruction. As General Franks pointed out, decisions and plans had to be made within the NSC in advance of the war to enable the full weight of U.S. government capabilities to be brought to bear quickly on stability and reconstruction tasks in Iraq. As will be discussed in Chapters Three through Six, prewar interagency planning and collaboration fell far short of what was necessary.

As depicted in Figure 2.1, the number of troops on the ground in Iraq, and in Baghdad, at the beginning of Phase IV was far less than those deployed in either Bosnia or Kosovo.[39] According to a RAND report published in 2003, if the levels of troops committed in Kosovo are used as a guide, 526,000 troops would have been needed to address immediate postwar security concerns in Iraq.[40] Drawing from his

Figure 2.1
Military Presence at Outset of Post-Conflict Operations

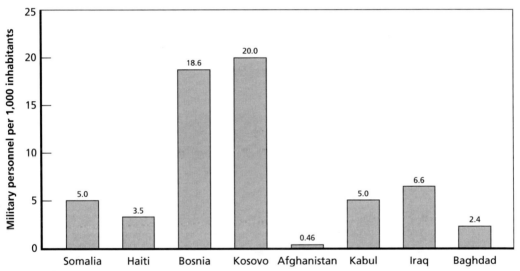

SOURCE: James Dobbins et al., *America's Role in Nation-Building*, Santa Monica, CA: RAND Corporation, MR-1753-RC, 2003.
RAND *MG642-2.1*

---

[39] The numbers used for Somalia, Haiti, Bosnia, and Kosovo are taken from another RAND study. Deviating from that study here, we have combined the total number of peacekeepers in Kabul with the 8,000 members of the U.S. military operating elsewhere in Afghanistan where they are conducting combat operations to arrive at the ratio of 0.46 (rather than the 0.18 one derives if one only counts the international peacekeeping forces). The ratio for Kabul is computed using the 5,000-man international peacekeeping force compared with an estimated population of 1 million. The ratio for Baghdad is based on a population figure of 6.2 million, and it assumes that six brigade equivalents could muster no more than 15,000 soldiers for active security responses.

[40] Dobbins et al., p. 197. For a broader discussion about required force sizes, see James T. Quinlivan, "Force Requirements in Stability Operations," *Parameters*, Winter 1995–96, pp. 59–69.

personal experience in the Balkans and a number of studies produced by and for the Army prior to the initiation of war with Iraq, then-Army Chief of Staff General Shinseki concluded that it would require approximately 400,000 soldiers to maintain security in the immediate aftermath of major combat operations.[41]

Anecdotal evidence suggests that other Army general officers shared General Shinseki's main concern, namely, that more troops would be needed immediately following combat with Iraqi forces than would be required to defeat those same forces; however, none spoke up publicly before the war.[42] What has become clear is that the military was unprepared for the immediate aftermath of the war. Perhaps the commander of V Corps, Lieutenant General William Wallace, summed it up best when he stated:

> The military did their job in three weeks. I give no credit to the politicians for detailed Phase Four planning. But I don't think that we, the military, did a very good job of anticipating [that] either. I don't think that any of us either could have or did anticipate the total collapse of this regime and the psychological impact it had on the entire nation. When we arrived in Baghdad, everybody had gone home. The regime officials were gone; the folks that provided security of the ministry buildings had gone; the folks that operated the water treatment plants and the electricity grid and the water purification plants were gone. There were no bus drivers, no taxi drivers; everybody just went home. I for one did not anticipate our presence being such a traumatic influence on the entire population. We expected there to be some degree of infrastructure left in the city, in terms of intellectual infrastructure, in terms of running the city infrastructure, in terms of running the government infrastructure. But what in fact happened, which was unanticipated at least in [my mind], is that when [we] decapitated the regime, everything below it fell apart.[43]

The sentiment expressed by both Generals Shinseki and Wallace has also been echoed by L. Paul Bremer, the administrator for the Coalition Provisional Authority (CPA). On October 4, 2004, Bremer stated that the United States made two major mistakes in Iraq: not deploying a sufficient number of troops and not adequately controlling the looting and lawlessness that ensued in the immediate aftermath of major combat operations. Bremer continued by saying, "We paid a big price for not stopping [the looting] because it established an atmosphere of lawlessness."[44] In an

---

[41] Interview with James Fallows for *Frontline: The Invasion of Iraq.* As of August 2007: http://www.pbs.org/wgbh/pages/frontline/shows/invasion/interviews/fallows.html

[42] For more information on this point, see Chapter Four.

[43] Interview with General William Scott Wallace for *Frontline: The Invasion of Iraq.* As of August 2007: http://www.pbs.org/wgbh/pages/frontline/shows/invasion/interviews/wallace.html

[44] Robin Wright and Thomas E. Ricks, "Bremer Criticizes Troop Levels: Ex-Overseer of Iraq Says U.S. Effort Was Hampered Early On," *Washington Post*, October 5, 2004.

earlier speech delivered on September 17, 2004, Bremer is reported as having said: "The single most important change—the one thing that would have improved the situation—would have been having more troops in Iraq at the beginning and throughout" the entire occupation of Iraq.[45]

The U.S. military experience in Iraq points to a shortcoming in U.S. military doctrine, which is predicated on existing military theory. Defeating enemy military forces must always remain a focus of military doctrine, but picking the right "end state" is crucial for planning. More emphasis needs to be placed on what it takes to "win the war"—from a broader context than simply defeating the enemy's military. The success of the ground campaign in OIF demonstrates the importance of theory as an effective guide to how battles should be prosecuted. However, the warfighting bias in existing theory leaves the military ill prepared to complete the overarching task of winning the peace—the ultimate objective of war.

---

[45] Wright and Ricks.

CHAPTER THREE

Civilian Planning Efforts

In addition to the U.S. military, several civilian government agencies invested time and effort in thinking about the challenges of postwar Iraq during 2001 and 2002. This chapter starts by examining the official interagency process that guided postwar planning for Iraq, which started in the summer of 2002. It then examines the work of the four government agencies that conducted the broadest planning efforts for postwar governance and reconstruction: the Office of the Secretary of Defense (OSD), the State Department, the U.S. Agency for International Development (USAID), and the National Security Council (NSC) staff. It concludes by examining several analyses of postwar requirements in Iraq, which were sponsored by think tanks and academic organizations in the months before the war and which were intended to inform and influence U.S. government planning efforts.

Interagency Planning: The ESG and the IPMC

During the spring and summer of 2002, CENTCOM and the Joint Staff ran a series of exercises as part of the war planning effort.[1] The Joint Staff wanted to include the interagency community in these exercises, but there was no standing interagency structure to plug into the process. Lieutenant General George Casey, then the Director for Strategic Plans and Policy (J5) on the Joint Staff, developed a structure for a new coordination body, called the Executive Steering Group (ESG), that would bring together representatives from the Joint Staff, the State Department's Bureau of Political-Military Affairs, the Office of the Under Secretary of Defense for Policy (OUSD(P)), the Office of the Vice President (OVP), and the Central Intelligence

---

[1] Prominent Hammer was one of the most important of these exercises; it was sponsored by CENTCOM in March 2002 to assess the feasibility of the emerging warplans. Bob Woodward, *Plan of Attack*, New York: Simon and Schuster, 2004, p. 114; Rowan Scarborough, *Rumsfeld's War*, Washington, D.C.: Regnery Publishers, 2004, pp. 45 and 175.

Agency (CIA).[2] The ESG started meeting in August 2002 under the chairmanship of Frank Miller, the NSC official to whom the Deputies Committee (see footnote 4) had delegated responsibility for Iraq policy.[3]

The ESG addressed a wide range of political-military planning issues, such as securing basing, access, and overflight rights, accelerating military construction, and identifying allied capabilities that could contribute to military operations. These issues formed the core of the interagency agenda, and if the ESG members could not reach agreement, the issue would be passed to the Deputies Committee (and on to the Principals Committee, if necessary) for resolution.[4] The ESG met three times a week from the time that it was formed until the war started in March 2003, and its membership grew as time went on.[5] By early 2003, its meetings regularly included representatives from other offices in the State Department (including the Bureau of European and Eurasian Affairs and the Bureau of Near Eastern Affairs), the Comptroller's office from OSD, OVP, the Joint Staff Directorate of Logistics (J4), and various NSC offices when specific agenda items addressed their areas of responsibility.[6] Such frequent meetings were needed to prepare for the twice-weekly Deputies Committee meeting and the weekly Principals Committee meeting on Iraq.[7]

The interagency structure proposed by Lieutenant General Casey also created several other organizations that supported the work of the ESG. To secure agreement on strategic planning guidance received from higher levels, the Iraq Political-Military Cell (IPMC) sat directly below the ESG and brought together working-level officials from each of the agencies participating in the ESG. It is important to note that the IPMC was never intended to do any independent planning; instead, it would enable agencies throughout the U.S. government to conduct their own planning efforts within a coherent strategic framework.[8] Unlike the ESG, the IPMC had a dedicated staff to oversee the coordination process. Normally the NSC would have been responsible for such a coordination function, but its small staff size prevented it from dedicating many personnel to this organization. Instead, the Joint Staff provided

---

[2] "Pre-war Planning for Post-war Iraq," information sheet published by the Office of Near East and South Asian Affairs within the Office of the Under Secretary of Defense for Policy. As of April 2004:
http://www.defenselink.mil/policy/sections/policy_offices/isa/nesa/postwar_iraq.html

[3] Peter Slevin and Dana Priest, "Wolfowitz Concedes Errors on Iraq," *Washington Post*, July 24, 2003; interview with U.S. government official, April 2004.

[4] The Principals Committee includes the Secretary of Defense, the Secretary of State, and the National Security Advisor; the Deputies Committee, as its name suggests, consists of their deputies.

[5] The ESG stopped meeting when OIF began, but after major combat operations ended it was reconstituted to address emerging postwar issues. Interview with NSC official, April 2004.

[6] Interview with NSC official, April 2004; "Pre-war Planning for Post-war Iraq."

[7] Interview with NSC official, April 2004, and OSD official, August 2004.

[8] Interview with NSC official, April 2004; "Pre-war Planning for Post-war Iraq."

most of the IPMC staff members, since it contained action officers who could be re-assigned relatively easily.[9]

Several working groups, reporting to the ESG, were established to focus on specific issues. The Iraq Relief and Reconstruction working group (IR+R) had primary responsibility for postwar issues, although some of the other working groups touched on these issues during their work. IR+R was co-chaired by Elliott Abrams from the NSC and Robin Cleveland from the Office of Management and Budget; it included representatives from the OSD Office of Stability Operations, the Joint Staff J5, the State Department's Bureau of Population, Refugees, and Migration (PRM), USAID, the CIA, and occasionally other agencies such as the Treasury and Justice Departments.[10] As its name suggests, IR+R focused on providing humanitarian relief in the immediate postwar period, as well as on reconstruction over the longer term.[11]

The ESG, IPMC, and the various working groups represented a fairly significant effort to reach interagency consensus on strategic guidance for Iraq. Nevertheless, this structure ended up being less effective than it might otherwise have been, for two main reasons. First, ESG meetings were not always attended by every agency, or by the same representatives of each agency. CIA representatives missed many meetings, for example; one official speculated that this might have stemmed from the distance of their headquarters in Langley.[12] Attendance from OSD offices was reportedly inconsistent, with different people representing organizations at different times and with representatives often lacking the seniority necessary to make decisions.[13] Interagency coordination suffered as a result, because OSD was the only organization that would link those planning the war with those responsible for issuing strategic guidance. Other DoD offices arrived at meetings with different positions, rather than one coordinated DoD policy.[14]

Second, postwar issues were not well coordinated through this structure. The IPMC spent much of its time working on postwar issues, using the work of the IR+R working group as a basis for its efforts. However, the ESG focused primarily on war planning issues and devoted much less attention to postwar planning.[15]

---

[9] Interviews with CENTCOM official, November 2003, and NSC official, April 2004.

[10] "Pre-war Planning for Post-war Iraq."

[11] Interviews with OSD officials, November 2003 and March 2004, and NSC official, April 2004.

[12] Interview with NSC official, April 2004.

[13] Interviews with NSC and Joint Staff officials, April 2004. The reasons for OSD's inconsistent representation remain unclear. One official noted that DoD officials generally displayed ambivalent feelings about the interagency process—that it provided a useful mechanism for gathering support from other agencies but required DoD to cede too much control—and speculated that this may have been one reason that OSD did not make a firmer commitment to participating in this process.

[14] Woodward, p. 321.

[15] Interview with NSC official, April 2004.

The Office of the Secretary of Defense

From the fall of 2001 through the spring of 2002, planning within the Department of Defense for possible combat and post-combat operations in Iraq was concentrated in the hands of the department's senior leadership. Secretary of Defense Donald Rumsfeld and Deputy Secretary of Defense Paul Wolfowitz were deeply involved in the development of the warplans during this time, but few other DoD officials were involved.[16] As the interagency process geared up during the summer of 2002, the planning process within the Pentagon expanded to include many other defense officials, including those within OSD. OSD representatives participated in the interagency meetings that started in July 2002, including the humanitarian relief planning efforts that are discussed in Chapter Six.

The Office of Special Plans

In August 2002, OSD renamed and expanded an existing office to handle the increased workload associated with potential military operations in Iraq. According to the DoD organizational structure depicted in Figure 3.1, the Office of the Under Secretary of Defense for Policy (OUSD(P)) is responsible for formulating national security and defense policy. Within OUSD(P), the Assistant Secretary of Defense for International Security Affairs (ISA) focuses on four regions of the world, including the Near East and South Asia (NESA).[17] Within NESA, the Office of Northern Gulf Affairs was responsible for coordinating policy toward Iraq and Iran. Northern Gulf Affairs had two people working on Iraq when the summer started, and it quickly became clear that additional personnel would be needed to handle the ever-expanding workload and interagency coordination. In August, the Office of Northern Gulf Affairs was renamed the Office of Special Plans (OSP) and augmented with more than a dozen temporary personnel.[18] Its responsibilities included developing policy recommendations for a wide range of issues, including coalition building, troop deployments, government reorganization, de-Ba'athification, maintaining the oil sector, training a police force, and war crimes prosecution.[19]

---

[16] Woodward.

[17] More information about the responsibilities of OUSD(P) and ISA can be found at the following web site. As of August 2004:
http://www.defenselink.mil/policy/#

[18] Interviews with OSD officials, August 2004.

[19] "Pre-war Planning for Post-war Iraq"; Dana Priest, "Feith's Analysts Given a Clean Bill of Health," *Washington Post*, March 14, 2004.

Figure 3.1
OUSD(P) and NESA Organization Chart

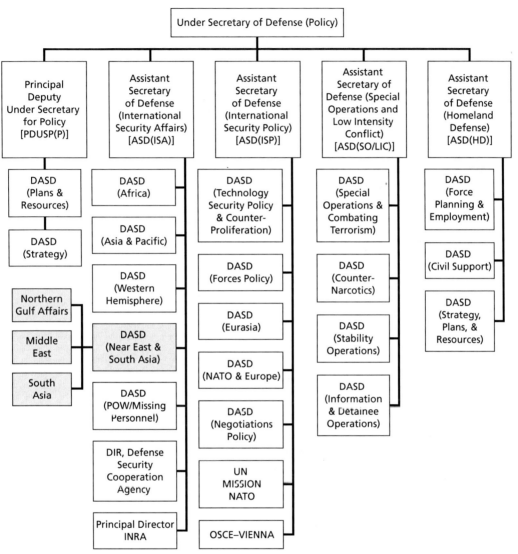

SOURCE: Department of Defense.
RAND MG642-3.1

Its nondescript name was chosen deliberately to avoid drawing attention to the fact that the Pentagon was considering the possibility of war and its aftermath in Iraq while simultaneously seeking international support at the United Nations.[20] As Douglas Feith, the Under Secretary of Defense for Policy, noted after the war, "The Special Plans Office was called Special Plans because, at the time, calling it Iraqi Planning Office might have undercut our diplomatic efforts."[21] The name provoked a lot of criticism from observers, who wondered why Iraq planning was occurring in such a seemingly secretive environment and what such an ambiguously named office was really working on. One OSD official later noted that the name caused such a stir that it might have been better to give the office a more substantive name, despite any possible diplomatic repercussions.[22]

OSP has been mistaken in the press for the Counter Terrorism Evaluation Group (CTEG), which was a two-person cell designed to review intelligence produced on terrorist networks.[23] The CTEG was established within OUSD(P) in October 2001, with the mission of investigating the linkages between international terrorism, state sponsors, and Iraq. CTEG proved to be a very controversial organization, with both the House Permanent Committee on Intelligence and the Senate Select Committee on Intelligence investigating its role in prewar intelligence.[24] For the purposes of this report, however, it is important to note that OSP was not linked to the CTEG, and was not involved in intelligence activities.

Policy Guidance

OSP developed policy guidance for the Secretary of Defense on a wide range of issues. One policy area that consumed a great deal of time and energy was postwar governance and, specifically, the process through which an interim government could be established. The key debate centered around the question of "externals versus internals"—whether the interim government should be constituted with exiles who had lived outside Iraq, or with people who had remained inside the country and endured Saddam's rule. OSP staff recommended that the interim government incorporate mostly externals because their objective was to turn power over to the Iraqis as

---

[20] Interviews with OSD officials, August 2004. The United Kingdom was also concerned about this tension. After major combat ended, the British government noted, "We felt that overt planning for the post-conflict [phase] would make it appear that military action was inevitable (which it was not) and could seriously prejudice ongoing attempts to reach a diplomatic solution." House of Commons Defence Committee, *Lessons of Iraq: Government Response to the Committee's Third Report of Session 2003–04*, May 26, 2004, p. 41. See also United Kingdom Ministry of Defence, *Operations in Iraq: Lessons for the Future*, December 2003, p. 61.

[21] Douglas Feith, quoted in Mark Fineman, Robin Wright, and Doyle McManus, "Preparing for War, Stumbling to Peace," *Los Angeles Times*, July 18, 2003.

[22] Interview with OSD official, August 2004.

[23] Articles that confuse these two offices include Paul Harris, Martin Bright, and Ed Helmore, "U.S. Rivals Turn on Each Other as Weapons Search Draws a Blank," *The Observer* (London), May 11, 2003; Seymour M. Hersh, "Selective Intelligence," *The New Yorker*, May 12, 2003; Julian Borger, "The Spies Who Pushed for War," *The Guardian* (London), July 17, 2003.

[24] Dana Priest, "Pentagon Shadow Loses Some Mystique," *Washington Post*, March 13, 2004.

quickly as possible. OSP staffers concluded that it would take a long time to identify and vet internals to ensure that no Ba'athists participated in the new government; they argued that it was better to stand up a government of externals quickly—possibly even in exile before the war started—rather than wait for this process to unfold after the war.[25] However, they were not able to secure interagency agreement on this point because agencies including State and the CIA argued that an interim government composed solely of externals would have no domestic legitimacy.[26]

As part of its focus on governance issues, OSP spent a lot of time working with exile and Kurdish opposition groups. Staff members participated in a number of meetings with such groups in the late summer and fall of 2002.[27] These meetings helped pave the way for the major conferences of opposition groups held in London in December 2002 and in northern Iraq at the end of February 2003.[28] The office was also involved in efforts to create the Free Iraqi Forces (FIF), a military force of exiles that would be under CENTCOM's command and trained to fight alongside U.S. forces during the war in Iraq. OSP officials thought that the FIF would be helpful on the battlefield practically, by serving as a focal point for Iraqi defections, and symbolically, showing Iraqis liberating their own country. Although several hundred potential FIF members were trained in Hungary, less than a hundred of them deployed to Iraq.[29] In addition to the lack of willing Iraqi volunteers, the FIF suffered

---

[25] Interviews with OSD officials, August 2004.

[26] OSD officials later expressed a great deal of frustration with this decision, arguing that the interim government that took power on June 28, 2004, was composed exclusively of exiles. They argued that the same government could have been in place in the fall of 2003 but for interagency opposition to their plans, and the faster transition would have resulted in much less opposition to the continuing U.S. presence in Iraq. Interviews with OSD officials, August 2004.

[27] Interviews with OSD officials, August 2004. Some of these meetings were co-sponsored with the State Department. See Peter Slevin, "Iraqi Opposition Pledges Anti-Hussein Unity," *Washington Post,* August 10, 2002; Michael R. Gordon, "Iraqi Opposition Groups Meet Bush Aides," *New York Times,* August 10, 2002; Michael R. Gordon, "Iraqi Opposition Gets U.S. Pledge to Oust Hussein for a Democracy," *New York Times,* August 11, 2002.

[28] The December and February meetings were officially organized by Zalmay Khalilzad, who was appointed Special Presidential Envoy and Ambassador at Large for the Free Iraqis on December 2, 2002. See Michael Howard, "Conference Delegates Vie for Political Role in New Iraq," *The Guardian* (London), December 16, 2002; Andrea Gerlin, "Iraqi Opposition Groups Agree on Power-Sharing Plan," *Philadelphia Inquirer,* December 18, 2002; Gareth Smyth, "Emboldened Opposition Gathers," *Financial Times* (London), February 27, 2003; Judith Miller, "Iraqi Leadership Team to Prepare for a Transition to Democracy," *New York Times,* February 28, 2003; Judith Miller, "Ending Conference, Iraqi Dissidents Insist on Self-Government," *New York Times,* March 3, 2003.

[29] After the fall of Baghdad, the U.S. military airlifted Ahmed Chalabi and several hundred armed personnel to the northern part of Iraq. These people were described as part of the FIF, but it is unclear whether they received official training in Hungary as part of that program or were more hastily assembled and designated as part of the FIF. On prewar FIF training, see Greg Jaffe, "On a Remote Base, U.S. Drill Sergeants Train Iraqi Exiles," *Wall Street Journal,* February 24, 2003; Adam LeBor, "Exiles Prepare for a Happy Return at Camp Freedom," *The Times* (London), February 24, 2003; Ian Traynor, "U.S. Closes Exiles Training Camp After Only 100 Show Up," *The Guardian* (London), April 2, 2003. On the airlift into northern Iraq, see Peter Finn, "New Force Moves to Gain Sway," *Washington Post,* April 12, 2003; Rupert Cornwell, "U.S.-Backed Iraqis Launch Bid for

from a lack of enthusiastic support within the Department of Defense. In particular, CENTCOM opposed relying on large numbers of Iraqis during the military campaign, reportedly because they would not add significant military capabilities and might increase risk to U.S. military personnel if they were not vetted properly.[30]

OSP also developed policy guidance on the future of the Iraqi army and the de-Ba'athification process, both of which became controversial issues after the Coalition Provisional Authority took power. They assumed that the army would continue to exist after the war ended and that it could be used for reconstruction projects while its personnel were being vetted and its units reconstituted.[31] OSP also helped develop a de-Ba'athification policy that would prevent senior Ba'ath Party officials from holding public positions in the new government.[32]

OSD's Role in Policymaking

Many of the people interviewed for this report claim that OSD officials were dictating policy to other U.S. government agencies.[33] The very existence of the Office of Special Plans—and the confusion surrounding its creation, purpose, and name—may have given its guidance disproportionate weight during the interagency planning process, thus leading many government officials to believe that OSP's "contributions" were de facto instructions. Clearly, frustration arose over the lack of coordination of OSP's guidance papers with other agencies before the meetings in which they were presented.[34] OSD officials counter that all of their planning efforts fed into the interagency process discussed at the beginning of this chapter, and that they lacked both the authority and intent to dictate policy to their counterparts in other agencies.[35] Furthermore, at least one senior State Department official reported receiving directions that OSD was in charge of both war and postwar planning efforts.[36] In any case, because the Department of Defense seems to have been perceived by others as

---

Power," *Independent on Sunday* (London), April 13, 2003; Peter Finn, "Pentagon's Iraqi Proteges Go Home to Mixed Welcome," *Washington Post*, April 17, 2004.

[30] Interviews with OSD officials, August 2004.

[31] This assumption was incorporated into the planning efforts of the Office for Reconstruction and Humanitarian Assistance (ORHA), as discussed in Chapter Five.

[32] Interviews with OSD official, August 2004.

[33] Interviews with State Department official, October 2003, and ORHA officials, November 2003 and January 2004.

[34] Interviews with State Department official, October 2003, ORHA officials, November 2003 and January 2004, and OSD official, August 2004.

[35] Interviews with OSD officials, August 2004.

[36] Greg Suchan, the Principal Deputy Assistant Secretary of State for Political-Military Affairs, has been quoted as saying that "the direction had come from the 7th floor [Secretary of State's office] that OSD was in charge and State was to help out in anyway [sic] possible, but not to rock the boat." See Donald R. Drechsler, "Reconstructing the Interagency Process After Iraq," *Journal of Strategic Studies*, Vol. 28, No. 1 (February 2005), p. 8.

first among equals within the interagency planning process, OSD's influence may indeed have been disproportionate.

This perception was likely exacerbated after the Department of Defense was officially designated as the lead agency for postwar planning. The President apparently reached this decision in October 2002, to ensure that there would be one U.S. government agency in charge of postwar Iraq. The President wanted to avoid some of the problems plaguing postwar Afghanistan, where responsibilities for reconstruction were divided among several different coalition partners and U.S. agencies.[37] All the postwar planning efforts that were under way would be centralized in a single office within DoD, though it would have extensive interagency representation.[38] OSD developed a few wiring diagrams to show how such an office might be organized.[39] A delay occurred after this decision was reached in October, for as discussed in Chapter Five, the Department of Defense was not designated as the postwar lead agency until January 2003. None of the officials interviewed for this report could explain that delay, though some speculated that senior officials may have had difficulty selecting a person to lead the new office or that naming a postwar coordinator in October might have undermined ongoing diplomatic efforts at the United Nations.[40]

State Department Planning

The State Department participated in all of the interagency planning processes described above, which formed the core of its contribution to the planning effort. In addition, State sponsored the broadest assessment of postwar requirements that would be conducted within the U.S. government, the Future of Iraq project, a process that correctly identified many of the reconstruction and governance problems that would be faced after the war ended. Yet the information and insights it generated were not well integrated into other U.S. government planning efforts. State Department officials believed that they were marginalized because State was seen as not supporting the possible war with Iraq.[41] Whatever the cause, State ultimately lost more battles than it won during the interagency planning process.

---

[37] Interview with OSD official, August 2004.

[38] As Douglas Feith later noted, "Bush gave Rumsfeld overall authority for the postwar plan, to maintain what he called 'a unity of concept and a unity of leadership' . . . Since so many of the responsibilities were military security responsibilities, the only person who could really do that was the secretary of defense." Fineman, Wright, and McManus.

[39] These organization charts are depicted in Figures 5.1 and 5.2.

[40] Interviews with OSD officials, August 2004.

[41] Interviews with State Department officials, October 2003 and July 2004. See also the discussion of Tom Warrick and ORHA in Chapter Five.

The State Department and Interagency Planning

Though the Future of Iraq project continued its work throughout 2002 and early 2003, the State Department as a whole remained on the margins of the postwar planning process. Representatives from various offices within State did participate in and influence the outcomes of the official interagency planning process described at the end of this chapter, but State as a whole never effectively influenced the planning process. Why was this the case? According to one official, one reason is that State never developed a single agency position. Tensions among the various bureaus within the State Department are a perennial problem, and the individuals who participated in the planning process reflected the views of their own bureau rather than a coherent position of the Department of State. These views emphasized different points at best, and directly contradicted each other at worst. This reduced State's ability to influence policy formation and frustrated many other agency representatives, who observed that interagency meetings work only when each agency shows up with a single coordinated position.[42]

A second, and perhaps more important, reason for State's lack of influence is that senior U.S. officials believed that State opposed the possibility of war with Iraq. Many State Department officials supported strengthening the United Nations sanctions regime before threatening regime change, while senior policymakers—apparently including Vice President Richard B. Cheney and Secretary of Defense Donald Rumsfeld—had reportedly concluded that diplomatic approaches, particularly through the UN, would be ineffective.[43] OSD officials remained concerned throughout the planning process that any criticism from State was ultimately designed to undermine the policies they supported. State Department officials report that representatives from the Bureau of Intelligence and Research (INR) were largely ignored during scenario planning meetings they attended and claim they were deliberately not invited to a DoD-sponsored wargame during the summer of 2002.[44]

These two factors combined to make the State Department a weak player in interagency preparations for postwar Iraq. Even though the Future of Iraq project was the most comprehensive effort within the U.S. government to examine the challenges and requirements of Iraq after Saddam, its insights and suggestions were not used as

---

[42] Interview with former State Department official, October 2003.

[43] The Bush administration opposed involving the United Nations or discussing sanctions until August 2002, when Secretary Powell convinced the President that a renewed diplomatic effort involving the United Nations was a necessary precondition for building international support for possible military operations against Iraq. Woodward describes part of this discussion as follows: "Powell believed he had Cheney boxed in, and to a lesser extent Rumsfeld. [Powell] argued that even if anyone felt that was the only solution, they could not get to war without first trying a diplomatic solution. It was the absolutely necessary first step . . . Powell believed he had them, although he sensed that Cheney was 'terrified' because once the diplomatic road was opened up, it might work." Woodward, pp. 154–157; quote from pp. 156–157.

[44] Interview with former State Department officials, October 2003 and March 2004.

a basis for postwar planning efforts within the interagency process. The project remained stovepiped within the State Department, and those who were not directly involved with it knew very little about its efforts or even its existence. In fact, the project was not allowed to be officially briefed to the interagency community until October 2002, reportedly because of concerns that it would be dismissed as another State Department effort to undermine support for the war by identifying postwar challenges.[45] Internal bureaucratic challenges and external suspicion about State's true motives, therefore, combined to marginalize State's influence on the postwar planning process, as well as to limit the dissemination of ideas and information from the Future of Iraq project.

The Future of Iraq Project

The Future of Iraq project was initially conceptualized during the fall of 2001, as a way to identify the challenges that might occur after Saddam Hussein was removed from power and to develop expertise among those who might serve in a new Iraqi government. On February 4, 2002, Secretary of State Colin Powell approved the formation of the project and the outline of subjects that it would contain. State Department officials soon started working on proposals for the various working groups within the project. They believed that the project should be sponsored by a neutral nongovernmental organization, and the project's initial planning meeting in April was sponsored by the Middle East Institute (MEI). However, senior U.S. government officials opposed the involvement of MEI, and by late April 2002, State decided to sponsor the project itself.[46] The project was coordinated by Tom Warrick, a foreign service officer from the State Department's Bureau of Near Eastern Affairs. By June, Congress approved $5 million in funding for the project, and the first working group met in July. Most of the working groups' meetings were held in the United States, though some were held in the United Kingdom and one was held in Italy. Each meeting lasted approximately a day and a half, including some informal gatherings to facilitate discussion and the exchange of information.

The project consisted of 17 working groups, across a wide range of functional areas.[47] The organizers knew that the quality of the working groups would be un-

---

[45] Interview with State Department official, July 2004.

[46] These senior officials reportedly opposed MEI sponsorship for two reasons. First, the president of MEI had publicly criticized the administration's policies on the Arab-Israeli conflict. Second, some members of the Iraqi political opposition reportedly opposed the existence of the project, since they were not invited to participate in most of the working groups and feared that the project would reduce their influence. Ahmed Chalabi reportedly asked senior U.S. officials to prevent MEI from sponsoring the project, in the hopes that it would not be able to find a different sponsor. Interview with a State Department official, July 2004, and David Phillips, *Losing Iraq*, New York: Westview Press, 2005, p. 37.

[47] The 17 working groups, in alphabetical order: Anti-Corruption Issues; Building a Free Media; Civil Society–Capacity Building; Defense Policy; Democratic Principles; Economy and Infrastructure; Education; Foreign Policy; Local Government; Oil and Energy; Preserving Iraq's Cultural Heritage; Public Finance; Public Health

even, but they thought that it was important to start dialogue in as many of these areas as possible. The State Department solicited as many nominations for Iraqi participants as possible, demonstrating its intent to be as inclusive as possible and to reach people who were not active in political opposition groups. Each Iraqi nomination was vetted by several U.S. government agencies, and the final participants were selected by State to ensure that a wide range of views was represented. Each meeting was chaired by a U.S. government official, usually from the State Department. The Iraqis conducted the bulk of the discussions at the main table while observers from other U.S. government agencies sat behind them, deliberately symbolizing their secondary role in these discussions.[48]

The working papers developed by the Future of Iraq project encompass more than 2,000 pages in 13 volumes.[49] The output of the various working groups was uneven, as expected, with some groups producing hundreds of pages of documents and others producing only a few pages.[50] Nevertheless, State Department officials believed that the process of holding the meetings was as important as their output. As one participant later noted, "It involved Iraqis coming together, in many cases for the first time, to discuss and try to forge a common vision of Iraq's future."[51]

Press reports have widely described the Future of Iraq project as a State Department "plan" for the reconstruction of Iraq. Such a characterization is unwarranted. Plans require a concrete set of prioritized steps that should be taken in a given situation, and a plan ideally assigns responsibility for each of those steps. The Future of Iraq project did not contain any such prioritization; it was not something that could be taken off the shelf and immediately executed.[52] The collected papers of the project fill 13 volumes, so it was clearly not something that Jay Garner (or, later, Paul Bremer) could easily digest. The project was designed not as a plan but as a *process*, which would help collect Iraqi expertise on the major reconstruction challenges and generate ideas on how to address those challenges. It was also designed to help prepare project members to participate directly in reconstruction efforts, and the fact that some project members became members of the new Iraqi government suggests

---

and Humanitarian Issues; Public Outreach; Return of Refugees and Internally Displaced Persons; Transitional Justice; and Water, Agriculture, and the Environment. See "Prepared Testimony of Marc Grossman before the Senate Foreign Relations Committee," February 11, 2003.

[48] Interview with State Department official, July 2004.

[49] Phillips, p. 37. Most of these papers were declassified in September 2006 and are available on a National Security Archive web site. As of February 2008:
http://www.gwu.edu/~nsarchiv/NSAEBB/NSAEBB198/index.htm

[50] James Fallows, "Blind into Baghdad," *The Atlantic Monthly*, January/February 2004, p. 57.

[51] David L. Phillips, quoted in David Rieff, "Blueprint for a Mess," *New York Times Magazine*, November 2, 2003.

[52] Under Secretary of Defense for Policy Douglas Feith reportedly described the Future of Iraq project as "a bunch of concept papers." Jeffrey Goldberg, "A Little Learning," *The New Yorker*, May 9, 2005.

that the project succeeded in this area.[53] In that sense, the project has had an important, though indirect, impact on the long-term reconstruction of Iraq.

Though the project was not a plan, its insights and suggestions could certainly have been used as the *basis* for a postwar plan. This did not occur because senior U.S. government officials marginalized the project's influence in the postwar planning process. As will be described in Chapter Five, Tom Warrick, the project's planning leader, briefly worked for the Office of Reconstruction and Humanitarian Assistance (ORHA) before Secretary of Defense Rumsfeld told Jay Garner to fire him. Had Warrick remained with ORHA, insights and expertise from the project might well have been used as the basis for some of ORHA's plans. But he did not, and ORHA planners had only a passing acquaintance with the Future of Iraq project and its contents.[54]

USAID Planning

USAID's planning for postwar Iraq began informally in the summer of 2002, although it did not receive formal tasking to begin planning until early 2003. The agency was instructed to develop plans for actions and responses in the event of a war in Iraq. USAID developed a plan in the fall of 2002, even though it—like other U.S. agencies—did not have personnel on the ground in Iraq at that time and lacked reliable information about conditions there. The plan included "branches" for several possible contingencies, such as the use of WMD, and was continually refined through the winter and early spring.[55]

Reconstruction Planning and Contracting

USAID's planning efforts drew on its considerable experience with post-conflict humanitarian relief and reconstruction activities throughout the world.[56] Its plans for Iraq focused on providing and reconstituting electricity, water and sewers, public health, education, local government, agriculture and food supply, and roads and buildings, among other issues. USAID tried to gather as much current information about Iraq as possible, but such information was spotty. For example, the CIA pro-

---

[53] Eric Schmitt and Joel Brinkley, "State Dept. Study Foresaw Trouble Now Plaguing Iraq," *New York Times*, October 19, 2003; interview with State Department official, July 2004.

[54] Interviews with ORHA officials, November 2003 and December 2003.

[55] Interviews with OSD official, November 2003, and USAID official, May 2004.

[56] USAID planners successfully argued that humanitarian relief and reconstruction were interrelated tasks and needed to be addressed together, rather than as separate steps in a process. USAID's planning for Iraq thus considered both relief and reconstruction to be part of a single operation. Interview with U.S. government official, May 2004. For more on U.S. government planning for humanitarian relief, see Chapter Six.

vided very good information about ports, airports, and targeting issues, but its information about infrastructure and basic services was thinner and less reliable because of Saddam's efforts to disguise their true level of degradation. CIA information did help USAID planners understand that they would probably need to import fuel because of the limitations of the Iraqi oil refineries. USAID was then able to contract with the Defense Logistics Agency (DLA) and others to provide fuel. Plans were also developed to increase electrical production and to connect Iraq to local and regional power grids.[57]

USAID carries out the vast majority of its projects by contracting with international organizations (IOs), nongovernmental organizations (NGOs), and private companies. USAID reportedly began assessments to support contract allocations as early as August 2002, and in mid-September it started holding weekly coordination meetings with relief agencies and NGOs.[58] On January 16, 2003, the director of USAID, Andrew Natsios, signed the order authorizing local procurement for Iraqi relief and reconstruction. A sum of $63 million was initially allocated for this mission, and USAID let 11 contracts almost immediately after the order was signed. Before the war commenced in March 2003, USAID had awarded numerous contracts and grants for key tasks in infrastructure reconstruction, education, and government support.[59] Some NGOs chose not to sign contracts with USAID, seeing U.S. government money as tainted by the continuing preparations for war.[60]

Coordination with Other Agencies

From the fall of 2002 through the spring of 2003, representatives from USAID participated in the interagency planning process described below as well as the CENTCOM planning efforts described above. USAID planners had participated in CENTCOM planning during the preparations for and execution of Operation ENDURING FREEDOM in Afghanistan, and that model was successfully continued in the buildup to OIF.[61] At these meetings, USAID representatives had the opportunity to voice their concerns and requirements for logistics, food, tents, generators, and other supplies. They urged CENTCOM to avoid bombing telephone

---

[57] Interview with USAID official, May 2004.

[58] Fallows, p. 62; interview with USAID official, May 2004.

[59] Contracting in the areas of health and economic growth reportedly lagged behind. Treasury was the lead agency for contracting on economic growth, which posed interagency coordination challenges, and USAID health personnel found that integrating a possible combat environment into their plans posed some unexpected challenges. Interviews with U.S. military officer, November 2003, and with U.S. government official, May 2004.

[60] Interview with USAID official, May 2004. For more on contracting issues, see Chapter Twelve.

[61] For more on the preparations for Operation ENDURING FREEDOM, see Olga Oliker et al., *Aid During Conflict: Interaction Between Military and Civilian Assistance Providers in Afghanistan, September 2001–June 2002*, Santa Monica, CA: RAND Corporation, MG-212-OSD, 2004.

exchanges and to maintain as much long-term electrical capacity as possible while targeting the power grid.[62]

USAID representatives also had opportunities to discuss these issues with General Tommy Franks, the CENTCOM commander, and the staff of Lieutenant General David McKiernan, the commander of land forces. During all of these meetings, USAID representatives noted consistently that CENTCOM personnel maintained that the military would not be responsible for public security and policing, humanitarian relief, any civilian effects of WMD use or unexploded ordnance, or civilian government. Responsibility for these tasks remained undefined, although the military assumed that USAID would be involved in at least some of them.[63]

USAID's Iraq Task Force Chief, Lewis Lucke, was one of the first civilians to arrive in theater, deploying to Kuwait in November 2002. He spent the balance of that year traveling between Kuwait and the United States, focusing on reconstruction planning. Humanitarian planning was outside of the USAID mission's scope, although some was considered in the assistance context. In Kuwait, coordination was with the U.S. military, with the World Food Program, and with the United Nations Development Program (UNDP).[64] Lucke returned to Kuwait early in 2003 as the reconstruction coordinator (in addition, George Ward served as humanitarian coordinator and Michael Mobbs as civil affairs coordinator), with Lucke's deputy, Chris Milligan, remaining as the senior USAID representative at ORHA in Washington.[65]

The National Security Council

The NSC staff coordinated the interagency process described earlier in this chapter. Additionally, the Deputies Committee tasked the NSC staff to pull together materials on likely postwar issues.[66] The work done by the IR+R and IPMC postwar efforts (discussed above) informed this NSC staff effort, although the ESG's failure to achieve consensus on these issues meant that the NSC staff had to secure interagency approval through the usual formal process. The NSC staff briefed President Bush on its recommendations for postwar reconstruction during March 2003, which he endorsed. Some of these recommendations were presented and endorsed after OIF had already begun. These recommendations did not constitute a plan for reconstruction, but instead provided strategic guidance for all U.S. agencies to follow as they started

---

[62] Interview with USAID official, May 2004.

[63] Interview with USAID official, May 2004.

[64] Interview with USAID official, July 2004.

[65] ORHA's structure is discussed at greater length in Chapter Five.

[66] Unless otherwise noted, the information in this section is drawn from an interview with an NSC official, April 2004.

engaging in reconstruction activities. Two of the most important substantive areas—both of which were to become highly controversial—were de-Ba'athification and the restructuring of Iraqi military and security institutions.

De-Ba'athification

The NSC staff recommendations called for 25,000 full members of the Ba'ath Party to be removed from their positions. This number represented just over 1 percent of the 2 million people on the government payroll, enabling government agencies to retain many people in managerial and technical leadership positions while removing the most senior Ba'athists.[67] The NSC staff guidance charged DoD and the State Department to develop a coordinated vetting policy that would apply to most government officials[68] and to generate a de-Ba'athification plan that could be implemented by ORHA. All personnel would have to be vetted for involvement in human rights abuses and war crimes, as well as whether it would be politically acceptable to let them retain their positions. Ba'ath Party membership would not constitute automatic grounds for removal, but the burden of proof would fall on the individual to demonstrate why he should retain his position. Coalition officials would not pursue any subsequent rounds of de-Ba'athification, but the NSC guidance stated that when Iraq was ready to hold elections, regime records should be made public so that the voters could decide on the candidates' suitability for office.

Restructuring Iraqi Military and Security Institutions

The NSC staff also established some guidelines for the process of disarmament, demobilization, and reintegration (DDR) of the Iraqi military. This guidance stated that the Iraqi military should become a de-politicized military force under civilian control that would no longer threaten Iraq's neighbors and could defend the territorial integrity of the country. Its composition would also reflect Iraq's population, including a diverse set of ethnic and sectarian groups. Existing army units should be told to stay in garrison, since immediately demobilizing hundreds of thousands of army personnel would only put them out on the street. Army units should be used to assist with reconstruction tasks in the immediate postwar period, even as their organizational structures were being reformed. Three to five regular army divisions would need to be established and trained during the transition period, in order to form the core of the new army. The Special Republican Guard, Republican Guard, and paramilitary forces would be disarmed and dismantled during this transition as well.

The guidance also emphasized the need to reform the government's security institutions as part of the transition process, particularly by inculcating the principle of

---

[67] Woodward, p. 339.

[68] Iraqi military and intelligence services would be vetted separately by DoD and the CIA.

civilian control over the military. An Iraqi national security council should be formed, controlled by the civilian commander-in-chief and various civilian deputies. The new civilian minister of defense would control all of Iraq's military forces, while the new minister of the interior would control the internal security institutions, including the national police and border guards.

Other Analyses of Postwar Requirements

In the months before the war, a number of think tanks and NGOs released reports on postwar requirements in Iraq. Even though many of these reports were not sponsored by the U.S. government, they were available to U.S. decisionmakers, and they demonstrate the range of issues that were identified before the war started. These reports are listed and discussed by date of release.

Council on Foreign Relations, *Guiding Principles for U.S. Post-Conflict Policy in Iraq.* The Council co-sponsored a working group on postwar policy in Iraq during the second half of 2002 and released a report of its findings that December. This report stresses a theme that emerged repeatedly in later reports: unless the U.S. government develops detailed plans for how to govern and reconstruct Iraq, "the United States may lose the peace, even if it wins the war."[69] The group distinguishes between short-term, medium-term, and long-term objectives. For each of these three phases, it identifies the type of leadership that should be in place as well as key security, economic, and governance objectives.[70] The report specifically notes that in the immediate postwar period, the United States will have to establish a single coordinator for Iraq within the U.S. government; establish law and order; preserve Iraq's territorial integrity; and distribute humanitarian assistance and reestablish vital services, among other tasks.[71]

Center for Strategic and International Studies, *A Wiser Peace.* The first paragraph of this report notes that the success of any war with Iraq "will be judged more by the commitment to rebuilding Iraq after a conflict than by the military phase of the war itself."[72] It argues that postwar planning efforts conducted to that point had

---

[69] Edward Djerejian and Frank G. Wisner, Co-Chairs, *Guiding Principles for U.S. Post-Conflict Policy in Iraq,* Washington, D.C.: Council on Foreign Relations and the James A. Baker III Institute for Public Policy, Rice University, December 2002, p. 1.

[70] Djerejian and Wisner, pp. 28–30.

[71] Djerejian and Wisner, pp. 4–9.

[72] Frederick D. Barton and Bathsheba N. Crocker, *A Wiser Peace: An Action Strategy for a Post-Conflict Iraq,* Washington, D.C.: Center for Strategic and International Studies, January 2003, p. 6. This study was released at the same time as a parallel study on general post-conflict government. See *Play to Win: Final Report of the Bi-Partisan Commission on Post-Conflict Reconstruction,* Washington, D.C.: Center for Strategic and International Studies and the Association of the U.S. Army, January 2003.

been incomplete and insufficient and identifies persistent mistakes that had occurred during recent post-conflict reconstruction efforts. The second part of the report includes the following ten recommendations to be undertaken to achieve success during the reconstruction period:

- Create a transitional security force that is effectively prepared, mandated, and staffed for post-conflict security needs, including the need for a constabulary force.

- Develop a comprehensive plan for securing and eliminating weapons of mass destruction.

- Plan and train for other post-conflict missions that create a foundation for a peaceful and secure Iraq that will enhance regional security.

- Establish an international transitional administration and name a transitional administrator.

- Develop a national dialogue process and recruit a national dialogue coordinator.

- Recruit a rapidly deployable team of international legal experts, judges, prosecutors, defense attorneys, corrections officers, and public information experts.

- Identify and recruit international civilian police officers.

- Call a debt-restructuring meeting and push the UN to review past war-related claims against Iraq.

- Immediately review sanctions against Iraq and prepare the documentation necessary to suspend them.

- Convene a donors' conference.[73]

The report concludes by noting that preparations for this wide range of reconstruction tasks must be undertaken before conflict starts in order to be effective after the conflict ends.[74]

The Atlantic Council of the United States and American University, *Winning the Peace.* During the second half of 2002, the Atlantic Council and American University co-sponsored a working group on postwar Iraq, and it published a policy paper containing the group's findings in January 2003. The paper emphasizes four main themes: developing a common vision of Iraq's future that is widely shared within the international community; U.S. power sharing with the United Nations

---

[73] Barton and Crocker, pp. 14–25.

[74] Barton and Crocker, p. 26. On March 25, 2003, the center issued a follow-up report that assessed the Bush administration's progress on implementing the ten recommendations listed above. It concluded that the administration earned "a mixed grade on its planning and preparations, which have been significant in certain areas but are still seriously lagging in others." Bathsheba N. Crocker, *Post-War Iraq: Are We Ready?* Washington, D.C.: Center for Strategic and International Studies, March 25, 2003.

and some sort of interim ruling council; investing in the Iraqi economy and oil sector; and ensuring that Iraq no longer poses a threat to its neighbors. It stresses that international cooperation will be critical to these fundamental reconstruction tasks and that a substantial peacekeeping force will be required after combat operations end.[75]

U.S. Army War College Strategic Studies Institute, *Reconstructing Iraq.* This report, released in February 2003, provides a historical review of past U.S. occupations and identifies challenges that the United States would be likely to face during a military occupation of Iraq. The main theme of the report:

> The possibility of the United States winning the war and losing the peace in Iraq is real and serious . . . Preparing for the postwar rehabilitation of the Iraqi political system will probably be more difficult and complex than planning for combat. Massive resources need to be focused on this effort well before the first shot is fired.[76]

This report notes that U.S. and allied forces would be required to occupy Iraq for an "extended period of time," but warns that "long-term gratitude is unlikely and suspicion of U.S. motives will increase as the occupation continues."[77] It expresses doubt that the exiles would be welcomed as the new leadership in Iraq, and it discusses the religious and tribal cleavages that might cause some political fragmentation. It also analyzes the size and costs of an occupation force, the potential for terrorism against U.S. forces, and requirements for economic assistance.[78]

The most important part of this report is the mission matrix for Iraq, which we reproduce here as an appendix. It provides the single most comprehensive list of postwar reconstruction tasks produced either inside or outside the U.S. government. It lists 135 tasks that must be accomplished throughout four phases of the reconstruction period: providing security (up to six months), stabilization (six months to one year), building institutions (one to two years), and handing power to local authorities (more than two years). Thirty-five of these tasks are considered critical for the military, meaning that there is a serious risk of mission failure if they are not immediately addressed and resourced. The matrix shows, however, that military forces will bear responsibility for the vast majority of reconstruction tasks during the immediate postwar period. During the first six months, the military will have to be in-

---

[75] Richard Murphy, Chair, *Winning the Peace: Managing a Successful Transition in Iraq*, Washington, D.C.: The Atlantic Council of the United States and the American University Center for Global Peace, January 2003.

[76] Conrad C. Crane and W. Andrew Terrill, *Reconstructing Iraq: Insights, Challenges, and Missions for Military Forces in a Post-Conflict Scenario*, Carlisle, PA: U.S. Army War College, Strategic Studies Institute, February 2003, p. 42.

[77] Crane and Terrill, p. 18.

[78] Crane and Terrill, pp. 23–42.

volved in almost three-quarters of all reconstruction tasks: it will bear sole responsibility for 70 of the 135 reconstruction tasks (52 percent) and will share responsibility with civilian authorities for another 28 tasks (21 percent). While it may be possible to dispute the authors' specific assignments of responsibility in individual cases, the unmistakable thrust of their work is that the military will bear heavy responsibilities during the first phase of any occupation, before civilian authorities are capable of taking on many of these responsibilities. The report concludes that military leaders, particularly within the U.S. Army, should conduct more detailed analyses of their responsibilities and missions in postwar Iraq.[79]

Council on Foreign Relations, *Iraq: The Day After.* In March 2003, the Council issued the findings of a second working group that picked up where the December 2002 report left off. This report contains specific recommendations about humanitarian issues, structuring a transitional administration immediately after the war, reconstructing the oil industry, and maintaining regional stability.[80]

---

[79] Crane and Terrill, p. 53.

[80] Thomas R. Pickering and James R. Schlesinger, *Iraq: The Day After*, Washington, D.C.: Council on Foreign Relations, 2003.

CHAPTER FOUR

Task Force IV

Establishing Task Force IV

In December 2002, CENTCOM sponsored a wargame called Internal Look, which exercised the plans for the invasion of Iraq.[1] The wargame focused almost exclusively on Phases I through III of the warplans involving major combat operations, with little attention paid to Phase IV, the postwar period.[2] Little consideration was given to ways in which postwar requirements might affect the conduct of combat operations.[3] There was general discussion about the postwar period at the end of the exercise, which included broad statements that the State Department would largely be responsible during that time but did not include any specifics. Retired General Gary Luck, who was serving as a senior advisor during the exercise, asked what the military's role would be during Phase IV, but no ready answers to his question were forthcoming. He commented that the military needed to stand up a planning cell to fill this gap in the planning, as well as start coordinating with the other U.S. government agencies, international organizations, and NGOs that would be involved in Phase IV.[4] In subsequent discussions, it was suggested that the Standing Joint Force Headquarters (SJFHQ) concept, developed by U.S Joint Forces Command (JFCOM), might serve as a good organizational model for such a planning cell.[5] Lieutenant General George

---

[1] Internal Look is an annual CENTCOM exercise that does not always focus on Iraq. The 2002 version, held between December 8 and 17, was used to exercise the developing warplans for Iraq. Daniel Williams and Vernon Loeb, "At Qatar Base, U.S. Begins a Test Run for War," *Washington Post,* December 10, 2002; Woodward, pp. 237 and 244; Scarborough, *Rumsfeld's War*, p. 176.

[2] As CENTCOM commander General Tommy Franks named the phases: "Phase I—Preparation. Phase II—Shape the Battlespace. Phase III—Decisive Operations. Phase IV—Post-Hostility Operations." Franks, p. 350.

[3] One observer noted that there was no interest in postwar civilian objectives during this exercise. Interviews with USAID official, February 2004.

[4] Interview with TFIV official, March 2004.

[5] For more on the SJFHQ concept, see United States Joint Forces Command, "Doctrinal Implications of the Standing Joint Force Headquarters (SJFHQ)," Joint Warfighting Center Joint Doctrine Series, Pamphlet 3, June 16, 2003. As of June 2004: http://www.dtic.mil/doctrine/jel/other_pubs/jwfcpam3.pdf
Also see Colonel Douglas K. Zimmerman, "Understanding the Standing Joint Force Headquarters," *Military Review*, July–August 2004, pp. 28–32.

Casey, then J5 Director on the Joint Staff, took the lead in developing the concept for the proposed planning cell.

The members of Task Force IV (TFIV) arrived in Tampa during the second week of January 2003 and quickly found themselves amidst much chaos and confusion. Their offices were still being wired, and their computer hardware and software systems were in the process of being installed. The new staff members spent several days meeting each other and trying to figure out what to do, because the TFIV commander, Brigadier General Steven Hawkins, was often away from the office receiving briefings on postwar planning efforts conducted by the U.S. government and other organizations. They also tried to learn more about the joint planning process and the SJFHQ concept, because this was the first time it had ever been put into practice.[6]

Task Force IV Planning

TFIV did not receive any formal planning guidance, so its members relied heavily on public statements by key administration officials to provide some broad principles on which to base their planning. In particular, they used the testimony that Douglas Feith, the Under Secretary of Defense for Policy, provided to the Senate Foreign Relations Committee on February 11, 2003.[7] In his testimony, Feith outlined five specific objectives of the postwar period:

- To "demonstrate to the Iraqi people and the world that the United States aspired to liberate, not occupy or control them or their economic resources";
- To "eliminate Iraq's chemical and biological weapons, its nuclear program, the related delivery systems, and the related research and production facilities";
- To "eliminate likewise Iraq's terrorist infrastructure";
- To "safeguard the territorial unity of Iraq";
- To "begin the process of economic and political reconstruction, working to put Iraq on a path to become a prosperous and free country."[8]

Feith also noted that if there were to be a war, the United States would need a "commitment to stay as long as required to achieve the objectives I have just listed,"

---

[6] Interviews with TFIV officials, March and April 2004.

[7] Interview with TFIV official, March 2004.

[8] Douglas J. Feith, prepared statement submitted to the U.S. Senate Committee on Foreign Relations, February 11, 2003.

but also "a commitment to leave as soon as possible, for Iraq belongs to the Iraqi people."[9]

TFIV conducted planning efforts in a number of important areas, though much of its work remains classified. Initially, TFIV planners focused on how they would organize themselves and how they would assess the power grid, water supplies, government structures, and security needs that would form the baselines for their work. TFIV then addressed many substantive topics, including food distribution, infrastructure maintenance, and what to do with the Iraqi army after the war.[10]

TFIV also planned extensively for the military headquarters that would conduct post-conflict operations. The plans called for TFIV to expand through augmentation into a fully functional coalition headquarters called the Combined Joint Task Force-Iraq (later renamed CJTF-7). TFIV planned for CJTF-Iraq to divide Iraq into seven multinational sectors, as shown in Figure 4.1, each commanded by a two-star general.

Based on a troops-to-task analysis, TFIV concluded that CJTF-Iraq would require at least 90 battalion equivalents, as detailed in Figure 4.2. When the requisite headquarters, maintenance units, hospitals, and other support structures were added to this combat force, TFIV estimated that CJTF-Iraq would require a minimum of approximately 200,000 U.S. military personnel and most likely more.[11] The CJTF-Iraq headquarters would have a relatively traditional military staff structure. It would consist of separate sections for personnel, intelligence, operations, logistics, planning, engineering, and so on, each commanded by a two-star general. The plan deliberately sought coalition participation in the command structure, with the personnel, intelligence, and logistics billets slated for U.S. officers and the rest slated for coalition officers.[12]

---

[9] Feith, prepared statement.

[10] TFIV's recommendations on the Iraqi army paralleled the guidance subsequently issued by the NSC. TFIV advocated removing the top two or three levels of the army command structure, leaving the rest of the soldiers in garrison, and gradually using the units to rebuild roads and to undertake other reconstruction tasks. Interview with TFIV official, April 2004.

[11] In late February 2003, General Eric Shinseki, then the Army Chief of Staff, asked TFIV for its estimates on how many troops would be required for postwar operations. Shinseki was told that depending on initial conditions, anywhere between 300,000 and 500,000 troops would be needed. These figures were similar to those generated by Shinseki's staff and helped form the basis for his congressional testimony that "something on the order of several hundred thousand soldiers" would be required for postwar operations in Iraq. Note that these figures were higher than the 250,000 troops that General Franks had estimated. Interviews with TFIV officials, March and June 2004; General Eric Shinseki, testimony to the U.S. Senate Armed Services Committee, February 25, 2003; Franks, pp. 366 and 393.

[12] The engineering billet was initially slated for a coalition officer, but Hawkins changed it to a U.S. officer. Interview with TFIV official, March 2004.

Figure 4.1
Planned Sectors for CJTF-Iraq

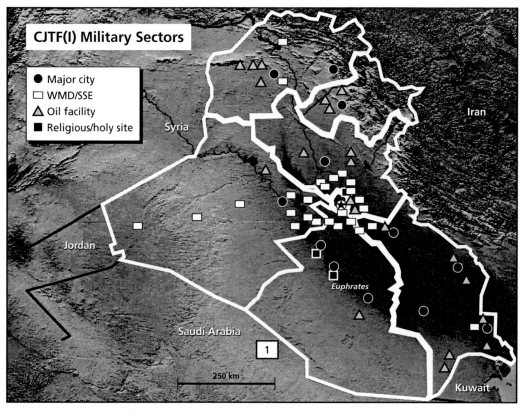

SOURCE: Task Force IV briefing to the CFLCC, February 21, 2003.
RAND *MG642-4.1*

As part of its planning efforts, TFIV identified ten key policy issues that remained unaddressed. On February 21, 2003, TFIV briefed the CFLCC on the areas for which it needed more interagency guidance, so that the request could be submitted up the chain. The ten issues were:

- Under what mandate will the transitional civil authority and coalition forces be operating? Is there a specific UN mandate for civil and military authority?

- What will be the coalition policy toward surrender and capitulation? Will there be a national ceasefire agreement?

- Who sets vetting policy? Who does the vetting? How quickly can it be done? What will be the criteria for criminal referral, government employment, police service, border guard service, and military service?

- What is coalition policy on disarming Iraqi military, security, and police forces? What is coalition policy on suspension and reactivation of the Iraqi military?

Figure 4.2
Proposed Force Structure for CJTF-Iraq

Ser	Capability	North Al Mawsil	West Ar Ramadi	Central Samarra	East Kirkuk	Baghdad	South-West Al Hillah	South-East Al Basrah	Total
1.	Command	2*	2*	2*	2*	2*	2*	2*	14 BCTs of
2.	Mech (wheeled)	2 x Bdes	1 x Bn	2 x Bn	2 x Bn	2 x Bde	2 x Bn	2 x Bde	25x Bns
3.	Light Inf	1 x Bde	1 x Bde	1 x Bn	2 x Bn	1 x Bde	1 x Bn	1 x Bde	16x Bns
4.	Light Artillery	1 x Bn	None	None	None	1 x Bn	None	1 x Bn	3x Bns
5.	Engineer (combat)	1 x Bn	1 x Co	1 x Co	1 x Co	1 x Bn	1 x Bn	1 x Bn	5x Bns
6.	Aviation	1 x Bde	1 x Bn	1 x Co	1 x Co	1 x Bn	1 x Co	1 x Bde	9x Bns
7.	SOF	1 x Co	1 x Co	1 x Co	1 x Co	1 x Co	1 x Co	1 x Co	1x Bn
8.	Log (combat)	1 x Bn	1 x Co	1 x Bn	1 x Bn	1 x Bn	1 x Bn	1 x Bn	8x Bns 1x Co
9.	Chemical	Corps asset	Corps asset	Corps asset	Corps asset	1 x Bn	Corps asset	Corps asset	
10.	Carabinieri	1 x Bn	1 x Co	1 x Co	1 x Co	1 x Bn	1 x Bn	1 x Bn	5x Bns
11.	MP (Arab/Muslim)	1 x Pl	None	None	None	1 x Co	None	1 x Pl	2x Cos
12.	Engineer (constr)	1 x Bn	1 x Co	1 x Co	1 x Co	1 x Bn	1 x Co	1 x Bn	4x Bns 1x Co
13.	Log (routine)	1 x Bn	1 x Bn	1 x Bn	1 x Bn	1 x Bn	1 x Bn	1 x Bn	7x Bns
14.	Civil Affairs	1 x Bn	1 x Co	1 x Co	1 x Co	1 x Bde	1 x Bn	1 x Bde	9x Bns

SOURCE: Task Force IV briefing to the CFLCC, February 21, 2003.
RAND MG642-4.2

- What is the policy on return of internally displaced persons (IDPs) and refugees?
- What Iraqi political, ethnic, and opposition groups will be allowed to participate in the interim government?
- What legal system will Iraq be under during coalition control?
- What currency or currencies will be legal tender?
- What is the funding mechanism? Who controls funds at the national, regional, and local levels?
- What should coalition forces do with Iraqi prisons and prisoners?[13]

TFIV did not receive any feedback on these questions; as the previous chapter notes, the interagency community was in the process of formulating guidance on many of these issues.

The lack of timely planning guidance was only one factor that hampered TFIV's ability to achieve its mission. TFIV's planning problems were exacerbated by

---

[13] "Resourcing Nation Building," Task Force IV briefing to the CFLCC, February 21, 2003.

a series of serious operational challenges that were encountered throughout the task force's existence.

Operational Challenges

During its ten-week life span, TFIV faced several bureaucratic and organizational challenges. These are easier to address functionally rather than chronologically. Three of the most important challenges were staffing issues; difficult relations with CENTCOM and CFLCC; and an unclear relationship with the Office of Reconstruction and Humanitarian Assistance (ORHA), which was established by DoD in January 2003 to plan and execute the postwar mission as well.[14]

Staffing Issues

No procedures existed to guide the process of staffing TFIV. It had slots for 58 personnel plus the commander, the same number envisaged in the SJFHQ concept. Yet SJFHQs were deliberately designed to be standing organizations staffed by permanently assigned personnel who worked together in peacetime as well as in times of crisis. TFIV, by contrast, had no standing organization upon which to draw, and it had to be assembled very quickly. Because the SJFHQ concept assumed that these organizations would be established well in advance of a crisis, it offered no guidance for TFIV.

Staffing for TFIV occurred on a completely ad hoc basis. The services were tasked to provide personnel to TFIV, and the 58 slots ended up being divided roughly evenly among them. The services scrambled to fill these slots quickly, though their procedures for identifying relevant personnel varied considerably. The Air Force, for example, searched through its personnel database to identify O-5s and O-4s who had completed both weapons school and their professional military education requirements, ideally at the School of Advanced Air and Space Studies. The Army, by contrast, sent out a general tasker through Forces Command (FORSCOM) without any specific requirements. Because most Army staffs were stretched thin as they prepared to execute the Iraq warplan on top of ongoing operational commitments, it was harder for the Army to identify any personnel who could be quickly reassigned to TFIV, much less those who had specific expertise or educational backgrounds. While such personnel constraints were understandable, there was a general belief—both inside and outside TFIV—that the Army did not send its best people to participate in this organization.[15]

---

[14] See Chapter Five for more on ORHA.

[15] Interviews with TFIV officials, March and April 2004.

As a whole, TFIV's staff lacked the specialized expertise needed to fulfill its mission. Few of the staff members had previous experience in a joint environment, and that lack of experience caused many coordination problems. Staff members frequently used terminology and concepts unfamiliar to staffers from other services. The planning process posed particular challenges in this regard. TFIV chose to use Army planning procedures, because the organization's senior leadership came from the Army, but these procedures sometimes confused TFIV personnel from other services.[16]

Detailed regional expertise was also lacking, because Middle East experts were in high demand throughout the military during the buildup to the war. For example, TFIV did have a slot for a political advisor (POLAD)—a critical function for reconstruction planning—yet the person assigned to fill that slot was a POLAD with expertise on China.[17] TFIV members tried to learn as much as they could about the region in a short amount of time, but their efforts could not substitute for true regional expertise.

Questions also arose about whether TFIV's leaders possessed the rank and expertise necessary for their positions. The December 2002 order from the CJCS specified that a one-star general would command TFIV, a rank that personnel both inside and outside the task force considered insufficient for an organization that had to cooperate closely with three- and four-star commanders.[18] Brigadier General Hawkins was reportedly picked by Lieutenant General Casey to lead TFIV because of his experience with reconstruction activities in the Balkans, but his largely tactical background was not as well suited to this assignment—nor were the background and experience of the people he picked to serve as his deputy and as his chief of staff.[19]

TFIV members were also frustrated by the funding arrangements for their organization. The Joint Staff did not provide funding for TFIV, which meant that all costs would have to be borne by the services through the component commands. TFIV staff members quickly became frustrated by this arrangement, because it effectively left them without a budget.[20]

Relations with CENTCOM and CFLCC

Task Force IV suffered from difficult relations with both CENTCOM and CFLCC throughout its existence. Part of the reason was structural: one observer described TFIV as an "orphan command," because no single organization took ownership of it

---

[16] Interview with TFIV official, November 2003.

[17] Interview with TFIV official, November 2003.

[18] Interviews with TFIV officials, November 2003, March 2004, and April 2004.

[19] Interview with TFIV official, November 2003.

[20] Interviews with TFIV officials, November 2003 and March 2004.

and managed and reviewed its work.[21] Bureaucratic politics compounded this problem, because both CENTCOM and CFLCC viewed TFIV as an outside organization that the Joint Staff had forced upon them.[22]

According to the orders issued by the Chairman of the Joint Chiefs of Staff, TFIV would be established by JFCOM and would then be placed under the operational control (OPCON) of CENTCOM. Yet TFIV and CENTCOM, particularly the CENTCOM J5, never developed a close working relationship. Most interaction between TFIV and the CENTCOM staff occurred at relatively senior levels, usually at the O-6 level and above. Hawkins met frequently with Generals John Abizaid, David McKiernan, and William Wallace, and with General Tommy Franks when he was in Tampa. At the working level, however, coordination consisted of little more than informal conversations among staff members. While this is common for subordinate staffs, it made formal coordination, when that was required, problematic in the absence of higher-level guidance from their leaders. Some CENTCOM officials resisted cooperation with TFIV, because they viewed the organization as an effort by the Joint Staff to interfere with the planning process.[23] Such bureaucratic frictions were exacerbated by TFIV's physical separation from the CENTCOM staff, operating from a building outside the main headquarters and lacking connectivity to the SIPRNET (the military's classified intranet) for much of its time in Tampa.[24] Although CENTCOM headquarters certainly faced space constraints during this time, the arrangements for locating TFIV in Tampa posed significant challenges for its staff.

In late January 2003, TFIV deployed from Tampa to Kuwait. Hawkins and Abizaid reportedly debated whether the task force should deploy to Qatar and collocate there with the CENTCOM forward headquarters, or to Kuwait, where it could collocate with CFLCC, so the staff could coordinate with those planning the land campaign and with the new DoD-sponsored Post-War Planning Office (which later became ORHA), which would spend time in Kuwait before the war started.[25] Hawkins arranged for CENTCOM to add another plane to the group of aircraft moving CENTCOM headquarters forward to Qatar at the end of January; the plane carrying TFIV continued on to Kuwait.[26] The Joint Staff was not involved in this decision, and it continued to assume that TFIV would remain in Tampa at least through

---

[21] Interview with TFIV official, November 2003.

[22] Interview with TFIV official, March 2004.

[23] Interviews with TFIV officials, November 2003 and June 2004.

[24] Interviews with CENTCOM official, January 2004, and TFIV official, April 2004.

[25] Interview with TFIV official, June 2004.

[26] Interview with TFIV official, March 2004.

March. Some members of the Joint Staff expressed a great deal of frustration at not knowing where TFIV was for almost two weeks.[27]

The decision to deploy forward at the end of January created two significant problems that hampered TFIV throughout its further existence. First, TFIV was not able to develop all the interagency relationships it needed to operate effectively. TFIV had only existed for two weeks before it deployed; this was barely enough time to get up and running, much less identify points of contact and coordinate with various U.S. government agencies. TFIV had started making these connections during the third week of January, during a CENTCOM meeting that involved interagency representatives as well as various NGOs and IOs, but the organization deployed forward before these connections had developed into working relationships. Connectivity challenges in Kuwait compounded the problem, since it took a couple of weeks for TFIV to establish telephone service and internet access from its new location. In retrospect, some members of TFIV thought that it might have been better for the organization to remain in Tampa, to facilitate cooperation with the other U.S. government agencies involved in postwar planning.[28]

Second, the decision to deploy forward created frictions with CFLCC that were never resolved. General Franks transferred OPCON of TFIV to General McKiernan once TFIV arrived in theater,[29] but the CFLCC staff generally viewed TFIV with suspicion. It did not understand the purpose of TFIV, or why an organization created by JFCOM, with ties to the joint community, was suddenly becoming involved in planning efforts that had been going on for months. Most of CFLCC's staff did not know that TFIV existed until it arrived at Camp Doha in Kuwait. It was further frustrated by Hawkins's continuing efforts to take charge of all postwar planning activities. Many members of TFIV believe, in retrospect, that Hawkins pushed it into a turf battle that it was destined to lose, since McKiernan had two more stars than Hawkins, and that the resultant friction made it impossible for them to fulfill their responsibilities.[30]

Relations with CFLCC did not improve once initial impressions had been set. For example, TFIV lacked access to the warplans until late February, which directly affected its ability to plan for the postwar period.[31] Also in February, McKiernan decided that TFIV should report to British Major General Albert Whitley, who filled a new position as the deputy commander for post-conflict operations, instead of directly to him. This decision was widely interpreted as an effort to constrain and fur-

---

[27] Interview with TFIV official, April 2004, and Joint Staff officials, April 2004.

[28] Interviews with TFIV officials, March 2004 and April 2004.

[29] Interview with TFIV officials, June 2004.

[30] Interviews with OSD official, November 2003, and TFIV officials, March 2004 and April 2004.

[31] Interview with TFIV official, April 2004.

ther marginalize Task Force IV. One observer noted that by this time, CFLCC viewed TFIV as a grain of sand in an oyster, posing constant irritations for the rest of the command.[32] Instead of working closely with CFLCC, as JFCOM and the Joint Staff had originally envisaged, Task Force IV never effectively coordinated its work on Phase IV with the CFLCC planners responsible for Phases I through III.

Relations with ORHA

The Office of Reconstruction and Humanitarian Assistance is discussed at length in the next chapter, but it is worth noting here that it and TFIV did not work closely together. The missions of both organizations were somewhat related: TFIV was intended to plan and lead military efforts in the immediate postwar period while ORHA focused on both short-term and long-term civilian reconstruction efforts. Members of TFIV and ORHA spent time trying to determine how they should relate to each other, but they were never able to figure out an effective working relationship.[33]

The two organizations also suffered from a competition for personnel and, in an environment that lacked robust communications, the fact that they were not collocated. Because both organizations shared a similar mission, they found themselves competing for trained planners and regional experts for their staffs, which meant that this valuable expertise was divided rather than concentrated in a single effort. The distance between the two headquarters compounded their problems. As discussed in the next chapter, ORHA set up its headquarters in the Kuwait Hilton, which meant that the two organizations were separated by a 45-minute drive and a security perimeter that ORHA personnel were not automatically cleared to enter. The physical distance, combined with both TFIV's and ORHA's lack of robust communications, made coordination difficult.[34] The staffs did make some adjustments to facilitate coordination; TFIV, for example, scheduled more of its personnel to work overnight shifts after realizing that ORHA worked late hours to facilitate coordination with Washington.[35] Yet most communication between the two organizations remained informal rather than institutionalized.

At several points, the head of ORHA—retired Army Lieutenant General Jay Garner—asked TFIV to come work for ORHA instead of working for CFLCC, so that all postwar planning could be consolidated into a single organization. Some of those involved in the discussion report that Hawkins supported this idea but that Franks, Abizaid, and McKiernan were all reluctant to do so; others report that Abi-

---

[32] Interview with OSD official, November 2003.

[33] Interviews with TFIV officials, November 2003, April 2004, and June 2004.

[34] Interviews with TFIV official, November 2003, and ORHA officials, November 2003 and December 2003.

[35] Interview with TFIV official, March 2004.

zaid had already approved the idea and that Hawkins resisted the move.[36] It is not clear why this did not occur, though placing a military organization such as TFIV under the civilian command of ORHA might have posed significant legal challenges. In any event, the confusing command relationships for oversight of TFIV made it difficult to determine who had final authority to make such a decision. CENTCOM and CFLCC argued that they did not have the proper authority to issue an order transferring TFIV to ORHA, because they only maintained operational control over TFIV. Yet the Joint Staff, which had established TFIV, does not maintain oversight of military organizations, since it is not in the chain of command. Because no one could figure out who had the ultimate authority to make this decision, TFIV and ORHA remained separate organizations. Many members of TFIV thought their failure to join ORHA was a crucial missed opportunity to consolidate planning efforts under the civilian organization that would ultimately oversee postwar reconstruction and assistance.[37]

TFIV became increasingly marginalized over time. As ORHA got up and running, it became the main point of contact for both military and civilian efforts to plan for the postwar period. ORHA could credibly claim to be in charge of these efforts because it received its mission directly from a presidential directive. Furthermore, Garner possessed much more influence throughout the chain of command than Hawkins. Garner was a retired three-star general whose subordinates, while he was on active duty, had included both Franks and McKiernan. Members of TFIV and ORHA both acknowledged that Garner "outranked" Hawkins, even in retirement, because of his wide network of contacts and direct access to these commanders and to Secretary Rumsfeld.[38]

The Dissolution of Task Force IV

By the end of March, it became clear that TFIV would have no direct role in postcombat operations. TFIV had nowhere near the number of personnel necessary to become the nucleus for the military headquarters overseeing post-conflict operations, as had originally been intended, and CFLCC continually argued that it was not responsible for running such a headquarters.[39] Because ORHA would be responsible for postwar civil assistance and administration, TFIV lacked any distinctive areas of responsibility. Members of TFIV started hearing rumors about the organization's

---

[36] Interviews with TFIV officials, November 2003, March 2004, and June 2004.

[37] Interviews with TFIV officials, November 2003, March 2004, and June 2004.

[38] Interviews with TFIV officials, November 2003, December 2003, and March 2004.

[39] When this headquarters (called CJTF-7) was eventually established, it was run by V Corps and not by CFLCC.

impending dissolution soon after combat operations started on March 19, and during the last week of March, CFLCC commander McKiernan announced that TFIV would be disbanded.[40] Hawkins protested that McKiernan did not have the authority to make that decision, because CFLCC only had operational control of TFIV. Yet McKiernan remained steadfast, and Hawkins did not further protest this decision.[41]

By this time, many members of TFIV had moved on. Some staff members were reassigned to other positions within CFLCC, and about half of the TFIV staff went to work for ORHA. The work that they had done as part of TFIV served as useful background for their new responsibilities, and some ideas and concepts from TFIV worked their way into ORHA documents. For example, the ORHA campaign plan drew heavily on the campaign plan that TFIV had developed. However, ORHA was so busy with the challenges of getting itself organized and deployed that it never systematically assessed or integrated the work that TFIV had done.[42]

Because combat operations had commenced by this point, the departing TFIV members who went to work for CFLCC were spread across the command into the areas where additional personnel were needed, rather than kept together to continue working on postwar issues. Some TFIV personnel went to the postwar planning cell that had been established within the CFLCC C5, but that organization focused exclusively on military planning and did not incorporate TFIV's broader focus on interagency coordination. Others went to work for the CFLCC C9, the office that planned civil-military operations. As a result, virtually no institutional memory survived about what TFIV had done. Some individual staff members took their hard drives with them when they left, but no single office took responsibility for maintaining or analyzing the gigabytes of documents that TFIV had produced.[43]

---

[40] Some members of TFIV suspect that British Major General Whitley, then McKiernan's deputy for postwar planning, was responsible for the decision to disband TFIV. Interview with TFIV official, June 2004.

[41] Interview with TFIV official, March 2004.

[42] Interviews with ORHA official, December 2003, and TFIV official, March 2004.

[43] Interview with TFIV official, March 2004.

The Office of Reconstruction and Humanitarian Assistance

Organization and Staffing

On January 9, Douglas Feith, the Under Secretary of Defense for Policy, notified retired Army Lieutenant General Jay Garner that the Department of Defense was about to be tasked to stand up a new office to plan for the postwar administration of Iraq, and that Secretary Rumsfeld wanted Garner to lead this office and assemble its staff.[1] Garner's background seemed ideally suited for this assignment: as a retired Army lieutenant general, he understood the military and the DoD bureaucracy, and as a former commander of Operation PROVIDE COMFORT, he knew a great deal about Iraq and maintained a network of contacts there. Feith told Garner that a wide range of planning efforts had already been conducted throughout the U.S. government and that his job would be to coordinate and integrate these previous efforts rather than to generate new plans.[2] This office, which became known as the Office of Reconstruction and Humanitarian Assistance (ORHA), would start work in Washington; in the event of war, ORHA would deploy to Iraq as the nucleus of the office that would administer the country after conflict ended. Garner was told that he would hand off responsibility to a new presidential envoy within a few months—perhaps even before the organization deployed forward.[3]

After agreeing to take the job, Garner started assembling his staff. He asked two of his former colleagues, retired Lieutenant General Ron Adams and retired Lieutenant General Jerry Bates, to be his deputy and his chief of staff respectively. He also brought in several other retired military personnel, leading one of them to quip that they were the Department of Defense's version of the *Space Cowboys*—referring to the Clint Eastwood movie in which aging astronauts return to space for one final

---

[1] Fineman, Wright, and McManus; Transcript of Jay Garner interview, dated July 17, 2003, for a *Frontline* episode called "Truth, War & Consequences," first aired October 9, 2003. As of October 2007: http://www.pbs.org/wgbh/pages/frontline/shows/truth/interviews/garner.html

[2] Garner interview with *Frontline*.

[3] Garner interview with *Frontline;* Woodward, p. 283.

mission.[4] Once the senior leadership was in place, staffing the rest of the organization proved to be an ongoing challenge. It took several weeks to get a core staff in place, and even then the staff was largely military. It was extremely difficult to get various U.S. government agencies to provide personnel for ORHA—including civilian representatives from the Department of Defense. In late January, Garner met with National Security Advisor Condoleezza Rice and other members of the National Security Council to discuss which federal agencies would provide personnel for specific positions within ORHA. The NSC staff verbally tasked the agencies to support ORHA, but few agencies followed through and provided the requested personnel. In mid-February, senior ORHA officials asked the NSC staff to press the agencies to provide their personnel as quickly as possible. Although the NSC staff agreed to do this, many interagency representatives did not join ORHA until after the organization moved to Baghdad—and some did not arrive until ORHA had been replaced by the Coalition Provisional Authority (CPA) in June.[5]

The initial organization chart for ORHA, shown in Figure 5.1, was drafted through the interagency process described in Chapter Three and was given to Garner when he arrived.[6] It established three separate pillars within ORHA: humanitarian affairs,[7] civil administration, and reconstruction.[8] Each pillar would be led by a senior U.S. government official and would be staffed by lower-level experts in each area. The deputy coordinators for each of the three pillars, as well as the chief of staff, were designated as nondeployable positions to ensure that ORHA maintained an organizational presence in Washington. The chart also contained a fourth branch, called Expeditionary Support Staff, to help get the organization up and running and prepare for its eventual deployment to the theater.

This initial organization scheme raised some significant concerns among the ORHA leadership. It envisaged each of the three pillars operating independently. It lacked some important functions, including political and legal advisors, security, a dedicated communications support team, and public affairs. It tasked OSD to provide the civil administration coordinator, a task that might have been more appropriate for State, and it was not clear why the deputy coordinators of the three pillars

---

[4] Brigadier General (ret.) Buck Walters, quoted in Richard Woods, Tony Allen-Mills, and Nicholas Rufford, "Painful Rebirth of Iraq in Cauldron of Defeat," *The Sunday Times* (London), April 13, 2003.

[5] Interviews with OSD official, November 2003, and ORHA officials, December 2003, March 2004, and April 2004.

[6] Interview with ORHA official, April 2004.

[7] ORHA's role in humanitarian relief is discussed in Chapter Six.

[8] The civil administration pillar started off as a pillar dedicated solely to oil reconstruction. Some ORHA staffers, however, expressed concern that a reconstruction pillar would imply control over all of the other government ministries, even those that did not include reconstruction tasks. The oil pillar's mandate was expanded as a result, and it was renamed civil administration. Interview with ORHA official, August 2004.

Figure 5.1
Initial ORHA Organization Chart

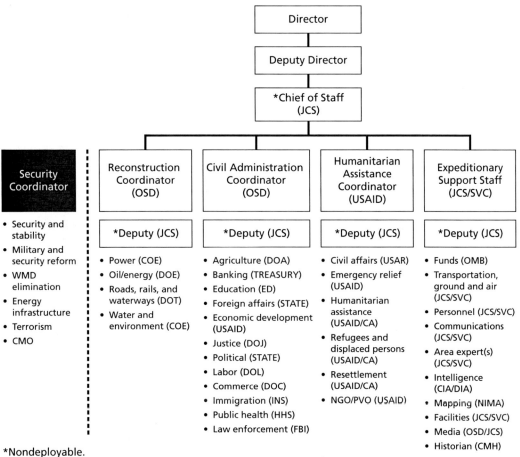

*Nondeployable.

SOURCE: Chart provided to ORHA, January 17, 2003.

RAND MG642-5.1

would be military personnel provided by the Joint Staff rather than civilian personnel from other U.S. government agencies. Perhaps most important, the organization chart contained a strict dividing line between ORHA and a separate Security Coordinator, who would have responsibility for all military, security, WMD, and terrorism issues within Iraq. This separation posed potential problems for ORHA's leadership; they believed that their reconstruction tasks would need to be closely coordinated, if not directly linked, with the security missions defined on the left side of the dotted line in Figure 5.1.[9]

---

[9] Interview with ORHA official, May 2004.

In late January, the ORHA leadership received a revised organization chart, shown in Figure 5.2, which included three major changes. First, the chief of staff and deputy coordinators were no longer described as nondeployable, meaning that ORHA could deploy to the theater as an intact organization. Second, the chart added three regional group directors[10]—one group for the northern part of Iraq (essentially the Kurdish areas), one group in the southern part of the country, and one for Baghdad and other cities in the center—to extend ORHA's reach throughout Iraq.[11] Third, it added details about the role of the new Security Coordinator. The Security Coordinator would now report to the CJTF Planner—intended to be General Hawkins of Task Force IV, though his name was not explicitly noted on the chart—who would report up to the three-star commander of CJTF-Iraq. The chart also added a liaison relationship between the ORHA director and the CJTF-Iraq commander. ORHA leaders thought that a liaison relationship would still be insufficient for the close coordination they anticipated would be needed between reconstruction and security activities. They were also highly concerned that the new structure did not provide sufficient unity of effort between its military and civilian elements. Both ORHA and CJTF-Iraq would ultimately report to the Secretary of Defense, but no single person in theater could coordinate and oversee the actions of both organizations.[12]

The ORHA staff started assembling and quickly found that the many administrative issues involved in setting up the organization left little time for substantive issues and long-term planning. Because it had only recently been established, the first members of the staff were crowded into a small space in the Pentagon that had few desks, phones, or computers. New staff members were arriving every day, which not only posed ever-increasing requirements for office space and supplies but also required a great deal of time for orientation, training, and personal introductions. ORHA also had to prepare for deployment to the theater on short notice, which involved medical exams, weapons training and certification, and many personal arrangements. The ORHA staff did manage to do some substantive work while juggling the many administrative demands but lacked the time, energy, and senior-level

[10] Later, some controversy arose over the decision to divide Iraq into three regions, because it seemed to reify some of the country's ethnic and sectarian differences. ORHA maintained that these divisions were based solely on geographic criteria, and that Iraq was divided into regions only to facilitate administrative management of the country. Interview with ORHA official, May 2004.

[11] The regional group director for the north was Bruce Moore, a retired Army major general; the director for the south was Roger "Buck" Walters, a retired Army brigadier general; and the director for the central region was Barbara Bodine, a former U.S. ambassador. David Phillips, *Losing Iraq*, New York: Westview Press, 2005, pp. 126–127.

[12] Garner reported directly to the Secretary of Defense, while the CJTF commander reported to the CENTCOM commander who then reported to the Secretary of Defense. Interview with ORHA official, May 2004.

Figure 5.2
Revised ORHA Organization Chart

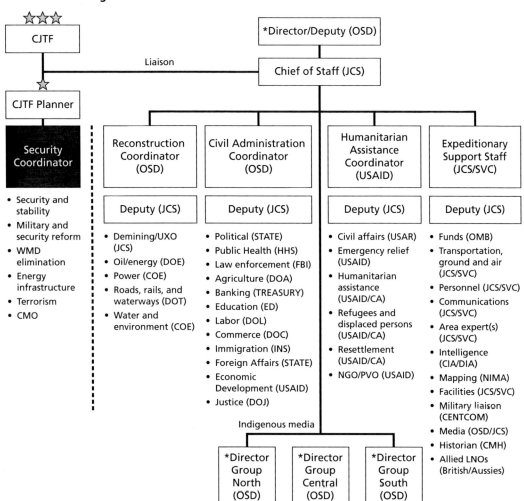

attention required for real strategic planning.[13] The staff nevertheless firmly believed in the mission they were being asked to conduct and developed such a strong esprit de corps that they soon started referring to themselves as "ORHAnians."[14]

Even though new personnel showed up regularly through February and March 2003, staffing remained an ongoing challenge for ORHA. Many U.S. government

---

[13] Interviews with ORHA officials, November 2003 and December 2003.

[14] Interviews with ORHA officials and U.S. military officers, November 2003, December 2003, April 2004, and May 2004.

agencies delayed or resisted providing personnel to ORHA, despite being directed to do so by the NSC staff. ORHA leaders were particularly frustrated by the failures of both civilian and military agencies within the Department of Defense to provide requested personnel. For example, when Garner asked the Joint Staff for the logistics, transportation, personnel, and communications experts it was tasked to provide for ORHA, the Joint Staff responded that most of these functions could not be provided until ORHA arrived in the theater—though none of these were in fact provided when ORHA arrived in Kuwait or Baghdad.[15] ORHA only deployed 151 staff members to Kuwait on March 16, even though Garner had estimated that ORHA would need a minimum of 300 people before it deployed.[16]

ORHA Planning

As ORHA got up and running, Garner identified several issues it would have to be prepared to address in the immediate aftermath of the war. First and foremost, Garner expressed concern that the war would trigger a major refugee crisis, particularly if Saddam Hussein used chemical weapons. Garner also worried that Saddam might choose to set the oil fields on fire and explode dams throughout the country, which would require extensive cleanup efforts; that epidemics of cholera and other diseases would erupt as sanitation and electrical systems failed; and that starvation would spread as the food distribution system broke.[17] In these areas, ORHA was able to draw on the extensive interagency planning efforts for humanitarian issues, which are discussed at length in the next chapter.

Garner also identified one other major problem: ORHA would have to ensure that the Iraqi ministries continued to function between the fall of Saddam Hussein and the establishment of a new permanent government.[18] ORHA, like many other organizations within the U.S. government, assumed that the Iraqi ministries would remain functional after the war. Although little was known about the inner workings of these ministries, Saddam's regime depended on a highly centralized government structure in which all important decisions were made in Baghdad.[19] It was therefore

---

[15] Interview with ORHA official, April 2004.

[16] Interview with ORHA official, May 2004.

[17] Garner interview with *Frontline*, July 2003; interview with former U.S. government official, March 2004.

[18] Garner interview with *Frontline*, July 2003.

[19] The U.S. intelligence community had great difficulty determining how the Iraqi government worked during Saddam's rule, because it was extremely hard to get reliable human intelligence (HUMINT) reporting from sources within such an absolutist regime. The intelligence community provided its best estimates of the strength and functioning of the Iraqi government, but after the war it became clear that much of its HUMINT was unreliable and inaccurate. Interviews with OSD official, November 2003, and ORHA official, November 2003.

assumed that the senior-most levels of ministry leadership—the minister and some senior Ba'athist officials—could be removed and replaced without substantially undermining the work of the ministries.[20] The large civil service staffs in the ministries would keep them running under new leadership.

How would this new leadership be determined? A Ministerial Advisory Team would be established for each ministry, and each team would be organized in the same way. The team would consist of a senior advisor, a senior government official from the United States or coalition partner who would guide ministry decisions but would retain veto authority during the transition period; expatriate Iraqi technocrats, who would provide technical expertise in the issues addressed by that ministry; and the "last Iraqi standing," who would be the most senior technocrat remaining in the ministry after the top-level Ba'athists were removed and would oversee the rest of the ministry personnel. Working together, the members of the team would serve as caretakers of the ministry until a permanent government was named.[21]

The February Rock Drill

On February 21 and 22, 2003, ORHA held a meeting at the National Defense University to vet the plans developed by each of ORHA's three substantive pillars. The meeting, which became known as the February rock drill,[22] included representatives from every U.S. government agency that would have a role in reconstruction, as well as from CENTCOM and TFIV.[23] Garner later noted that this "standing-room only" meeting involved several hundred people.[24] A large amount of time was spent discussing the Ministerial Advisory Teams that would be established for 23 Iraqi ministries.[25] As summarized in Table 5.1, the discussion of each ministry included an overview of its specific functions, the U.S. lead agency in charge of reconstruction, the name of the senior advisor, and the Iraqi expatriates who would be involved.

---

[20] One source reports that the number 50 came up frequently during discussions of how many senior officials would need to be removed. Gordon Corera, "Iraq Provides Lessons in Nation Building," *Jane's Intelligence Review*, January 2004, p. 31.

[21] Interviews with OSD official, November 2003, and ORHA official, November 2003.

[22] As Garner later explained, the meeting was called a rock drill because "you turn over all the rocks" to identify any problems or shortcomings with the plan. Garner interview with *Frontline*, July 2003.

[23] This meeting marked the first time that ORHA had any official contact with TFIV. Interview with ORHA official, March 2004.

[24] Garner interview with *Frontline*, July 2003. One participant quipped that the meeting was so big that it "was like a cast call for the remaking of *Ben-Hur*." Fineman, Wright, and McManus.

[25] Upon arriving in Baghdad, ORHA officials were quite surprised to learn about the existence of a few other ministries that the intelligence community had not previously known about, including the Youth Ministry. They had also been told that the Atomic Energy Ministry had been disbanded, but they discovered that it had been demoted from a ministry to an office and was still fully functional. Interview with ORHA official, November 2003.

The February rock drill was supposed to ensure that all interagency members knew what their responsibilities were and to help them define timetables, identify shortcomings, and generate personnel requirements. However, as Garner later noted, the meeting exposed "tons of problems" with planning and coordination efforts.[26] Two particularly serious shortcomings were identified, both of which foreshadowed problems that later plagued ORHA and, ultimately, CPA. First, it proved difficult to get various organizations within the U.S. government to provide personnel for the

Table 5.1
Planned Ministerial Advisory Teams, February 2003

Ministry	Functions	U.S. Lead Agency	Senior Advisor	Iraqi Expatriates
Agriculture	Land ownership Price and marketing supports Animal/crops production support Iraqi grain board	USAID	Lee Schatz, U.S. Department of Agriculture	Jamal Fuad Awshalim Lazar Khamu Zuher Mushan Sarbagh Saleh
Central Bank	Bank supervision Currency issuance Exchange rate management Monetary policy Trade finance	Treasury	George Mullinax, Treasury	N/A
Culture	Preservation of cultural sites and museums Oversight of all publications and artistic expression Propaganda Tourism	State	Ambassador Lange Schermerhorn Interim: Ambassador John Limbert	N/A
Defense	Border control National defense	DoD	Walter Slocombe Interim: George Oliver	N/A
Education	Primary and secondary education Study abroad programs	USAID	Interim: Dorothy Mazaka, USAID	Munaf Al-Tikriti Fuad Hussein Hind Makiya
Electricity Commission	Generation, transmission, distribution of electricity	U.S. Army Corps of Engineers	Interim: Andy Backus	N/A
Finance	Budget Customs duties collection Debts management Taxation	Treasury, USAID	David Nummy, Treasury	Sinan Al-Shabibi
Foreign Affairs	Diplomatic representation Diplomatic - public policy Consular affairs Protocol Research and analysis	State	Ambassador Keith Kenton Interim: Drew Erdmann, State	Mowafak Abboud Kais Al-Fahed

---

[26] Fineman, Wright, and McManus.

Table 5.1 (continued)

Ministry	Functions	U.S. Lead Agency	Senior Advisor	Iraqi Expatriates
Health	Hospital management Child malnutrition programs Medicine production, import, distribution Social welfare programs	USAID	Skip Burkle, USAID	Said Hakky Abdul Khaliq Hussein
Higher Education and Scientific Research	Supervision and administration of universities Iraqi national computer center Diplomatic – public policy Training/testing facilities of WMD Recruitment of scientific personnel for WMD pursuits	DoD	DoD, Vacant Interim: Mike Ayoub, State	Mohammed Al-Najim Muhammed Al-Ruubai Mustaft Al-Shawi Farouk Darweesh Khidhir Hamza
Housing and Construction	Construction of bridges, roads, mosques, and buildings Land surveying Urban planning Repair of civil airports and seaports Operations and maintenance of government facilities	USAID	Dan Hitchings, U.S. Army Corps of Engineers Interim: Albert Bleakley	Suher Al-Musely Usef Aziz Seherety
Industry and Materials	Production of military equipment (conventional) Generation and distribution of electricity Treatment and distribution of water	DoD	Ambassador Tim Carney	Ramsey Jiddou
Information	Management and operation of internet/Iraqi web sites Control of print media, radio, TV Monitoring of foreign journalists	DoD	James Woolsey Interim: Bob Reilly, DoD	Sam Alaskary
Interior	Administration of provinces (local government) Census Sanitation Internal security, law enforcement, prisons Border and immigration control Oversight of president's farms	Justice	Robert Gifford Interim: Dick Mayer, Justice	Ibraham Mostafa Jamal Ildin Sami Izara Al Majoun Sadoon Al-Dulaimi Brig. Gen. Tawfeek Al-Yasiri
Irrigation	Dams and reservoirs Irrigation projects River maintenance Water use/exploration	U.S. Army Corps of Engineers (USACE)	Gene Stakhiv, USACE Interim: Steve Browning, USACE	Hassan Janabi
Justice	Judicial appointments Supreme law college Administration of public prosecution Administration of court system	Justice, State/INL	Clint Williamson Interim: Bill Lantz, Justice	Riyadh Abdel Majid Prof. Munther Al Fadhal Salem Cheleby Hon. Zakia Hakki

Table 5.1 (continued)

Ministry	Functions	U.S. Lead Agency	Senior Advisor	Iraqi Expatriates
Labor and Social Affairs	Labor laws/labor conditions Social security Social welfare program Women's issues	USAID	Interim: Karen Walsh	Lori Latif Shak Bernard Hanish Khalid Hassan
Military Industriali-zation	Foreign procurement Military production (WMD) Arms development Sanctions evasion (through front companies)	DoD	DoD – TBD Interim: Tom Gross	Garabeet Ishaqian
Oil	Exploration, production, distribution, and marketing of crude oil Refining and distribution of petroleum products	USACE	DoD – TBD Interim: Gary Vogler, Dept. of Energy	Ibrahim Allolum Ali Katham Fathel Uthman Muhammed Ali Zainy
Planning	Census and statistical recording Development, investment plans, economic reporting Telecommunications networks (intercepts)	State	Ambassador Dave Dunford Interim: Bob Macleod	Isam al Khafaji Mohamed Alhakim Osama Al-Naeb
Religious Affairs	Religious endowment Supervising mosques, Islamic schools and Islamic charities Oversight of imams/ Friday prayers Collection/distribution of zakat	State	Ambassador John Limbert	Ali Al-Qizweeni Bassim Abbas Hilmi Wallid Saleh
Trade	Administration of Oil for Food program (contracting) Iraq-Jordan trade protocol, other trade pacts Computing center	State	Ambassador Robin Raphel Interim: Frank Ostrander	Bassim Abbas Hilmi
Transport and Communica-tions	Air traffic control system, maritime traffic Management of bridges, roads, civil airports, seaports, telecommunications networks, broadcast media, and telephones Distribution of food from Oil for Food program Transportation of weapons Management and operation of internet Network construction (telecommunications and posts)	USAID, State/EB, USACE	Dick Beard, State Interim: Stephen Browning	Saleh Almumayiz Najah Kadhim Majed Hanun Sam Kareem

SOURCE: ORHA briefing slides, February 21 and 22, 2003.

ministerial teams, particularly given the very short timeframes involved. ORHA had existed for barely a month by the time of the February rock drill, which did not allow much time to identify appropriate personnel and have them arrange for leaves of absence from their current positions.[27] This posed a particular problem for those designated as senior advisors, since most of these people held high-level positions within the U.S. government and could not quickly transfer their responsibilities to a successor. As Table 5.1 shows, interim senior advisors were named for 17 out of the 23 ministries. Few of the permanent senior advisors joined ORHA before it deployed to Kuwait in March 2003, and some had still not arrived in theater by the time that ORHA was replaced by CPA in June.[28]

But time was not the only reason why ORHA had difficulty getting secure commitments from U.S. government personnel. Bureaucratic resistance also played a role, with many agencies wary of putting their personnel under the control of a new organization within the Department of Defense. Furthermore, senior U.S. government officials often reviewed the personnel selected to work for ORHA (and later CPA) and occasionally intervened in the personnel process. In one well-known case, Garner was overruled when he attempted to get Tom Warrick, the director of the Future of Iraq project at the State Department, to join ORHA. Warrick had briefed some of the findings of this project at the February rock drill, which was the first time Garner had heard of it. Garner asked Warrick to work for ORHA, and Warrick readily accepted. Three days later, however, Secretary Rumsfeld told Garner that Warrick would have to be removed from the team. Garner protested the decision, arguing that ORHA needed Warrick's expertise,[29] but Rumsfeld responded that the decision had come from "such a high level [of the U.S. government] that I can't overturn it."[30] According to Bob Woodward, Rumsfeld later explained this decision to Secretary of State Powell by saying that work on postwar planning "had to be done by those who were truly committed to this and supportive of the change and

---

[27] Knowing that it was critical to their efforts, ORHA officials had hoped to hold such a coordination conference even earlier, but they were unable to make the arrangements in time. As it turned out, some of them were frustrated with the level of coordination that did occur as a result of this conference. One participant described the rock drill as a BOGSAT—a derogatory acronym for "bunch of guys sitting around a table"—rather than a real analysis of the problems that lay ahead. Interview with ORHA official, April 2004.

[28] Some of those named in the table never went to Iraq at all. Interview with State Department official, January 2004; interview with ORHA officials, April 2004 and August 2004.

[29] As George Ward, the ORHA official responsible for humanitarian operations, later noted, ORHA saw Warrick's knowledge as an important asset, because "we had few experts on Iraq on the staff." Eric Schmitt and Joel Brinkley, "State Dept. Study Foresaw Trouble Now Plaguing Iraq," *New York Times*, October 19, 2003.

[30] Corera, pp. 30–31. Garner later speculated that "people in the executive branch" did not like some of the work that Warrick had done, and that Warrick "just wasn't acceptable" to them. Garner interview with *Frontline*, July 2003. Rumsfeld's comment has led to speculation that the decision was made by Vice President Cheney or someone in his office. Peter Slevin and Dana Priest, "Wolfowitz Concedes Errors on Iraq," *Washington Post*, July 24, 2003; John Barry and Evan Thomas, "The Unbuilding of Iraq," *Newsweek*, October 6, 2003; Fallows, p. 72; Woodward, pp. 283–284.

not those who have written or said things that were not supportive."[31] Despite continuing appeals, Warrick never returned to ORHA. ORHA staffers remained largely unaware of the details contained in the Future of Iraq project's report; those who were aware of it noted that they were not sure what happened to the project or its findings.[32]

The February meeting also revealed a second serious shortcoming: several critical issues remained unaddressed, including the crucial question of who would provide security in postwar Iraq. When this question arose during the meeting, CENTCOM's representatives made clear that the military would only be responsible for removing Saddam Hussein from power and defeating his forces, not for providing security after the fall of the regime.[33] It was not clear, however, who would be responsible for ensuring law, order, and stability if the military did not take on this mission.[34] Other unresolved issues that surfaced during the meeting included how the Iraqi military would be restructured; court reform; the role of indigenous media; and salaries for civil servants, police officers, and military personnel. ORHA officials raised these unresolved issues many times during the subsequent months, particularly during videoconferences with General Franks, Secretary Rumsfeld, and NSC officials. Yet few of these issues were addressed until ORHA was replaced by CPA in June 2003.[35]

ORHA's Role in Reconstruction

ORHA's primary mission was to facilitate reconstruction efforts after the fall of the regime rather than directly conduct reconstruction activities itself.[36] USAID would be responsible for letting most of the major reconstruction contracts, as discussed in Chapter Three, and it was in the process of picking contractors for more than $1 billion worth of contracts when the war started. Yet the USAID personnel in theater

---

[31] Woodward, p. 284. David Phillips suggests that one reason for this decision may have been the fact that Warrick had criticized the Defense Department during a meeting with Iraqi exiles in Michigan. According to Phillips, one exile recalled that Warrick told those assembled at the meeting that "if you work with Paul Wolfowitz, the State Department will not give you anything." See Phillips, *Losing Iraq*, New York: Westview Press, 2005, p. 127.

[32] Stephen J. Glain, "Chaos Thrives in Baghdad Despite Prewar Planning," *Boston Globe*, June 10, 2003.

[33] ORHA officials later put together what they described as the "cop on the beat" briefing, which showed that coalition military forces would have to be prepared to provide security if the Iraqi police forces were not up to the task. The briefing went at least as high as the Deputies Committee. The consistent response to this briefing was that the military was not going to undertake that task, with no clarification about who would be responsible. Interview with ORHA official, April 2004.

[34] ORHA officials were particularly concerned about their own security, since they had been explicitly told that security would be provided for them when they arrived in Iraq. ORHA ended up hiring 100 Ghurkas and a personal security team from South Africa to provide security for the organization. Interviews with ORHA officials, January 2004, March 2004, and April 2004.

[35] Interviews with ORHA officials, November 2003, January 2004, and March 2004.

[36] Its mission also included facilitating humanitarian relief assistance, as discussed in Chapter Six.

reported to Garner, as did all other U.S. personnel involved in reconstruction efforts.[37]

The contracts that were let during this time included some fairly ambitious plans and timetables for reconstruction activities. According to one report, ORHA and USAID plans for rebuilding the Iraqi school system called for the following tasks to be accomplished within one year: launching a new nationwide accelerated learning program; getting "all children back in school," since only one-third of Iraqi children had been attending secondary school before the war; providing books and other supplies to all of Iraq's students; and ensuring that 25,000 schools would be able "to function at a standard level of quality."[38] At the end of the year, responsibility for the school system would be returned to the Ministry of Education. Similarly, USAID sought a private contractor to run the Iraqi health system, which consisted of 270 hospitals and 1,000 clinics. The USAID contract proposal in this sector stated that after a year, the Ministry of Health "will be reformed and prepared to take over operation of the health care system."[39]

Deploying to Kuwait

After the February rock drill, ORHA continued planning and absorbing new staff members at a rapid rate. At the end of the first week of March, a senior U.S. Department of Defense official told ORHA that it should deploy to Kuwait immediately—a move that many at ORHA interpreted as a signal that war with Iraq would most likely commence soon.[40] Since ORHA had been focusing on establishing itself in Washington, the members of ORHA spent a week scrambling to get ready for the deployment—getting inoculations, processing forms, receiving training in weapons handling and chemical protection, and so on—while still absorbing new staff members.[41] The organization grew from 90 to 151 people during this time, and all of the new staff members had to prepare for deployment as well. Many members of ORHA thought that they needed more time in Washington to solidify their relationships with other U.S. government agencies, especially since the revised ORHA structure (shown in Figure 5.2) called for the entire organization to deploy to the theater with-

---

[37] Neil King, Jr., "Bush Has an Audacious Plan to Rebuild Iraq Within a Year," *Wall Street Journal*, March 17, 2003; Karen DeYoung and Dan Morgan, "U.S. Plan for Iraq's Future Is Challenged," *Washington Post*, April 6, 2003.

[38] Quotes taken from the USAID contract proposal. Also see King.

[39] King.

[40] Interviews with OSD official, November 2003, and ORHA official, April 2004.

[41] The majority of ORHA staff members were civilians rather than military personnel, which meant that many people were receiving this kind of training for the first time. Interview with ORHA official, April 2004.

out leaving a back office in Washington.[42] Garner told his staff that he thought the deployment would serve as a forcing mechanism: once ORHA got to Kuwait, agencies would have no choice but to support ORHA by breaking through bureaucratic and logistics logjams.[43]

ORHA staff members started deploying to Kuwait as soon as they were ready, and once Garner arrived on March 16, the deployment was complete. However, ORHA discovered that no one in the theater had been expecting them. General McKiernan, the CFLCC commander, did not want ORHA to collocate with his staff at Camp Doha, because he worried about the security implications of allowing a mostly civilian organization within the perimeter of his military headquarters.[44] In the absence of any alternatives, ORHA started setting up its headquarters at the Kuwait Hilton, where it remained for several weeks. The Hilton was approximately 45 minutes away from Camp Doha, and the normal coordination problems caused by that distance were exacerbated by the lack of functioning communications in theater and the requirement that ORHA personnel go through extremely lengthy security checks before being allowed inside the Camp Doha perimeter.[45] Furthermore, ORHA had to build an office infrastructure from scratch, since it had never planned to locate at the Hilton. It had no desks, chairs, telephones, computers, internet access, or office supplies; thus, it had to scavenge for all these resources for the second time since its inception.[46]

ORHA faced many challenges while in Kuwait. One of the most important was that it never developed an effective working relationship with CFLCC. The exact command relations between the two organizations were a source of confusion at the outset,[47] but it was clear that the two organizations would need to work together closely so that ORHA could access the land assets that CFLCC controlled. In practice, however, CFLCC's warfighting needs always had a higher priority than ORHA's reconstruction needs. Whenever ORHA requested the security, communications, and transportation support that the Joint Staff had promised would be available in theater, CFLCC responded that those assets were not available because they were involved in combat tasks. While no one at ORHA wanted to undermine the

---

[42] Interviews with OSD official, November 2003, and ORHA officials, December 2003 and May 2004.

[43] Interview with OSD official, November 2003.

[44] Interview with ORHA official, April 2004.

[45] Interviews with ORHA official, November 2003, and TFIV official, March 2004.

[46] Interviews with ORHA officials, November 2003, December 2003, March 2004, and April 2004.

[47] Garner and many other ORHA staff reportedly told the CFLCC staff that ORHA worked directly for Secretary of Defense Rumsfeld. CENTCOM and CFLCC, by contrast, believed that ORHA had been put under the OPCON of CFLCC during Phase IVa. Donald R. Drechsler, "Reconstructing the Interagency Process After Iraq," *Journal of Strategic Studies*, Vol. 28, No. 1, February 2005, p. 20.

war effort, they grew frustrated that reconstruction needs were always subordinated to combat needs.[48]

Furthermore, ORHA's planning efforts were limited by its lack of knowledge of the content of the warplans until shortly before the war started. Garner's original concept of operations called for ORHA to go into Basra and start reconstruction efforts as soon as U.S. and coalition military forces secured that city. During the second week of March, Garner reportedly had a videoconference with General McKiernan, who informed him that the warplans called for most military forces to go straight into Baghdad. McKiernan told Garner that he should not expect to go into Basra, immediately rendering many of ORHA's plans obsolete. The Phase IV plan envisaged ORHA entering Baghdad after 120 days, by which time most of the country would be pacified and the military would have the necessary resources available to support this civilian organization.[49] This came as quite a shock to ORHA, especially coming just a few days before the onset of major combat operations. Garner had briefed his concept of operations to senior civilian and military officials for several weeks, without ever being informed that the warplans might require him to operate differently. ORHA could have developed an alternative plan if it had known crucial components of the warplans from the outset of its efforts.[50] Furthermore, ORHA was directed to remain in Kuwait during major combat operations. General McKiernan argued that he did not have enough forces to provide security for ORHA, and that ORHA would therefore have to wait in Kuwait until combat ended and the security situation stabilized.[51] Not only did this render ORHA's plans ineffective, but once Baghdad fell and the looting started, it exposed ORHA to the charge that it was doing nothing to stop the destruction of the country.

While in Kuwait, ORHA continually refined its reconstruction plans. The ORHA staff held meetings almost every day to posit different scenarios and determine how they would provide humanitarian assistance, rebuild infrastructure, and develop local governance in each one.[52] They developed detailed timetables and checklists of issues that would need to be addressed during the initial transition period. Garner repeatedly emphasized to his staff—and later to the Iraqi people—that ORHA's mission remained limited and that they should be prepared to "work their way out of a job within 90 days."[53] This led some observers to question whether

---

[48] Interviews with ORHA officials, November 2003, December 2003, and April 2004.

[49] Drechsler, p. 20.

[50] Interviews with ORHA officials, December 2003 and April 2004.

[51] Interviews with ORHA officials, December 2003 and April 2004.

[52] Interview with ORHA official, April 2004.

[53] Susan B. Glasser and Rajiv Chandrasekaran, "Reconstruction Planners Worry, Wait and Reevaluate," *Washington Post,* April 2, 2003.

ORHA's plans would adequately address the complexity of the postwar situation in Iraq. One participant likened a briefing in one of ORHA's planning sessions to "a Boston Consulting Group presentation to IBM. It was so different than what the situation really is in Iraq. That is going to be a big, big shock to them."[54]

Arriving in Baghdad

Garner entered Iraq for the first time on April 11, 2003—two days after the toppling of the statue of Saddam Hussein—on a day-long visit to the southern port of Umm Qasr.[55] He traveled to Nasiriya a few days later but again returned to Kuwait at the end of the visit.[56] Garner understood that CENTCOM did not want ORHA to go into Baghdad while the security situation remained unstable, but he grew increasingly concerned that they were losing valuable time for the reconstruction process. On April 17, Garner met with General Franks in Qatar and later described the meeting as follows:

> [I] said, "You got to get me into Baghdad." He said, "You know, it's really hot there right now, it's really going to be hard to protect you." I said, "I think we'll take our chance." He said, "Well, let me talk to the military commanders." It was either the night of the 17th, [or] the night of the 18th, he called and said, "Go ahead, and we'll give you all the support we can."[57]

An ORHA advance party went into Baghdad on April 19. Garner arrived in Baghdad on April 21, and the majority of the ORHA staff followed on April 24.[58]

ORHA quickly discovered that conditions in Iraq were quite different from the conditions they had anticipated. They had expected and planned for an extensive humanitarian crisis, but as discussed in the next chapter, that crisis never materialized. As Garner later noted, "I thought the first 30 days were going to be all food, water, and medicines."[59] Instead, they were confronted by two major reconstruction challenges that they had not expected. First, looting destroyed far more of Baghdad's infrastructure than ORHA had ever anticipated. The war itself caused little damage

---

[54] Glasser and Chandrasekaran.

[55] Garner had hoped to travel farther north to Basra, but his aides reportedly said that the longer trip would not be safe. Jane Perlez, "Aid Groups Urging Military to Protect Essential Services," *New York Times*, April 12, 2003.

[56] See Chapter Ten for more on Garner's meetings in Nasiriya.

[57] Garner interview with *Frontline*, July 2003.

[58] Monte Reel, "Garner Arrives in Iraq to Begin Reconstruction," *Washington Post*, April 22, 2003; interview with ORHA official, March 2004.

[59] David Luhnow, "Shortages, Lack of Time to Plan Bedevil New U.S. Agency in Iraq," *Wall Street Journal*, June 5, 2003.

to infrastructure, largely because of heavy reliance on precision weaponry, but the looting destroyed almost three-quarters of the government ministries and many other structures and facilities.[60] As Garner explained,

> What happened in Baghdad is not only did they take everything out of the build-ings, but then they pulled all the wiring out of the buildings, they pulled all the plumbing out of the buildings, and they set it on fire. So the buildings were not useable at all. In fact, some of them probably are not structurally sound enough to ever be used—they'll have to be torn down and rebuilt . . . I knew that there would be looting. I think all of us knew that. But I never anticipated we would not be able to use the buildings, unless they were destroyed by the military.[61]

The destruction of the ministries also made it very difficult for ORHA to iden-tify ministry personnel, since they had nowhere left to work. The ministerial senior advisors had to rely on word of mouth to identify these people. As Garner recalls, "we literally had to put our people on the streets of Baghdad walking around and asking if they knew anyone" who worked in the various ministries.[62] The ministries set up temporary offices within the palace that ORHA used as its own headquarters, with each ministry identified by a piece of paper taped on the outside of the door.[63] These conditions meant that most of the ministerial advisory teams could not take over functioning ministries as planned but, instead, had to start with the very chal-lenging task of rebuilding the ministries, often from scratch.

Continuing combat operations posed the second unanticipated challenge to ORHA. All the military plans for Iraq envisaged that combat operations would come to a definitive end, to be followed by post-conflict stability operations in a relatively benign environment. Instead, both kinds of operations ended up overlapping to a great extent. The fall of Saddam's regime created a power vacuum that enabled for-mer regime loyalists, insurgents, and foreign fighters to compete with one another for power, as well as organize and execute attacks on coalition forces. The unsettled secu-rity environment slowed reconstruction efforts because of the ongoing looting and sabotage and the increasing fearfulness of the Iraqi population.

ORHA therefore found itself operating in a risk-filled environment, in which U.S. and coalition military forces continued to conduct military operations against the growing insurgency. As ORHA grew increasingly dependent on U.S. military forces to provide security, it became more difficult for ORHA to establish effective working relationships with Iraqis, which were essential for reconstruction. Garner

---

[60] The headquarters of 17 out of the 23 government ministries were destroyed. Garner interview with *Frontline*, July 2003.

[61] Garner interview with *Frontline*, July 2003.

[62] Corera, p. 31. See also Garner interview with *Frontline*, July 2003.

[63] Luhnow.

told Lieutenant General William Wallace, the commander of V Corps, that ORHA needed security provided in 30 different locations around the city. Wallace responded that he did not have enough forces to fill all the requests for security he was receiving and that he could only provide ORHA with some, not all, of the security forces Garner had requested. This limited the ability of ORHA personnel to move around Baghdad and get their work done. In some cases, ORHA personnel were stranded at their ministries because they were not supposed to leave without security escorts.[64] This problem was never satisfactorily resolved, and it limited the amount of contact that ORHA (and later CPA) officials could have with the Iraqi people. As one British ORHA official later noted, "at the U.S. military's insistence, we traveled out from our fortified headquarters in Saddam's old Republican Palace in armored vehicles, wearing helmets and flak jackets, trying to convince Iraqis that peace was at hand, and that they were safe."[65]

The Transition to CPA

Unbeknownst to most of its staff, ORHA was already being phased out of existence within days of arriving in Baghdad. On the evening of April 24, Secretary Rumsfeld called Garner and told him that President Bush planned to appoint L. Paul Bremer as his permanent envoy to Iraq.[66] The reasons for the timing of this decision remain unclear. From the outset, the plan was to replace Garner with a presidential envoy. Yet the fact that Garner was notified of this transition almost as soon as he arrived in Baghdad led to much speculation that he was being held responsible for the looting and the chaos that emerged in his absence—despite the fact that the military commanders had prevented him from entering the country any earlier.[67] Press reports stated that the decision to replace Garner with Bremer was reached hastily, after President Bush and his senior advisors decided that both the perception and the reality of reconstruction efforts were deteriorating.[68]

---

[64] Corera, p. 32.

[65] Larry Hollingworth, quoted in David Rieff, "Blueprint for a Mess," *New York Times Magazine*, November 2, 2003.

[66] Garner interview with *Frontline*, July 2003; Fineman, Wright, and McManus.

[67] One source speculated that the decision may have been influenced by U.S. special envoy Zalmay Khalilzad and Assistant Secretary of State Ryan Crocker, who visited Baghdad in late April and were appalled at conditions in the city. They reportedly returned to Washington with a warning that civil order had collapsed and that local warlords or Iranian agents might move to fill the vacuum. Joshua Hammer and Colin Soloway, "Who's in Charge Here?" *Newsweek*, May 26, 2003, pp. 28–29.

[68] Karen DeYoung, "U.S. Sped Bremer to Iraq Post," *Washington Post*, May 24, 2003.

U.S. officials first discussed Bremer's appointment with the press on May 1[69] and officially announced the appointment on May 6.[70] It immediately set off speculation that Garner had been fired because ORHA was unable to address the postwar challenges.[71] Garner has explicitly denied that he was fired, stating that the plan had always been for him to stand up the organization and then transfer power to a permanent presidential envoy after a few months.[72] Others, both inside and outside the administration, interpret these events less charitably, arguing that their timing was designed to convey the impression that Bremer was making a clean start, even though Garner did exactly what senior U.S. officials had told him to do. Critics expressed great frustration that no one in the Bush administration said anything to support Garner after this dismissal or to correct the impression that he had been fired.[73]

Bremer arrived in Baghdad on May 12, 2003,[74] with a mandate to create a new Coalition Provisional Authority (CPA). Whatever the exact circumstances of Garner's removal, CPA represented a fundamental shift in U.S. policy. Whereas ORHA was a temporary organization designed to assist a new Iraqi government during a short transitional period, CPA would possess all the powers of an occupation authority. It therefore signified an implicit acknowledgement that the reconstruction of Iraq would be a much longer and more complicated endeavor than U.S. policymakers had anticipated. Bremer arrived from Washington with what Garner later referred to as "a suitcase full of directives,"[75] which represented major changes in U.S. policy.[76] The two decisions that proved to be the most controversial later on—disbanding the Iraqi military and de-Ba'athification—deliberately changed the policies that had been approved through the interagency process earlier in the year and implemented by ORHA.

The transfer of power from ORHA to CPA did not progress smoothly. Bremer arrived in Baghdad with a staff that he had assembled in Washington and started

---

[69] Mike Allen, "Expert on Terrorism to Direct Rebuilding," *Washington Post*, May 2, 2003; Steven R. Weisman, "U.S. Set to Name Civilian to Oversee Iraq," *New York Times*, May 2, 2003.

[70] "President Names Envoy to Iraq," White House, Office of the Press Secretary, May 6, 2003; James Dao and Eric Schmitt, "President Picks a Special Envoy to Rebuild Iraq," *New York Times*, May 7, 2003.

[71] Hammer and Soloway, pp. 28–29.

[72] Garner argues that the misperceptions surrounding this decision resulted from "DoD or the administration or whoever was in charge [doing] a very poor job of prepping the press on what the plan was." Garner interview with *Frontline*, July 2003.

[73] Interviews with ORHA officials, November 2003, December 2003, and April 2004, and with State Department official, January 2004.

[74] Patrick E. Tyler, "New Overseer Arrives in Iraq," *New York Times*, May 13, 2003.

[75] Corera, p. 32.

[76] Although RAND researchers could not definitively determine who was involved in making these decisions, the wide range of officials interviewed for this report unanimously agreed that they were made in Washington before Bremer arrived in Iraq.

building the new organization largely from scratch. Although Garner and Bremer worked together closely, their staffs did not. Many ORHA officials grew frustrated that Bremer's staff not only assumed control so quickly, but also largely dismissed the work and expertise that ORHA had generated during the past six months as not relevant to the new mission.[77] As CPA staff started to grow, the ORHA staff started to shrink. Some ORHA staff members stayed on to work for CPA, but most of them decided to return home. Garner left Iraq on June 1, almost two weeks after ORHA was replaced by CPA.

---

[77] Interviews with U.S. military officers, November 2003 and December 2003, and with former U.S. government officials, March 2004 and April 2004.

CHAPTER SIX

Humanitarian Planning

The U.S. government conducted extensive planning for the humanitarian relief efforts that any war with Iraq might require. An interagency planning team started meeting in the fall of 2002 and worked with international organizations (IOs) and nongovernmental organizations (NGOs) to generate detailed humanitarian relief plans for a wide range of possible scenarios. As it turned out, the war in Iraq did not generate a significant humanitarian emergency, largely because of the ways in which the war was fought. This does not mean that the humanitarian planning efforts focused on the wrong areas: they helped ensure that humanitarian concerns were factored into the warplans, and worst-case scenario planning for any potential humanitarian disaster is prudent. However, it is important to note that humanitarian relief plans are *not* the same as reconstruction plans. Relief requires intensive short-term efforts, while reconstruction requires steady efforts over a prolonged period. The preparations for providing food, water, and shelter to needy Iraqis stand in contrast to the problems with reconstruction planning noted in previous chapters.

Interagency Humanitarian Planning

During the summer of 2002, representatives from the Office of the Secretary of Defense, CENTCOM, and the Joint Staff started meeting to develop humanitarian relief plans for a possible war in Iraq.[1] They soon expanded to form a Humanitarian Planning Team (HPT), which brought in representatives from the NSC staff, USAID, and the State Department.[2] In late September 2002, the HPT started working with the Iraq Relief and Reconstruction working group (IR+R), one of the inter-

---

[1] The OSD representatives came from the Stability Operations office within the office for Special Operations and Low Intensity Conflict; the Joint Staff representatives came from the J7, Operational Plans and Interoperability. Interview with OSD official, November 2003.

[2] Interview with CENTCOM official, January 2004.

agency groups discussed in Chapter Three.[3] Humanitarian operations were seen as a critical component of operations in Iraq, with President Bush stressing that humanitarian assistance needed to be part of the operation from the very beginning.[4]

The HPT and IR+R working group assessed potential humanitarian relief needs in a wide range of areas, including relocating refugees as well as providing food, water, and electrical power.[5] They were well aware that significant humanitarian challenges already existed in Iraq. Approximately 60 percent of the population depended on food distributions from the United Nations Oil for Food (OFF) Program, and the United Nations estimated that 800,000 people were internally displaced within Iraq and an additional 740,000 Iraqis were refugees in neighboring countries.[6] Although exact scenarios were hard to predict, it was expected that any war with Iraq would displace as many as 2 million additional people; disrupt and perhaps destroy key parts of the food distribution network; disrupt the electrical supply, which would directly affect water and health services; and cause many UN and NGO personnel to leave the country, thus reducing the services they provided.[7] These interagency planners, therefore, assumed that the Iraqi population would require extensive humanitarian assistance both during and after combat operations.

Although the detailed plans remain classified, they were based on six main principles, as follows.[8]

1. **Minimize displacement, infrastructure damage, and disruption of services.** The military campaign was designed to minimize the number of civilian casualties, through careful vetting of targets and high reliance on precision munitions. The humanitarian mapping program was a key element of this process, which involved U.S. military outreach to humanitarian IOs and NGOs to identify important humanitarian facilities, key infrastructure, and cultural and historical sites so that they would not be unintentionally targeted during combat operations. USAID started holding weekly meetings with humanitarian organizations in late 2002 in order to facilitate the exchange of information.[9] Some IOs and NGOs were reluctant to share this information with the U.S. military, lest they

---

[3] Remarks of Andrew Natsios, Administrator of USAID, "Briefing on Humanitarian Reconstruction Issues," White House Office of the Press Secretary, February 24, 2003.

[4] Woodward, p. 147.

[5] Interview with OSD official, November 2003.

[6] Remarks of Elliott Abrams, NSC Senior Director for Near East and North Africa, "Briefing on Humanitarian Reconstruction Issues," February 24, 2003.

[7] United Nations, *Likely Humanitarian Operations*, December 10, 2002, paragraph 17; Joseph J. Collins, Deputy Assistant Secretary of Defense for Stability Operations, "DoD News Briefing," February 25, 2003.

[8] Abrams, "Briefing on Humanitarian Reconstruction Issues," February 24, 2003.

[9] Abrams and Natsios, "Briefing on Humanitarian Reconstruction Issues," February 24, 2003; Collins, "DoD News Briefing," February 25, 2003.

be seen as supporting the war. Some organizations refused to cooperate officially but quietly provided the requested information through informal channels or during several highly confidential, off-the-record meetings with U.S. government representatives.[10] The U.S. government also provided IOs and NGOs with a phone number and a web site where they could nominate targets to be placed on the no-strike list.[11]

2. **Rely primarily on civilian relief agencies.** IOs and NGOs, and not the U.S. military, would take the lead in providing humanitarian assistance. The U.S. military would focus on providing a secure environment in which those civilian and international agencies could do their work. It would provide limited humanitarian relief in cases where these agencies did not have the presence or capability to act on their own—such as during the early phases of a war—but the planners anticipated that this would occur for only a short time before the relief agencies would be capable of operating on their own. In short, no one expected serious security problems to arise.

3. **Develop effective civil-military relations.** USAID assembled and trained a Disaster Assistance Response Team (DART), which was designed to enter Iraq alongside U.S. military forces. Its tasks would be to assess immediate humanitarian needs; coordinate the provision of relief with the military, IOs, NGOs, and other donors; and make local grants for immediate relief projects. It was the largest DART ever assembled, composed of more than 60 humanitarian relief experts from USAID; the State Department's Bureau for Population, Migration, and Refugees (PRM); and the Department of Health and Human Services' Public Health Service. The DART would consist of a core team, with three mobile field offices that would report to the DART leader.[12] The interagency team also prepared to establish a Humanitarian Operations Center (HOC) in Kuwait, so that U.S. military, U.S. civilian, UN, IO, NGO, and coalition representatives would all be located in one place and be able to coordinate their relief efforts.[13]

4. **Facilitating IO and NGO operations.** In addition to establishing coordination centers such as the HOC, the U.S. government also supported IO and NGO

---

[10] Fallows, p. 68; Steve Stecklow, "Before Iraq War, U.N., U.S. Hatched Plan to Feed Nation," *Wall Street Journal*, September 26, 2003; interviews with U.S. military officers, October and November 2003.

[11] Woodward, p. 277.

[12] The DART also included a representative from the Centers for Disease Control and an active duty military officer. "USAID Contingency Plans for Humanitarian Assistance to Iraq," U.S. Agency for International Development Fact Sheet, February 24, 2003; remarks of Bernd McConnell, Director, Office of Foreign Disaster Assistance, USAID, "U.S. Humanitarian Planning and Relief Efforts," February 25, 2003.

[13] The HOC would supplement, but not replace, the UN's Office for the Coordination of Humanitarian Affairs (OCHA). Abrams, "Briefing on Humanitarian Reconstruction Issues," February 24, 2003.

operations in a number of ways. The State Department's PRM gave more than $15 million to international agencies, particularly the UN High Commissioner for Refugees (UNHCR), for prepositioning supplies and contingency planning. USAID gave more than $9 million to UNICEF, the World Food Program, and other agencies for contingency planning efforts. Much U.S. financial assistance was provided very quietly—and in some cases was disguised as general contributions—because these organizations did not want to be seen as supporting the war.[14] The U.S. government also provided indirect assistance, such as simplifying the licensing process for them to operate inside Iraq and funding the Joint NGO Emergency Preparedness Initiative (JNEPI), which provided coordination and support to NGOs preparing to work in Iraq.[15] Other countries also contributed to these efforts. The UK's Department for International Development, for example, contributed £16.5 million for the prepositioning of humanitarian supplies, largely to UN agencies, NGOs, and the International Red Cross and Red Crescent.[16]

5. **Prepositioning U.S. government relief supplies and response mechanisms.** The United States stockpiled more than $26 million worth of supplies in the region, including water, medicine, shelter supplies, and blankets, that would serve an estimated one million people.[17] It prepositioned more than three million humanitarian daily rations (HDRs), each of which contains enough calories for one person for one day.[18] The United States also had plans to quickly restore essential services, such as water supplies, sanitation systems, and health services. All of these depended on the quick restoration of electrical power throughout the country.[19]

6. **Restoring the food distribution system.** As noted above, more than 60 percent of Iraqis depended on the rations provided by the OFF program. It was a complex food distribution system that involved more than 55,000 ration agents throughout the country. U.S. planners aimed to disrupt that system as little as

---

[14] Woodward, pp. 276–277.

[15] Abrams, "Briefing on Humanitarian Reconstruction Issues," February 24, 2003; "USAID Contingency Plans for Humanitarian Assistance to Iraq."

[16] *Operations in Iraq: Lessons for the Future*, United Kingdom Ministry of Defence, December 2003, p. 62.

[17] When these numbers were released in February 2003, USAID was in the process of reallocating an additional $56 million of its own funds to preparations for humanitarian relief in Iraq. Natsios, "Briefing on Humanitarian Reconstruction Issues," February 24, 2003; and Natsios, "U.S. Humanitarian Planning and Relief Efforts," February 25, 2003.

[18] U.S. planners deliberately decided that these HDRs would be provided only through handouts on the ground, in the wake of the significant controversy surrounding the airdrops of HDRs during the war in Afghanistan. For more on this controversy, see Oliker et al.

[19] Abrams, "Briefing on Humanitarian Reconstruction Issues," February 24, 2003. Efforts to provide electricity ended up being much more difficult than anticipated before the war. See Chapter Twelve.

possible during the war and to reestablish it as quickly as possible after the conclusion of hostilities.[20]

On February 25, 2003, USAID officials estimated that they had already spent $26 million on preparing for possible humanitarian relief efforts in Iraq, and noted that they were in the process of reallocating another $56 million of USAID funds toward this mission.[21]

ORHA was also involved in efforts to prepare for humanitarian relief during this time, since it was one of the organization's three core pillars. ORHA would not become directly involved in the provision of relief, since that would be done mostly by various IOs and NGOs, but it would assist the many humanitarian coordination efforts described above. The interagency planning team and ORHA officials particularly sought to engage the United Nations in cooperative planning efforts. While the various UN technical agencies (including UNHCR and the World Food Program) worked closely with ORHA and other U.S. government organizations, relations with the UN Secretariat were reportedly much more problematic, largely due to the ongoing political debates at the Security Council about whether or not to go to war.[22]

IO and NGO Frustrations

Despite the coordination efforts described above, many IOs and NGOs nevertheless remained frustrated by the perceived lack of support from the U.S. government. In one important example, the existing sanctions regime prevented U.S. humanitarian relief organizations from operating inside Iraq. During the fall of 2002, representatives from U.S. NGOs asked for a presidential directive that would exempt them from the sanctions, so that they could start operating inside Iraq and developing the infrastructure necessary for relief efforts. They were told that a decision would be made by December, but no such decision was made. The point was soon moot, since by early 2003 the NGOs had come to fear that the situation on the ground was too dangerous. Moreover, the U.S. government feared that relief workers might be taken hostage in the event of war; but the NGOs' frustrations lingered.[23] The frustrations were exacerbated by the fact that senior representatives from these NGOs regularly interacted with working-level U.S. government officials on humanitarian relief issues. In January 2003, the humanitarian groups that had been meeting with USAID asked to meet with either Secretary Rumsfeld or Paul Wolfowitz to discuss their concerns.

---

[20] Abrams, "Briefing on Humanitarian Reconstruction Issues," February 24, 2003.

[21] McConnell and Natsios, "U.S. Humanitarian Planning and Relief Efforts," February 25, 2003.

[22] Interview with ORHA official, January 2004.

[23] Fallows, p. 63.

Such a meeting was never scheduled, and they never met with anyone more senior than a deputy assistant secretary.[24]

Furthermore, some NGOs remained skeptical that relief plans were being based on relevant scenarios. They expressed concern that demands for humanitarian assistance in Iraq might dwarf the plans that had been developed to provide relief assistance. As the head of Interaction, an umbrella organization representing more than 100 U.S. NGOs, said two days before the war started, "We don't think the relief and reconstruction needs of the Iraqi people will be adequately met, based on the overly optimistic scenarios we understand the U.S. government is using."[25]

Actual Humanitarian Requirements

The vast majority of these plans were never needed, because the humanitarian situation turned out to be much better than expected during major combat. The conflict produced few refugees and displaced persons, essential services remained largely intact, and no significant food shortages occurred throughout the country. Why was there no humanitarian crisis? The answer falls into two broad categories: factors related to the conduct of the war itself, and the extensive preparations for humanitarian assistance.

First, the rapid speed of military operations and the swift fall of the Iraqi regime prevented major humanitarian needs from developing throughout the country. On the ground, U.S. military forces moved quickly toward Baghdad and avoided major combat in most of the population centers in southern and central Iraq. Combat operations in Baghdad caused the fall of the regime far faster than either U.S. military or civilian decisionmakers had expected.[26] In the air, the extensive reliance on precision munitions and the focus on leadership targets minimized the damage to Iraq's infrastructure. For both of these reasons, the vast majority of Iraqis chose to stay in their homes during the war rather than flee to other parts of the country or across the border. If combat operations had inflicted more damage on civilian residences and infrastructure—i.e., during a prolonged urban siege, through less discriminate target-

---

[24] They did meet with Joseph Collins, the Deputy Assistant Secretary of Defense for Stability Operations. Fallows, p. 69; Hammer and Soloway.

[25] Mary McClymont, quoted in Neil King, Jr., "Bush Has an Audacious Plan to Rebuild Iraq Within a Year," *Wall Street Journal*, March 17, 2003. These sentiments were echoed by Sandra Mitchell, the vice president of the International Rescue Committee, who testified to Congress that "The logistical and operational framework is not in place to support a humanitarian response." Barbara Slavin, "Rebuilding Iraq to Start Quickly," *USA Today*, March 20, 2003.

[26] U.S. officials were extremely concerned that Saddam would pursue a "Fortress Baghdad" strategy, digging into the city and causing prolonged and highly destructive urban warfare. Woodward, pp. 64, 126, 134–135, 147, 174, and 205–206; Franks, pp. 349 and 391.

ing, or because of a decision by Saddam to use chemical weapons, as many feared before the war—the number of refugees could easily have approached or even surpassed the estimated one million mark.

Second, food and supply distribution networks remained largely intact. Before the war, Saddam ordered the distribution of three months' worth of food rations, as opposed to the usual one-month supply. Humanitarian planners feared that many Iraqis might have sold some of their rations in a desperate move to get cash before the conflict, but this ended up not being the case. Furthermore, the humanitarian mapping effort proved quite successful. No major incidents were reported of U.S. military forces mistakenly targeting humanitarian facilities. The food distribution network resumed its full functioning in June 2003,[27] which obviated the need for the large relief stockpiles that had been prepositioned throughout the region.[28]

As a result, the expected humanitarian crisis never materialized. Relief agencies did enter Iraq after the war, but few emergency services were required. The DARTs did go into the country, but because humanitarian aid was not needed, they focused on facilitating reconstruction by assessing needs, working with military civil affairs personnel, and contracting for community projects.[29] ORHA did not need to implement its humanitarian plans, and its humanitarian assistance pillar was disestablished during the transition to CPA in June 2003.[30]

Assessing Humanitarian Planning

After the war, ORHA was heavily criticized for having planned for the wrong events. Its extensive humanitarian plans were not needed, and it remained unprepared to address the looting, lawlessness, and eventually the insurgency that emerged after Saddam was removed from power. These criticisms are largely unfair, however. That

---

[27] One observer reported that the distribution of rations for June 2003 was delayed by only one day and was on time after that. Interview with ORHA official, January 2004.

[28] Many of the supplies contained in these stockpiles could be reallocated to relief efforts in different parts of the world. As one U.S. official noted, the government was trying to maximize the number of nonconsumables in the stockpiles, because "if you don't use a water bottle in the Middle East, you can use a water bottle somewhere else." McConnell, "U.S. Humanitarian Planning and Relief Efforts," February 25, 2003.

[29] There was a fair amount of dissatisfaction with the DARTs in Iraq. Both military and civilian personnel note that the DARTs were slow to arrive in country because they were accustomed to operating in a permissive environment, and the opposite situation in Iraq hampered their ability to work. ORHA officials grew frustrated with their slow deployment, because they expected that the teams would conduct most of the assessments of assistance needs throughout the country. Bureaucratic and organizational tensions also caused some friction with ORHA, because the DARTs were accustomed to operating independently and were reluctant to follow Garner's directions. Interviews with OSD official, November 2003; with ORHA official, January 2004; with State Department official, January 2004; and with civil affairs officer, June 2004.

[30] CPA did retain ORHA's mandate for humanitarian assistance, which had grown to encompass human rights and a few other issues by that time. Interview with ORHA official, January 2004.

humanitarian plans were not needed does not mean they were faulty, and the humanitarian planners maintain that they *did* focus on the right areas. They argue that massive refugee flows and food shortages would have been catastrophic—both for the local populations and for the progress of the military campaign—if they had occurred and no plans had existed to address them.[31] The plans that were implemented, particularly the stockpiling of food and other supplies in the region, were executed effectively. It is always prudent to plan for worst-case scenarios. These planners clearly identified refugee flows and mass starvation as the major challenges they might face, and they developed potentially workable plans to address them.

It is true that they did not prepare for the looting and chaos that occurred. However, it is not clear that humanitarian relief planners *should* have been planning to address looting and other public security issues. Humanitarian relief and post-conflict stabilization are not the same things, though both tend to get lumped together in the category of post-conflict reconstruction. In the months before the war with Iraq, humanitarian relief planners correctly identified the major problems that would be faced in this area, and through a relatively effective interagency process they developed plans to deal with these contingencies. Responsibility for post-conflict stabilization requirements lay with the government agencies and organizations addressed in the previous chapters, not with those that were working on humanitarian relief issues.

---

[31] Interviews with OSD officials, November 2003.

CHAPTER SEVEN

Combat Operations During Phase IV

Military plans for Phase IV included three different stages. Phase IVa would commence immediately after major combat operations and would focus on stability, security, and the emergency restoration of essential services. Combined Forces Land Component Command (CFLCC) would be the supported military headquarters during this stage, which was envisioned to last approximately 90 days. Phase IVb, reconstruction, would be marked by a handover of military control from CFLCC to Combined Joint Task Force-Iraq, commanded by the U.S. Army's V Corps. During this stage, military forces would provide support to a coalition-led transitional civilian agency. Finally, Phase IVc would be initiated when an Iraqi interim government was established. Control of military operations would then transfer from CJTF-Iraq to a new coalition headquarters to be named the Multinational Force-Iraq (MNF-I). During this final stage of Phase IV, large-scale military operations were to be conducted with the consent of a sovereign Iraqi interim government. This chapter examines Phases IVa and IVb; Phase IVc did not commence until after the transfer of authority on June 28, 2004, which is the cutoff date for this report.

Phase IVa: Stability Operations

Although CFLCC anticipated a number of specific post-conflict security challenges, most did not materialize. Rather, wide-scale looting and civil violence became the primary security concern in the first few months of the post-combat phase of operations. This civil unrest would later be eclipsed by organized terrorist activities and attacks conducted against coalition forces, the Iraqi National Guard, Iraqi police, the new Iraqi government, Iraqi infrastructure, and Iraqi civilians by a growing insurgency.

Anticipated Security Challenges

U.S. and coalition forces had prepared for four primary security problems. First, drawing on lessons from the 1991 Persian Gulf War, coalition forces had planned to secure Iraq's oil fields and protect them from sabotage. To counter this threat, the

United States deployed special operations forces (SOF) in the north near Kirkuk and in the south near the Al-Faw Peninsula to protect the oil fields, critical pipelines, and offshore oil terminals.

Second, coalition forces anticipated large numbers of refugees fleeing combat zones, creating the possibility of a massive humanitarian crisis.[1] This expectation was drawn from CENTCOM's experience during Operation DESERT STORM, during which an estimated 5 million people fled their homes and sought protection in both Iran and Turkey.[2] To prepare for this possibility, the United States deployed civil affairs units near the Iraqi border and prepositioned food and other provisions to prevent the expected flow of refugees from turning into a humanitarian crisis.[3] These prepositioned resources would also be available in the event that Saddam Hussein used WMD to attack Iraqi citizens as he had in the aftermath of DESERT STORM.

Third, CFLCC was concerned about the challenges associated with securing and managing a large number of surrendering Iraqi prisoners of war (POWs). Consequently, coalition forces planned to have the capability to detain and process large numbers of POWs. Coalition forces anticipated using a portion of surrendering Iraqi forces to form an interim Iraqi National Guard that could be used to help local Iraqi police maintain law and order and prevent widespread civil unrest during Phase IV.[4]

Finally, coalition forces anticipated violence occurring within the Iraqi population, particularly ethnic violence between different factions in Iraqi society and reprisal attacks against Ba'ath Party members loyal to Saddam Hussein. CFLCC was prepared to dispatch civil affairs units to contact and work with local leaders to stem ethnic, religious, and political violence.[5]

Security Challenges During the Transition to Phase IVa

As coalition forces stormed across Iraq and pushed toward Baghdad, a plan of action for policing the country was publicly released. The plan specified that, immediately following major combat operations, U.S. forces would maintain security and continue to root out resistance fighters. During the initial stage of Phase IV, coalition forces would work with international law enforcement advisors to train an Iraqi police force, a task to be completed "as quickly as possible." Finally, the coalition would take action to prevent riots and restore security.[6]

---

[1] For more on humanitarian planning assumptions, see Chapter Six.

[2] Judith Miller, "Displaced by the Gulf War: Five Million Refugees," *New York Times*, June 16, 1991.

[3] Joel Brinkley and Eric Schmitt, "The Struggle for Iraq: Postwar Planning," *New York Times*, November 30, 2003; Bathsheba N. Crocker, *Post-War Iraq: Are We Ready?* Washington, D.C.: Center for Strategic and International Studies, March 25, 2003, p. 2.

[4] Eric Schmitt, "Plans Made for Policing Postwar Iraq," *New York Times*, April 9, 2003.

[5] Peter Baker, "Top Officers Fear Wide Civil Unrest," *Washington Post*, March 19, 2003.

[6] Schmitt, "Plans Made for Policing Postwar Iraq."

Baghdad fell to U.S. forces on April 10, 2003, with little resistance from Iraqi troops. Soon afterward, on April 15, coalition forces secured Tikrit, bringing the entire country under coalition control. Most of the post-conflict challenges for which coalition forces had planned did not arise. Iraqi civilians did not flee their homes, and most had stockpiled food. Moreover, the rapid collapse of the Iraqi military prevented the type of extended combat that might have caused large numbers of civilians to flee combat zones. Therefore, the predicted exodus of refugees, with the associated need for refugee camps near Iraq's borders, did not materialize.[7]

Similarly, while extensive planning had occurred to ensure that coalition forces could accommodate a large number of Iraqi POWs, this capability was never needed. Rather than being destroyed or surrendering, the Iraqi military generally melted away into the civilian population. Coalition forces reportedly were prepared to accommodate 50,000 POWs at a detention center in the port city of Umm Qasr, but only 6,200 Iraq soldiers were captured.[8] One official stated that there was no need to disband the Iraqi army because "[t]hey simply laid down their arms and went home."[9]

The protection of Iraq's oil infrastructure was one major success in the immediate aftermath of major combat operations. Although the Iraqi regime was in the process of implementing plans to sabotage oil fields and related pipeline and transfer facilities, the unexpectedly fast pace of operations left the regime ill prepared to initiate such operations. Moreover, the protection of the oil infrastructure had been both planned and resourced to facilitate success. For example, 5,300 U.S. and coalition special operations forces were deployed to northern Iraq in part to prevent destruction of the oil-producing facilities located in and around the oil-rich city of Kirkuk.[10] Following the start of major ground combat, 2,300 soldiers from the Army's 173rd Airborne Brigade were also deployed to the theater and given the primary mission of securing Kirkuk's oil fields. Likewise, a smaller number of Navy SEALs and Polish special forces seized key facilities along the Al-Faw Peninsula in southern Iraq.

A second major success story in the transition from Phase III to Phase IV operations was the protection of major dams and hydroelectric facilities. For example, the destruction of the Hadithah Dam—located on the Euphrates River approximately 125 miles northwest of Karbala—could have had catastrophic effects throughout the country. In addition to the adverse effect that flooding would cause for the U.S. mili-

---

[7] See Chapter Six for more details.

[8] Jonathan Weisman, "Iraq Chaos No Surprise, but Too Few Troops to Quell It," *Washington Post*, April 14, 2003.

[9] Esther Schrader, "Attack Puts Emphasis on Recruits," *Los Angeles Times*, November 3, 2003.

[10] U.S. Army Special Operations Command, "Global Scouts," a briefing presented at the NDIA Symposium, February 5, 2004. Working alongside coalition SOF were 69,000 militia controlled by the Kurdistan Democratic Party (KDP) and the Patriotic Union of Kurdistan (PUK). These units were under the operational control of the U.S. special operations commander in northern Iraq.

tary's march to Baghdad, the resultant water shortage was likely to cause severe hardships for the Iraqi people during the summer months.[11] On April 1, soldiers from the 3rd battalion, 75th Infantry (Ranger) conducted an airborne insertion to seize the Hadithah Dam during the hours of darkness. The raid took the small number of Iraqi defenders by surprise, and Rangers quickly gained control of the structure and the adjacent hydroelectric generating facility. The Rangers continued to hold this position until April 19, when the 1-502 Infantry Battalion from the 101st Airborne Division relieved them.[12] Navy SEALS and Polish special forces conducted a similar action when they seized the Mukarayin Dam, located approximately 57 miles northeast of Baghdad along the Diyala River, during the last week of April.

Initial Response to Civil Unrest

The planning documents for post-combat operations specified that Phase IVa operations would be conducted with forces deployed for Phase III. But because Phase III was supposed to take significantly longer than it did, initial security operations had to be undertaken with two full Army divisions less than had been anticipated.[13] In addition to having fewer forces for Phase IVa than originally planned, the forces that seized Baghdad on April 9 were armored and mechanized infantry units, primarily trained and equipped for combat against other military forces. These forces were employed in Baghdad because a major military confrontation was expected between Iraq's best forces, its Republican Guard, and U.S. troops. While armored and mechanized units served U.S. forces well in combat, they were not ideal for confronting challenges encountered in the post-combat phase of the operation.

On April 10, while Baghdad and other cities across the country were spiraling into looting, vandalism, and civil unrest, U.S. government officials responded to criticisms about the lack of law and order by suggesting that this was a normal turn of events that would play itself out in short order. For example, Deputy Secretary of State Richard L. Armitage argued that while coalition forces in Iraq were equipped for combat, "tens of thousands of U.S. troops" intended to stabilize the country were on the way. Armitage concluded this briefing by stating: "[w]e think the fractious behavior that you witnessed in Basra, and to some extent in Baghdad, will settle

---

[11] Debbie Quimby, "ERDC Contributes to Operation Iraqi Freedom from the Home Front," *ERDC Information Bulletin*, December 5, 2003.

[12] Colonel Gregory Fontenot (USA, ret.) and Lieutenant Colonel Edmund J. Degen, *On Point: The United States Army in Operation Iraqi Freedom*, Fort Leavenworth, Kansas, 2004, p. 310.

[13] The 4th Infantry Division was still offloading in Kuwait when major combat operations ended. In addition, the 1st Cavalry Division was in the deployment process at the end of April 2003 when it was told it would not be deployed for OIF.

down."[14] In a similar manner, Secretary of Defense Donald Rumsfeld described Iraq's lawlessness as "untidiness" that would soon subside.[15]

The expectations expressed by Armitage and Rumsfeld did not materialize; rather, the country continued to experience wide-scale violence and destruction, particularly in Kirkuk, Mosul, and Baghdad. In Kirkuk, Kurdish peshmerga troops succeeded in driving out Saddam's forces at the beginning of the war. However, these forces failed to prevent the city from falling into chaos. In addition to wide-scale looting, the city's substantial Arab population came under attack, forcing many to flee south to Sunni Arab cities such as Tikrit.[16]

In the northern city of Mosul, Saddam's troops and police fled on April 11, leaving the city's residents without any local or coalition security forces to keep the peace.[17] The city erupted into looting and ethnic violence, leading to 31 deaths and 150 wounded in the first day of unrest.[18] In an effort to control the outbreak of violence, 200 special operations soldiers were dispatched to the city, where they encountered hostility and gunfire. To help restore law and order to a city that the Commander of Joint Special Operations Task Force–North (JSOTF-N) considered "reminiscent of Dodge City and the Wild West," CENTCOM directed that the 26th Marine Expeditionary Unit (MEU), which was afloat in the Mediterranean, immediately deploy to northern Iraq.[19] The 26th MEU was placed under the operational control of JSOTF-N on April 12 when it landed at the Irbil airfield in northern Iraq. At the direction of JSOTF-N, the 26th MEU deployed to Mosul, with its lead elements arriving on the morning of April 13. After securing the airfield, the Marines began to establish a presence in the city by setting up roadblocks, conducting urban patrols, ensuring that roads remained open and passable for both military and civilian vehicles, and searching for weapons caches.[20] The 26th MEU remained in Mosul until the 101st Airborne Division assumed responsibility for the city on April 22. Marine commanders also met with the tribal leaders of Mosul to gain their support in establishing local control of the population.[21]

---

[14] Peter Slevin and Bradley Graham, "U.S. Military Spurns Postwar Police Role," *Washington Post,* April 10, 2003.

[15] Daniel Williams, "Rampant Looting Sweeps Iraq," *Washington Post,* April 12, 2003.

[16] C.J. Chives, "Marines Keep Order in Saddam Hussein's Hometown," *New York Times*, April 16, 2003.

[17] Williams, "Rampant Looting Sweeps Iraq."

[18] David Rhode, "Deadly Unrest Leaves a Town in Northern Iraq Bitter at U.S.," *New York Times*, April 20, 2003.

[19] Interview with JSOTF-N personnel, February 2004.

[20] 1st Lieutenant (USMC) Christopher Mercer, "Charlie Company Leads BLT into Northern Iraq," 26 MEU Public Affairs Office, 2003, pp. 1–2.

[21] Interview with JSOTF-N personnel, February 2004.

Although the United States deployed a force of around 2,300 paratroopers from the 173rd Airborne Brigade to the north under the operational control of JSOTF-N, this force was not scheduled to arrive in Kirkuk until April 15. But when violence erupted in Kirkuk on April 10 when the Iraqi military collapsed, the ground movement of the 173rd Airborne Brigade was expedited, with the lead elements arriving at Kirkuk's airfield on April 13.[22] Working with the SOF units that had arrived days earlier, the 173rd Airborne Brigade assumed responsibility for maintaining law and order within the city. At April's end, the newly arrived 4th Infantry Division (ID) moved into Tikrit and conducted a link-up with the 173rd Airborne Brigade. Once this link had been established, operational control of the 173rd Airborne Brigade was transferred from JSOTF-N to the 4th ID. The force began working with local leaders to address grievances, restore order, and build rapport within the region. By May, northern Iraq was relatively stable, particularly in comparison with Baghdad.[23]

As anticipated by CFLCC planners in February 2003, maneuver forces were not in optimal positions when major combat operations ended to provide security for a subsequent occupation. The 1st Marine Expeditionary Force had forces as far north as Tikrit even though its occupation zone was in the south. The 4th ID had to move north rapidly from Kuwait to relieve Task Force Tripoli in the Tikrit area. Similarly, elements of the 3rd Infantry Division had to move into the eastern and northern portions of the Baghdad area to relieve the 1st Marine Division. In turn, the 1st Marine Expeditionary Force relieved the 82nd Airborne Division in southern Iraq. The 173rd Airborne Brigade secured Kirkuk and linked with the 4th Infantry Division. The 101st Airborne conducted a long-distance air assault to secure Mosul, where it assumed operational control of the 26th Marine Expeditionary Unit and expanded its operations to include all of northern Iraq. The 3rd Armored Cavalry Regiment took responsibility for western Iraq, including the upper reaches of the Euphrates and the entire western desert to the borders with Jordan and Syria. Thus, by early May, coalition forces were conducting stability and support operations in newly assigned sectors as depicted in Figure 7.1. Within weeks or days, these forces were in position to take more forceful stands to maintain security, but as will be discussed below, substantial damage to government facilities and infrastructure had occurred during the intervening period.

Stabilization Efforts, March to June 2003

Coalition forces, including British troops and U.S. Marines, took the port city of Umm Qasr in the first days of ground combat and immediately began working with

---

[22] The 173rd first arrived in northern Iraq through an airborne insertion but then moved to Kirkuk over land.

[23] Marni McEntee, "173rd Airborne Brigade Takes to the Streets of Kirkuk," *European Stars and Stripes*, May 7, 2003; Scott Wilson, "A Mix of 'President . . . and Pope,'" *Washington Post*, May 16, 2003.

Figure 7.1
Phase IVa Occupation of Iraq

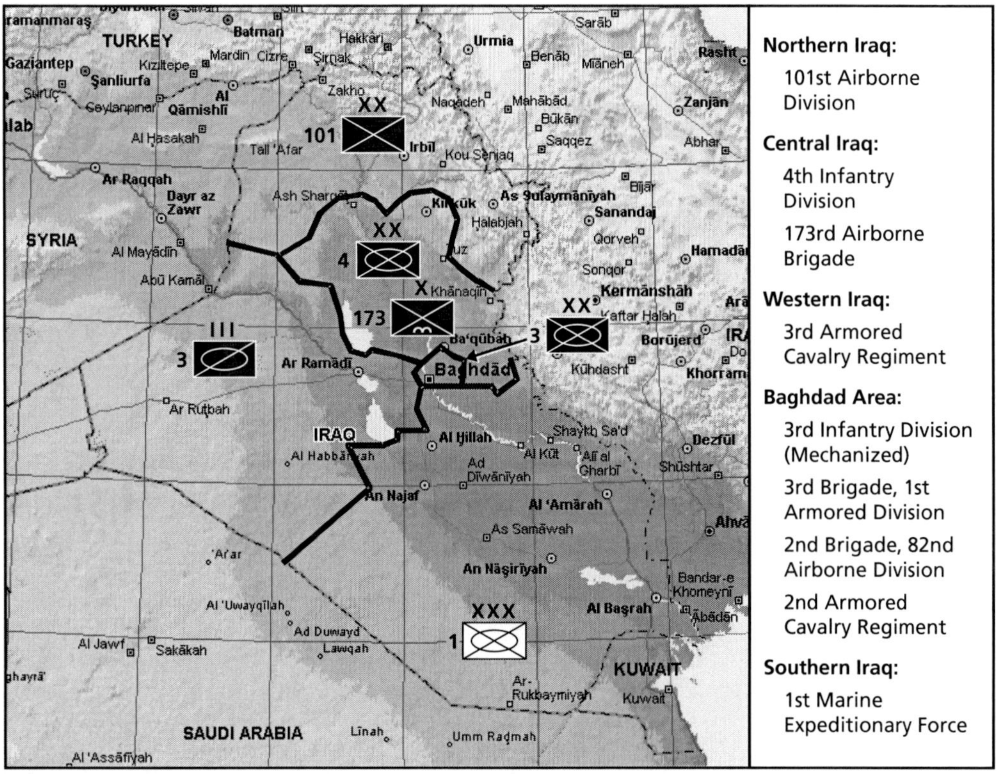
TURKEY
SYRIA
IRAQ
SAUDI ARABIA
KUWAIT
XX
101
XX
4
X
173
III
3
XX
3
XXX
1
Baghdad
Irbil
Kirkuk
As Sulaymaniyah
Ar Ramadi
Al Hillah
An Najaf
Al 'Amarah
An Nasiriyah
Al Basrah
Kermanshah
Ar Rutbah
Abu Kamal
Northern Iraq:
101st Airborne Division
Central Iraq:
4th Infantry Division
173rd Airborne Brigade
Western Iraq:
3rd Armored Cavalry Regiment
Baghdad Area:
3rd Infantry Division (Mechanized)
3rd Brigade, 1st Armored Division
2nd Brigade, 82nd Airborne Division
2nd Armored Cavalry Regiment
Southern Iraq:
1st Marine Expeditionary Force

SOURCE: V Corps, "The Road to Victory in Operation Iraqi Freedom," briefing, Heidelberg, Germany, 2003.
RAND *MG642-7.1*

local leaders to restore order and provide basic services for its citizens. The United States deployed civil affairs units to Umm Qasr, and they worked alongside British civil affairs soldiers to rebuild the city's infrastructure and government. In addition, the State Department dispatched a permanent team of USAID officers to help with reconstruction and, specifically, the reopening of the port.[24] Coalition forces succeeded in coordinating efforts with local leaders, mostly Shi'ite Muslim clerics, to restore order and services to Umm Qasr. On May 15, British forces handed over the southern port city to a 12-member council, followed by the transfer of the port to Iraqi leaders on May 22.[25]

---

[24] Thanassis Cambanis, "Forces Get First Test of Building New Order," *Boston Globe*, April 4, 2003; Glenn Frankel, "British Troops Bring Their Brand of Civic-Minded Peacekeeping to Iraq," *Washington Post*, April 4, 2003.

[25] Tini Tran, "First Iraqi City Handed Over to Civilian Government," *Washington Times*, May 16, 2003; Marc Lacey, "British Give Port Control to the Iraqis," *New York Times*, May 23, 2003; Michael Smith, "British Patrols Vital for Law and Order," *Daily Telegraph* (London), May 26, 2003.

Baghdad, in contrast, continued to experience wide-scale crime and unrest throughout the reconstruction phase. In Baghdad, chaos erupted in most neighborhoods after the fall of Saddam's regime. Initially, U.S. forces did little to stop the looting. Major General David H. Petraeus, commander of the 101st Airborne Division and the first officer in charge of central Iraq, told reporters that "we should discourage looting, but we're not going to stand between a crowd and a bunch of mattresses."[26] Within days virtually every government building was sacked and burned along with most of the city's hotels, department stores, museums, schools, universities, hospitals, and Saddam's palaces.[27] A notable exception was the Ministry of Oil, which was heavily guarded by a company of Marines. One reporter in Baghdad observed that in the neighborhoods where U.S. forces were positioned, looters were deterred, "but there were far too few American soldiers to make a difference in most neighborhoods."[28]

On April 11, following nearly a week of looting and lawlessness in the capital, the State Department announced that it would send a team of 26 police and judicial officers to Iraq to prepare for the deployment of law and order experts.[29] On April 12, the *Washington Post* reported that U.S. Marines were not endowed with "policing authority" and could only detain individuals directly hostile toward U.S. forces.[30] The following day, the *New York Times* reported that policing the streets and restoring law and order had become the priority of the Marines in Baghdad.[31] However, this proved to be a difficult task, and one for which many of the troops in Baghdad had not been trained.[32] In addition to policing, Marines were assigned as mayors of certain neighborhoods in Baghdad and tasked with hearing complaints from local Iraqis and restoring law and order. One Marine stated, "It's one of the hardest things I've ever been asked to do, because I have almost no training to fall back on."[33]

In addition to coalition efforts to curb looting, Iraqi citizens banded together to begin policing their neighborhoods and cities. Shortly after chaos erupted in Baghdad, a cleric in the Shi'ite section of the city issued a fatwa banning looting and

---

[26] William Branigin and Rick Atkinson, "Anything, and Everything, Goes," *Washington Post*, April 12, 2003.

[27] Dexter Filkins, "In Baghdad, Free of Hussein, a Day of Mayhem," *New York Times*, April 12, 2003; Rajiv Chandrasekaran, "'Our Heritage Is Finished,'" *Washington Post*, April 13, 2003.

[28] Chandrasekaran, "'Our Heritage Is Finished.'"

[29] Filkins, "In Baghdad, Free of Hussein, a Day of Mayhem."

[30] Jonathan Finer, "Marines Get New Mission: Restoring Law and Order," *Washington Post*, April 12, 2003.

[31] Dexter Filkins and John Kiefner, "U.S. Troops Move to Restore Order in Edgy Baghdad," *New York Times*, April 13, 2003.

[32] Brett Lieberman, "Marines Uneasy About Shift to Humanitarian Mission," *Newhouse.com*, April 15, 2003.

[33] Jonathan Finer, "Marines Get Hands-On Civics Lesson," *Washington Post*, April 14, 2003; Filkins and Kiefner, "U.S. Troops Move to Restore Order in Edgy Baghdad."

helped organize joint civilian and military patrols.[34] These ad hoc forces put up road-blocks and searched cars for looted materials, taking the confiscated goods to a nearby mosque where they were returned to their rightful owners. Items whose ownership could not be determined were distributed to the poor.[35] Before long, soldiers, Marines, MPs, and other U.S. security forces in Baghdad patrolled alongside civilian groups to stabilize the city.[36]

As Baghdad and other major cities in Iraq continued to experience violence and unrest, coalition forces came under increasing international criticism for failing to establish law and order. They were particularly criticized for their failure to uphold provisions of the Geneva Conventions, which require that an occupying force establish law and order following the end of major combat.[37] In addition, coalition forces were criticized for failing to return basic services to Iraq's cities, particularly Baghdad, which regularly lacked potable water, electricity, garbage collection, and a functioning phone system.[38]

Officials in Washington responded to these charges by claiming that the speed with which coalition forces had succeeded in toppling Saddam had left them without sufficient forces in the theater to stabilize Iraq. They claimed that other troops had been slated to handle the stabilization of the country, such as the 4th Infantry Division and the 1st Armored Cavalry Division, both of which contained MPs and civil affairs units.[39] Furthermore, restoring law and order was complicated because coalition troops on the ground lacked proper equipment to act as policing forces. As the *New York Times* reported,

> The armored forces stationed in the city [Baghdad] have found it difficult to maneuver through the streets in their M1 Abrams tanks and Bradley fighting vehicles. Many tank crews cannot even carry out foot patrols, because it was never envisioned that they would play a peacekeeping role and they were never issued the heavy body armor to operate outside their vehicles.[40]

---

[34] Chandrasekaran, "'Our Heritage Is Finished.'"

[35] Alissa J. Rubin, "Crowds Seize Loot to Return to Owners," *Los Angeles Times*, April 17, 2003.

[36] Esther Schrader, "Fighting Force Is Giving Way to Police Force," *Los Angeles Times*, April 16, 2003. This article reported that U.S. forces have 2,000 "civil affairs and military police specialists attached to forces in Iraq."

[37] Schrader, "Fighting Force Is Giving Way to Police Force"; Weisman, "Iraq Chaos No Surprise, but Too Few Troops to Quell It." One article noted that applying these laws is not always clear-cut. In the Iraqi case, portions of the country were still involved in major combat, thus making it difficult to categorize the country as "post-combat," which would require the application of the Fourth Geneva Convention's laws. Seth Stern, "U.S. Struggles with New Rules as War Turns to Occupation," *Christian Science Monitor*, April 21, 2003.

[38] Marcus Stern, "Uncertain Road to Rebuilding," *San Diego Union-Tribune*, April 21, 2003.

[39] Weisman, "Iraq Chaos No Surprise, but Too Few Troops to Quell It."

[40] Michael R. Gordon, "Baghdad's Power Vacuum Is Drawing Only Dissent," *New York Times*, April 21, 2003.

In the middle of April, Washington announced plans for a "rolling Phase IV" in which coalition forces would move toward greater restoration of law and order in addition to performing other tasks such as finding Saddam Hussein, rounding up leading Ba'ath Party members, and searching for Iraq's purported WMD.[41] To this end, Washington also announced that it would deploy 4,000 more troops to Baghdad—raising the number of U.S. forces in the city from 12,000 to 16,000—in addition to sending 2,000 more MPs and extra civil affairs units.[42] It further proclaimed that Iraq would be divided into three zones: north, central (including Baghdad), and south.[43] By mid-May, the number of coalition troops in Baghdad had risen from 16,000 to a reported 25,000.[44]

Along with the increase in U.S. troops to Baghdad, coalition commanders also reviewed their strategies for mitigating civil unrest and looting throughout the country. On May 1, the CFLCC Commander, Lieutenant General David McKiernan, issued a proclamation forbidding "looting, reprisals and criminal activity."[45] At the same time, coalition forces also began using lethal force against looters raiding a munitions warehouse in Tikrit, killing 14 Iraqis.[46] In mid-May, the newly appointed civilian administrator of Iraq, L. Paul Bremer, publicly announced that coalition forces would begin shooting looters in Baghdad.[47] This was backed by a statement from Secretary of Defense Donald Rumsfeld, who announced that "The forces there will be using muscle to see that the people who are trying to disrupt what's taking place in that city are stopped and either captured or killed."[48] To this end, they shot and killed a looter in one of Baghdad's Shi'ite slums, Sadr City, which prompted

---

[41] Thom Shanker, "Allied Forces Now Face Both Combat and Occupation Duty," *New York Times*, April 13, 2003.

[42] Scott Peterson, "In Peacekeeping Mode, U.S. Troops Tested," *Christian Science Monitor*, April 30, 2003; Michael R. Gordon, "U.S. Planning to Regroup Armed Forces in Baghdad, Adding to Military Police," *New York Times*, April 30, 2003; Christine Spolar, "Baghdad's Disorder Gets in Way of Reconstruction," *Chicago Tribune*, May 1, 2003.

[43] Vernon Loeb, "Occupation of Iraq Has No Time Limit," *Washington Post*, May 10, 2003; Peter Slevin, "Baghdad Anarchy Spurs Call for Help," *Washington Post*, May 13, 2003.

[44] Patrick Healy, "Security Proves a Daunting Task for U.S. in Iraq," *Boston Globe*, May 18, 2003. Some estimates were even higher: General Peter Pace, Vice Chairman of the Joint Chiefs of Staff, testified before the Senate Appropriations Subcommittee that the number was 49,000. See Tom Hundley and E.A. Torriero, "U.S. Boosts Troop Strength in Baghdad to Help Fight Crime," *Chicago Tribune*, May 18, 2003.

[45] Alissa J. Rubin, "U.S. Struggles in Quicksand of Iraq," *Los Angeles Times*, May 5, 2003.

[46] Michael R. Gordon, "Between War and Peace," *New York Times*, May 2, 2003.

[47] Patrick E. Tyler, "New Policy in Iraq to Authorize G.I.'s to Shoot Looters," *New York Times*, May 14, 2003.

[48] Edmund L. Andrews, "U.S. Military Chief Vows More Troops to Quell Iraqi Looting," *New York Times*, May 15, 2003.

demonstrations in protest.[49] Coalition forces also stepped up patrols across the city, arresting hundreds of looters and detaining them for a standardized 21 days.[50]

On May 20, in another effort to stem violence in the capital city and elsewhere, coalition forces announced that they would require Iraqi citizens to hand over their automatic and heavy weapons by June 14. Those who refused would be subject to arrest.[51] No incentives, such as payment for each forfeited weapon, were given for Iraqis to comply with this order.[52] After an initial outcry from Iraqi citizens who complained that they needed these weapons in their homes to protect against looters, the coalition announced that citizens would be allowed to keep small firearms and AK-47s but would require a permit to carry them outside the home.[53] Coalition forces estimated that Iraq contained more than 24 million firearms in addition to an enormous cache of light and heavy military armament, such as rocket-propelled grenades, anti-aircraft guns, and surface-to-air missiles; at the time of the proclamation, AK-47s were selling on the streets of Baghdad for a reported $10.[54]

Coalition forces had hoped that requiring Iraqis to hand over their automatic and heavy weapons would reduce attacks on coalition forces, slow the spread of weapons to neighboring countries, and curb rampant looting. However, the amnesty program proved to be a failure. By June 14, Iraqis had forfeited "123 pistols, 76 semiautomatic rifles and shotguns, 435 automatic rifles, 46 machine guns, 162 anti-tank rocket-propelled grenade launchers, 11 antiaircraft weapons, and 381 hand grenades."[55] One U.S. soldier noted that Iraq had a gun culture not unlike certain regions in the United States and that disarming the country "would be like trying to disarm Texas. It isn't going to happen."[56] The weapons amnesty concluded with "Operation MARKET SWEEP," which attempted to crack down on black-market arms sales.[57]

---

[49] Peter Ford, "U.S. Troops in Perilous Police Work," *Christian Science Monitor*, May 19, 2003.

[50] Patrick E. Tyler, "U.S. Steps Up Efforts to Curb Baghdad Crime," *New York Times*, May 16, 2003.

[51] Michael R. Gordon, "Allies to Begin Seizing Weapons from Most Iraqis," *New York Times*, May 21, 2003.

[52] The stated punishment for failing to hand in illegal weapons was one year in jail and a $1,000 fine. Gary Strauss, "Amnesty Begins for Automatic Weapons," *USA Today*, June 2, 2003; E.A. Torriero, "U.S. Aims to Tame Gun Riddled Iraq," *Chicago Tribune*, May 31, 2003.

[53] Strauss, "Amnesty Begins for Automatic Weapons"; Torriero, "U.S. Aims to Tame Gun Riddled Iraq"; Edmond L. Andrews, "Allied Officials Now Allow Iraqi Civilians to Keep Assault Rifles," *New York Times*, June 1, 2003.

[54] Rachel Stohl, "War Ends, but Iraq Battle over Small Arms Just Begins," *Christian Science Monitor*, May 28, 2003.

[55] Daniel Williams, "Few Iraqis Meet the Deadline for Turning in Their Guns," *Washington Post*, June 15, 2003.

[56] Sandra Jontz, "Soldiers Face Dangerous Situation at Weapons Turn-In Stations in Iraq," *European Stars and Stripes*, June 3, 2003.

[57] Sandra Jontz, "Weapons Sweep Through Baghdad Marketplace Not Very Productive," *European Stars and Stripes*, June 13, 2003.

In addition to reducing arms among the public, coalition officials had to contend with an array of militias and private security forces that had emerged in the postwar chaos. Several political organizations—most notably the Patriotic Union of Kurdistan (PUK), the Kurdistan Democratic Party (KDP), the Iraqi National Congress (INC), and the Supreme Council for the Islamic Revolution in Iraq (SCIRI)—had their own militias, some of which had been supported by the U.S. government prior to the war.[58] Alongside these militias were private security forces and neighborhood watch groups, most of which were armed and wore some sort of uniform.[59] Initially, coalition authorities issued identity cards to many of these security forces to distinguish them from insurgents.[60] Later, ORHA and coalition forces announced that some of these forces would be disarmed, disbanded, or absorbed into the newly emerging Iraqi security forces.[61] Implementing this policy, however, proved to be a sensitive task. At the end of May 2003, coalition authorities announced that they would disarm the INC's "free Iraqi fighters" but allow the Kurds' peshmerga to keep their weapons. Although the coalition succeeded in disarming the INC, the policy produced bitter feelings over perceived favoritism toward the Kurds.[62]

Alongside issues of disarmament, coalition forces, their civil affairs units, and nongovernmental organizations (NGOs) also began working to resolve Iraq's water and electrical shortages.[63] Coalition forces also worked with local leaders, most notably with clerics and members of mosques, to identify needs and begin improving the social welfare of Iraqis.[64] Despite improvements, citizens still complained about the lack of resources available to them in the new Iraq, particularly jobs. To counter this problem, coalition forces began to distribute salaries to former government employees, and this included erstwhile members of the army, which had employed over 300,000 Iraqi men.[65] However, Iraq's postwar economy stalled, and high unemployment rates continued to be a problem, breeding discontent within the popula-

---

[58] Karen DeYoung and Daniel Williams, "Training of Iraqi Exiles Authorized," *Washington Post*, October 19, 2002; Daniel Williams, "Iraqi Exiles Mass for Training," *Washington Post*, February 1, 2003.

[59] Yochi J. Dreazen, "Disarming Iraq's Many Militias Poses Major Challenge for U.S.," *Wall Street Journal*, April 28, 2003.

[60] Dexter Filkins and Ian Fisher, "U.S. Is Now in Battle for Peace After Winning the War in Iraq," *New York Times*, May 3, 2003.

[61] See Chapter Nine of this report.

[62] Torriero, "U.S. Aims to Tame Gun Riddled Iraq"; Gareth Smyth, "Weapons Ultimatum Issued in Northern Iraq," *Financial Times* (London), June 18, 2003.

[63] See Chapter Twelve of this report.

[64] Anthony Shadid, "Troops Test Cooperation with Clerics," *Washington Post*, May 23, 2003.

[65] Peter Slevin, "A Sense of Limbo in the South," *Washington Post*, May 6, 2003; Wilson, "A Mix of 'President... and Pope.'"

tion.[66] ORHA and other agencies in Iraq implemented projects aimed at generating jobs and income, but progress was slow.[67]

Phase IVb: Recovery Operations

While the end-state conditions established in CENTCOM planning documents had not been fully realized, especially in terms of the security situation, General Franks decided to transition to Phase IVb upon the establishment of the Coalition Provisional Authority. Thus, shortly after ORHA transferred political control to CPA in May 2003, the control of coalition forces was transferred from CFLCC to a newly established Combined Joint Task Force 7 (CJTF-7). The formal transfer of authority to CJTF-7 took place on June 12, and the headquarters for the new command was collocated in Baghdad with CPA.

Organization for Phase IVb

As portrayed in Figure 7.2, CJTF-7 consisted of the former V Corps headquarters and all ground forces that served under the operational control of CFLCC during Phase IVa—including the 1st Marine Division, the 3rd Armoured Division (United Kingdom), and a Polish mechanized infantry division. In addition, Joint Special Operations Task Force—Arabian Peninsula (JSOTF-AP) was established and placed under the operational control of CJTF-7. In essence, CJTF-7 exercised operational control of all conventional and special operations forces remaining in theater with the exception of JTF-123, which consisted of special operations units associated with the Joint Special Operations Command (JSOC).

Along with the establishment of CJTF-7 came new geographic responsibilities. Rather than the three geographic areas of operation established by CFLCC for Phase IVa, Iraq was now divided into four geographic regions, with responsibility for security being assigned as depicted in Figure 7.3. The 3rd Armoured Division (UK) was designated the multinational division south (MND-S), and the Polish mechanized infantry division was designated the multinational division south central.

---

[66] Paul Weisman and Vivienne Walt, "Hostility Towards U.S. Troops Is Running High in Baghdad," *USA Today*, May 7, 2003.

[67] Warren Richey, "Efforts to Make Progress Tangible to Iraqis," *Christian Science Monitor*, May 12, 2003.

Figure 7.2
CJTF-7 Task Organization, June 2003

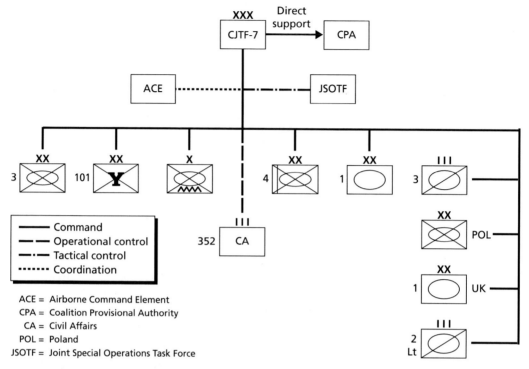

SOURCE: "V Corps: The Road to Victory in OIF," V Corps briefing.
RAND MG642-7.2

The Growing Insurgency

Coalition forces continued to come under attack from terrorist organizations, foreign fighters, and residual forces or individuals loyal to Saddam Hussein well after the collapse of the Iraqi military. In the first week of May 2003, an Iraqi was shot and killed as he attempted to run over two Marines with his vehicle at a checkpoint in the city of Kut. That same week, grenades were thrown at U.S. soldiers in Fallujah. Four members of a civil affairs unit were shot dead in Baghdad.[68] In addition to these attacks against coalition forces, Iraqis organized several anti-occupation protests. One protest west of Baghdad ended with U.S. troops shooting and killing 18 protesters after individuals in the crowd opened fire on U.S. forces.[69]

---

[68] Gordon, "Between War and Peace."

[69] Filkins and Fisher, "U.S. Is Now in Battle for Peace After Winning the War in Iraq."

Figure 7.3
CJTF-7 Areas of Operations, June 2003

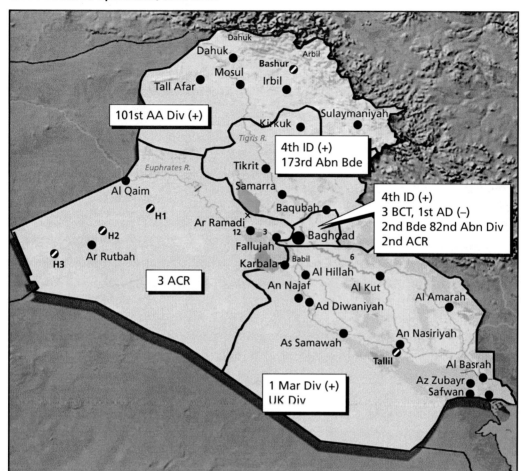

SOURCE: "V Corps: The Road to Victory in OIF," V Corps briefing.
RAND *MG642-7.3*

Initially, attacks against coalition forces appeared to be isolated incidents. As the summer of 2003 progressed, however, the attacks became more frequent and more sophisticated. In June, soldiers in Baghdad came under increasing "hit and run" attacks throughout the city, including an intense gun battle that resulted in the downing of an Apache helicopter.[70] In response, U.S. forces launched Operation DESERT SCORPION to uproot Saddam loyalists and thwart resistance fighters.[71] In July, the *New York Times* reported that attacks against U.S. forces in Fallujah had increased in

---

[70] Edmund L. Andrews, "Why Wheels of Recovery Are Spinning in Iraq," *New York Times*, June 15, 2003.

[71] David Wood, "U.S. Forces Scour Iraq in Attempt to Disarm Guerrillas," *Newhouse.com*, June 16, 2003.

sophistication from "light arms to rocket-propelled grenades to improvised explosives to mortars, which require training to use."[72] Attacks against coalition forces reached a new level in the fall of 2003, as CJTF-7 increased its offensive operations; a U.S. helicopter was shot down, killing 16 soldiers on board, and a car bomb exploded outside Italian military headquarters in Nasiriya, killing more than 20 people.[73] With the exception of a small spike in attacks that corresponded to the capture of Saddam Hussein in December 2003, the average weekly number of attacks against coalition forces in the winter of 2003–2004 was lower than that in the fall of 2003.

In the spring of 2004, one year after OIF began, attacks against coalition forces spiked once again. In April, two conflicts heated up simultaneously. First, more than 1,000 U.S. Marines surrounded and attacked insurgents in Fallujah following the brutal murder and public mutilation of four U.S contractors.[74] This battle lasted for a month before coalition forces negotiated a ceasefire with the insurgents. Almost simultaneously, coalition forces confronted the radical Shi'ite cleric Moqtada al-Sadr by closing his newspaper on March 28, claiming that it was inciting violence against the occupation and calling for al-Sadr's arrest in connection with the murder of a Shi'ite cleric in April 2003.[75] These acts prompted al-Sadr's militia, the Mahdi Army, to attack coalition forces, forcing a military confrontation that continued throughout April and most of May. These two battles made April 2004 an especially deadly month in terms of both U.S. and Iraqi casualties.[76]

Throughout Phase IVb, coalition forces did succeed in working with local leaders to improve law and order while reducing attacks against their own forces. In June 2003, U.S. forces called on local Sunni tribal leaders in the Sunni city of Hit to help restore order and crack down on insurgents in exchange for pulling out of the city's center and ending house-to-house searches.[77] This agreement resulted in the cessation of nearly all attacks against U.S. forces in this city.[78] U.S. troops also experimented with the controversial policy of paying "blood money" for Iraqis killed or wounded by coalition forces in raids and patrols, a practice designed to end reprisal attacks and

---

[72] Shaila K. Dewan, "G.I.'s Turn Over Policing of Iraqi Town to Local Force," *New York Times*, July 12, 2003.

[73] Anthony Shadid, "Iraqi Parties Move to Fill Security Role," *Washington Post*, November 14, 2003.

[74] Anthony Shadid and Sewell Chan, "Iraqi Militia Provokes More Clashes," *Washington Post*, April 6, 2004.

[75] Jeffrey Gettleman, "G.I.'s Padlock Baghdad Paper Accused of Lies," *New York Times*, March 29, 2004.

[76] Gregg Zoroya, "Iraq's Deadliest Month," *USA Today*, April 30, 2004.

[77] Hit is approximately 95 miles northwest of Baghdad, along the Euphrates River, with a population of 45,000.

[78] Yaroslav Trofimov, "U.S. Gains Relative Sense of Calm with Quiet Presence in Iraqi Town," *Wall Street Journal*, June 13, 2003.

customary throughout the Arab world.[79] This policy appeared to have reduced violence against coalition forces in and around Fallujah as well as in the north.[80]

In addition to targeting coalition forces, insurgents also attacked nonmilitary international organizations in the country. On August 8, 2003, a suicide bomber struck the Jordanian embassy in Baghdad, killing more than 15 people.[81] The blast was a setback for the newly emerging Iraqi police force, which had been assigned to protect the embassy.[82] A little over a week later, on August 19, a massive suicide bomb exploded at the UN headquarters in Baghdad. This blast claimed more than 20 lives, including the UN's special envoy to Iraq, Sergio Vieira de Mello.[83] These incidents angered the newly formed Iraqi Governing Council (IGC), which blamed the United States for not providing adequate security in the country and for failing to allow Iraqi security forces a greater role in protecting their country against insurgents.[84]

During the late summer and fall of 2003, insurgents began to target the IGC, the newly emerging Iraqi security forces, and Iraqi civilians. City council members, judges, and other Iraqi officials also came under increasing attack from insurgents, who accused these leaders of collaborating with the United States and compromising Iraq's sovereignty.[85] On September 25, a member of the IGC, Aqila al-Hashimi, was gunned down in her car while in Baghdad. A week later, on November 4, a judge was shot in Mosul after being abducted.[86] On November 18, an Iraqi education official tasked with training the Iraqi police force was killed in the city of Latifiyah.[87] In May 2004, in the face of a growing insurgency, the Pentagon and White House discussed transferring the Iraq Survey Group—a body of 1,400 military and civilian personnel tasked to search for WMDs—to counterinsurgency intelligence gather-

---

[79] Rajiv Chandrasekaran, "In Iraqi City, a New Battle Plan," *Washington Post*, July 29, 2003.

[80] Chandrasekaran, "In Iraqi City, a New Battle Plan."

[81] Michael R. Gordon, "One Hundred and Twenty Degrees in the Shade," *New York Times*, August 8, 2003.

[82] James Hider, "One Local Police Officer Has Job of Finding Embassy Bombers," *The Times* (London), August 9, 2003.

[83] Dexter Filkins and Richard A. Oppel, Jr., "Truck Bombing; Huge Suicide Blast Demolishes U.N. Headquarters in Baghdad," *New York Times*, August 20, 2003. The number of dead was later raised to 25.

[84] See Chapter Ten for more details.

[85] Yochi J. Dreazen, "Insurgents Turn Guns on Iraqis Backing Democracy," *Wall Street Journal*, December 10, 2003.

[86] Dreazen, "Insurgents Turn Guns on Iraqis Backing Democracy."

[87] Gregg Zoroya, "Danger Puts Distance Between Council, People," *USA Today*, October 21, 2003.

ing.[88] Death threats against the evolving Iraqi government persisted throughout the transfer of authority from CPA to an independent Iraqi government.

Iraq's fledgling security forces also came under an increasing number of attacks from insurgents. On October 27, 2003, suicide bombers attacked three police stations in Baghdad, killing 15 officers and wounding over 65.[89] On October 30, a suicide bomber disguised as a repairman detonated a car full of explosives at a police station in the capital.[90] In addition to attacks on police stations, Iraqi officers received death threats and threats against their families.[91] By the end of November, at least five Iraqi police chiefs had been assassinated.[92] Also in November, suicide bombers targeted the police station in Baquba, killing nearly 20 people.[93] By the end of 2003, attacks on Iraqi police stations had become a weekly occurrence, and an estimated 208 Iraq police had been killed since the country's liberation, 60 in Baghdad alone.[94]

In addition to attacking the Iraqi government and the fledgling security forces, the insurgency also targeted Iraqi citizens. The Shi'ite holy sites in Karbala and Najaf became the scene of several bloody attacks. On August 29, 2003, a car bomb exploded outside the Imam Ali shrine in Najaf, killing at least 85 people including the Ayatollah Mohammed Baqir al-Hakim, a prominent leader of the Shi'ite political party SCIRI.[95] Again on March 2, 2004—during the observation of the Ashura, the holiest day in Shi'ite Islam—at least nine simultaneous bombs were detonated among thousands of pilgrims in Karbala and Baghdad, killing over 200 people.[96] It is important to note that Shi'ite militias, volunteer "shrine guards," and Iraqi police guarded these holy sites. While U.S. forces were also in the area, they remained at a distance on the exterior perimeter in deference to the religious importance of the day. Despite these layers of security, the attackers succeeded in smuggling multiple bombs into the city and wreaking havoc.

---

[88] Eric Schmitt and Douglas Jehl, "Weapons Searchers May Switch to Security," *New York Times*, October 29, 2003; Andrew Koch, "Training More Locals Is Key to U.S. Strategy for Iraq," *Jane's Defence Weekly*, November 5, 2003.

[89] Tyler Marshall, "Iraqi Police Struggle to Hold Back a Crime Wave," *Los Angeles Times*, October 28, 2003.

[90] Mike Marshall, "Training of Iraqi Police Seen as Key to U.S. Success," *Newhouse.com*, December 2, 2003.

[91] Rajiv Chandrasekaran, "Iraqi Police Now Targets of Choice," *Washington Post*, November 2, 2003; Alex Berenson, "For Iraqi Police, a Bigger Task but More Risk," *New York Times*, November 9, 2003.

[92] Patrick J. McDonnell, "Iraqi Police Chief Paid for Friendship with U.S.," *Los Angeles Times*, November 26, 2003.

[93] Nicholas Blanford, "Iraqi Police Walk Most Perilous Beat," *Christian Science Monitor*, November 25, 2003.

[94] Andrew Higgins, "As It Wields Power Abroad, U.S. Outsources Law and Order Work," *Wall Street Journal*, February 2, 2004.

[95] "Shiites Report Top Leader One of Bombing Victims," *CNN.com*, August 29, 2003.

[96] Rajiv Chandrasekaran, "Shiites Massacred in Iraqi Blasts," *Washington Post*, March 3, 2004.

In response to the growing insurgency, the IGC called for Iraqi forces to play a greater role in policing and providing security for the country, arguing that the highly visible presence of coalition forces was fueling anger among Iraqis. Specifically, certain members of the IGC proposed deploying their private militias, such as the Kurdish peshmerga and the Shi'ite Badr Corps, as an immediate solution to the lack of trained Iraqi police and other security forces.[97] CPA, however, criticized this proposal, arguing that these forces were not trained to work together as one unit and that it could not endorse the introduction of ethnic or religious biases into Iraq's security forces.[98]

---

[97] Patrick E. Tyler, "Iraqi Factions Seek to Take Over Security Duties," *New York Times*, September 19, 2003.

[98] Tyler, "Iraqi Factions Seek to Take Over Security Duties"; Alex Berenson, "Security: Use of Private Militias in Iraq Is Not Likely Soon, U.S. Says," *New York Times*, November 6, 2003.

CHAPTER EIGHT

The Coalition Provisional Authority

In May 2003, the Office for Reconstruction and Humanitarian Assistance (ORHA) was officially replaced by the new Coalition Provisional Authority (CPA). CPA had a much wider mandate than ORHA. Under the direction of Ambassador L. Paul Bremer, CPA was tasked with creating new and democratic institutions throughout Iraq. Before discussing the substantive activities of CPA—which are addressed in the next four chapters of this report—it is important to start by examining the origins, goals, structure, and functioning of CPA itself. This chapter also addresses the relationship between CPA and CJTF-7, the coalition military authority in Iraq during the occupation.[1]

The Origins and Authorities of CPA

President George W. Bush appointed Bremer to the position of Presidential Envoy to Iraq on May 6, 2003.[2] The White House press release announcing the appointment noted that Bremer would oversee reconstruction and the creation of the institutions that would guide a democratic Iraq. CPA took ORHA's place just before Bremer's arrival in Baghdad on May 12 and his official designation as the administrator of CPA on May 13, 2003.[3] With this appointment, Bremer became the senior civilian official in charge of all policy efforts in Iraq.[4] As a result, retired Lieutenant General Jay M. Garner was no longer in charge of civilian operations in Iraq, although he remained in country until June 1, 2003.[5]

---

[1] This chapter was written by RAND analysts Keith Crane, Olga Oliker, and Andrew Rathmell, each of whom worked for CPA in Baghdad. All information in this chapter, including that not otherwise attributed, is informed by their personal experiences.

[2] White House, Office of the Press Secretary, "President Names Envoy to Iraq," May 6, 2003.

[3] L. Elaine Halchin, "The Coalition Provisional Authority: Origin, Characteristics, and Institutional Authorities," CRS Report for Congress, RL32370, Congressional Research Service, Washington, D.C., April 29, 2004.

[4] Military forces remained separate, under the command of CJTF-7.

[5] For more on Garner and ORHA, see Chapter Five of this report.

The legal basis of CPA was rather unusual, and it created some confusion. The organization was simultaneously an international organization and a U.S. government entity, enabling it to take advantage of some of the benefits of each. Legally, UN Security Council Resolutions 1483 and 1511 recognized the United States and the United Kingdom as occupying powers, and CPA as their instrument for conducting the occupation.[6] While a Congressional Research Service (CRS) report assessed this dual structure as confusing, there was reportedly little debate within the U.S. government or among allies about it.[7] Still, as the CRS report notes, although UN Security Council resolutions recognize CPA as an entity, the United Nations did not in any way authorize its creation. A classified National Security Presidential Directive issued in May 2003 by order of the President laid out the authorities of CPA as a U.S. government agency. It was further recognized as a U.S. government entity by the Emergency Supplemental Appropriations Act for Defense and for the Reconstruction of Iraq and Afghanistan, 2004 (November 6, 2003).[8]

As the administrator, Bremer was the official representative of the occupying powers and, regardless of the CPA's legal status, the proconsul of Iraq. CPA and CJTF-7 accepted their role as the occupying powers of Iraq (although their relations with each other were sometimes more cooperative and sometimes less, as is discussed below) and hence their responsibility to treat both former Iraqi military and Iraqi civilians in accordance with the requirements of international law (including the Geneva Conventions).[9] Moreover, CPA believed that its role as the political authority, when combined with the language in UN Security Council Resolution (UNSCR) 1483, allowed for a wide range of policy changes. These included the replacement of the Saddam-era Iraqi dinar with a new currency, the dissolution of the Iraqi army, and the creation of a new police force. As the administrator, Bremer ruled in part by issuing orders, decrees that had the force of law in Iraq. These decrees could modify or replace existing Iraqi laws.

There was no question that international law permitted an occupying power to delegate authority to CPA to issue orders and take actions in the interest of public order and safety. Many of the orders issued early on were fully in line with such "housekeeping" requirements, addressing media law, the penal code, and so forth. However, as the CPA agenda began to incorporate more fundamental changes in Iraq's governing structures, there was increasing debate among CPA staff and among

---

[6] Interview with CPA officials, August 2004.

[7] Halchin, pp. 4–5; interviews with CPA officials, August 2004.

[8] Halchin, pp. 4–5; interviews with CPA officials, August 2004. The legislation, H.R. 3289 [108th], can be examined via the following web site. As of October 2007:
http://www.govtrack.us/congress/bill.xpd?bill=h108-3289

[9] The Fourth Geneva Convention governs the protection of civilians in wartime. A guide to the Geneva Conventions, with full text, is available at www.genevaconventions.org.

the coalition capitals regarding the legitimacy of the authority to take such steps. The language in UNSCR 1483, seen by some as the source of this authority, states that the United Nations Security Council:

> Requests the Secretary-General to appoint a Special Representative for Iraq whose independent responsibilities shall involve reporting regularly to the Council on his activities under this resolution, coordinating activities of the United Nations in post-conflict processes in Iraq, coordinating among United Nations and international agencies engaged in humanitarian assistance and reconstruction activities in Iraq, and, in coordination with the Authority, assisting the people of Iraq through:
>
> (a) coordinating humanitarian and reconstruction assistance by United Nations agencies and between United Nations agencies and nongovernmental organizations;
>
> (b) promoting the safe, orderly, and voluntary return of refugees and displaced persons;
>
> (c) working intensively with the Authority, the people of Iraq, and others concerned to advance efforts to restore and establish national and local institutions for representative governance, including by working together to facilitate a process leading to an internationally recognized, representative government of Iraq;
>
> (d) facilitating the reconstruction of key infrastructure, in cooperation with other international organizations;
>
> (e) promoting economic reconstruction and the conditions for sustainable development, including through coordination with national and regional organizations, as appropriate, civil society, donors, and the international financial institutions;
>
> (f) encouraging international efforts to contribute to basic civilian administration functions;
>
> (g) promoting the protection of human rights;
>
> (h) encouraging international efforts to rebuild the capacity of the Iraqi civilian police force; and
>
> (i) encouraging international efforts to promote legal and judicial reform.[10]

The language in subparagraphs (c), (e), and (f) do suggest an expectation that CPA ("the Authority") will play a role in, among other things, building representative governance and economic and government reform in Iraq. However, this language also calls for extensive coordination with a variety of international bodies and structures, as well as a sizeable United Nations role. When that United Nations role shrank substantially after the August 19, 2003 attack on its headquarters in Baghdad,

---

[10] All UN Security Council Resolutions for 2003 can be found at the following web site, as of October 2007: http://www.un.org/Docs/sc/unsc_resolutions03.html

coordination became increasingly difficult. CPA officials coordinated their orders with the capitals of the three core coalition members (Washington, London, and Canberra),[11] international financial institutions, the United Nations, and others. However, the lack of a UN presence in Baghdad, combined with initial reluctance in Washington to give the UN a prominent role, made coordination difficult.

Furthermore, from the outset, there was a problem regarding the limited involvement of Iraqis in CPA decisionmaking. Although CPA was able over time to increase Iraqi participation through the IGC and the development of ministries, and there were a number of Iraqi staff in some offices that had involvement in decisionmaking (for instance, Iraqi staff in the CPA office of General Counsel assisted in the development of some orders), outreach to stakeholders remained uneven throughout CPA's tenure.[12]

Goals

The CPA vision statement, developed after the organization was fully up and running in July 2003, stated that its goal was:

> A durable peace for a unified and stable, democratic Iraq that provides effective and representative government for the Iraqi people; is underpinned by new and protected freedoms and a growing market economy; is able to defend itself but no longer poses a threat to its neighbors or international security.[13]

To achieve this vision, CPA set for itself four principal objectives, which it characterized as four "core foundations":

1. Security: establishing a secure and safe environment.
2. Governance: enabling the transition to transparent and inclusive democratic governance.
3. Economy: creating the conditions for economic growth.
4. Essential Services: restoring basic services to an acceptable standard.[14]

---

[11] Coordination with the United Kingdom was required because of the UK's status as a legal occupier of Iraq, but this was not always the case in practice. CPA officials also tried to coordinate many of their actions with Australia, despite the fact that Australia was not a legal occupier, because of its military commitment to the coalition.

[12] Interviews with CPA officials, August 2004.

[13] Coalition Provisional Authority, *Vision for Iraq*, July 11, 2003.

[14] Coalition Provisional Authority, *Achieving the Vision: Taking Forward the CPA Strategic Plan for Iraq*, July 18, 2003, p. 1. A fifth pillar, strategic communications, was subsequently added.

Throughout its existence, CPA focused with varying intensity on achieving these objectives. The inherent dual roles of governing and institution building proved difficult to reconcile. Moreover, CPA was constrained by the limits of its capacity and authority within the U.S. government. Security policy was a clear example: CPA's capacity to establish a secure and safe environment for Iraqis was constrained by its staffing and tasks. Clearly it had the role of developing Iraqi internal security forces and establishing a legal framework, which are tasks of institution building. With regard to day-to-day security, CPA also had a role in efforts to direct Iraqi police. These were variably effective, and at times CPA advisors were themselves involved in responding to crimes and criminal activity. They lacked, however, the numbers and capacity to do this on any large scale, even as these efforts sometimes distracted from their ability to focus effort on building Iraqi capacity.

Moreover, CJTF-7's role in this effort was unclear. As the occupying military power, it had the forces and, many felt, the mission of creating a stable and secure environment. It was not, however, prepared or configured for this role. Initially unclear about the extent of its policing role, as evidenced by its noninvolvement in the massive looting that followed the initial campaign, CJTF-7 focused on active military counterinsurgency operations as the insurgencies gathered pace. Thus, despite some policing activities, CJTF-7 consistently left a policing vacuum in which the average Iraqi citizen was faced with growing concerns about crime and lawlessness. CJTF-7 also had a role in the training and creation of certain security institutions: the Iraqi Civil Defense Corps (ICDC) initially, and, by the end of CPA's tenure, all Defense Ministry and Interior Ministry forces, as will be discussed in Chapter Nine. CJTF-7 was not under CPA's chain of command.

In the essential services sphere, priority was given to ensuring that oil and electricity were up and running, with attempts to restore power generation and distribution to prewar levels as quickly as possible. Water supplies were another focus. Communications, municipal services, health care, transport, housing, education, social welfare, and a whole range of central and local governance services were also run by CPA, which sought both to reconstruct them and to reform them in line with its perceptions of international best practices. While managing and reforming these services, CPA also, if late in the day, began efforts to build Iraq's capacity to manage these sectors on its own.[15]

The goal of economic policy was to create the institutions of a market economy while stimulating economic growth. The security situation made this a challenging proposition.[16] Finally, CPA was intent on creating the foundations for a democratic government, the rule of law, and respect for human rights in Iraq. There was, of

---

[15] See Chapter Twelve for more detail on restoring essential services.

[16] See Chapter Eleven for more detail on economic policy.

course, a tension inherent in attempting to do this under conditions of occupation marked by a limited Iraqi voice in government.[17]

Other Governing Institutions

Several Iraqi organizations functioned alongside CPA and under its direction. These included the government ministries, both those that had existed under Saddam Hussein and those set up by CPA. The ministries implemented policy and were staffed by Iraqis, who were in turn advised by CPA officials. In addition, a structure of local councils in various cities and the Iraqi Governing Council for the country as a whole were created to advise coalition personnel and to provide various representative bodies of Iraqis who could work with CPA to define and implement policy.

Iraqi Governing Council (IGC)

The IGC and its structures are discussed in Chapter Ten. From the perspective of CPA, however, the IGC functioned largely as an advisory group and, over time, a vetting agency of sorts for orders and appointments. Bremer did not formally transfer any standing decisionmaking authority to the IGC, although certain authorities were delegated to it for specific actions, such as the creation of an Iraqi Special Tribunal and responsibility for implementing de-Ba'athification. However, given the absence of alternative institutions and the need to put an Iraqi "face" on decisions and hear an Iraqi "voice" in making those decisions, the IGC quickly became CPA's primary Iraqi interlocutor and partner in governance. Bremer held regular meetings with the IGC covering a wide range of issues.

CPA also used the IGC to develop agreed-on policy programs. For instance, the IGC Security Committee met regularly with the CJTF-7 leadership and senior CPA security advisors. Although CPA advisors were also involved in the drafting effort, the IGC was primarily responsible for initial, albeit abortive, work on a full constitution in fall 2003 and for drafting and approving the Transitional Administrative Law (TAL), the interim constitution under which the interim government was to operate, signed in March 2004. As the June 30, 2004 transfer of authority got closer, the IGC's de facto power to veto or shape policies grew.

The appropriateness of the IGC as the primary Iraqi voice in governing occupied Iraq was highlighted by the final approval process for the TAL, which demonstrated that other political forces in Iraq wielded significant power.[18] Moreover, there were continuing concerns that the IGC, its membership entirely appointed by coali-

---

[17] For a broader discussion of the dilemmas posed by instilling democracy through authoritarian means, see Simon Chesterman, *You, The People*, Oxford: Oxford University Press, 2004.

[18] For more on the adoption of the TAL, see Chapter Ten.

tion personnel, was not in any real way representative of Iraq, especially because it failed to include a range of important actors—particularly the Sunni elite and religious and tribal leaders. Among the Iraqi public, the IGC was never particularly popular or trusted; it was generally seen as composed of expatriates who owed fealty to occupation forces. Ironically, the IGC was not, in fact, particularly acquiescent to CPA. Over time, it became increasingly assertive in taking positions different from those of Bremer and his staff.

Ministries

One senior advisor was assigned to each Iraqi ministry.[19] The senior advisors had existed under ORHA, as discussed in Chapter Five, and their fundamental roles and structures continued under CPA. Through August 2003, the senior advisor to each ministry functioned as the de facto minister; all the Iraqi ministers under Saddam had been dismissed. After Iraqi ministers were appointed and approved by the IGC, starting in late August 2003, the senior advisors still held veto authority over all decisions until the transfer of authority in June 2004.

The senior advisors worked to administer their portfolios by relying to a large extent on senior Iraqi civil servants who had met the de-Ba'athification criteria. After Iraqi ministers were named, the division of authority between the ministers and their senior advisors varied significantly from ministry to ministry. Stripped of senior Ba'athist officials, most ministries retained many of the structures and staff that had existed under Saddam. The exceptions were the Ministries of Displacement and Migration, Environment, Science and Technology, Human Rights, and Defense. The Ministry of Defense was dismantled in May 2003 and a new one created by CPA order in March 2004. The other new ministries were created by CPA order, in the case of the Ministry of Science and Technology, to replace the dissolved Ministry of Atomic Energy.[20]

As with CPA as a whole, the senior advisors had dual roles: the operations of their ministries and the building of institutions and capacity within them. For example, the senior advisor to the Ministry of Labor and Social Affairs was responsible for ensuring that his ministry paid pensions and disability payments on time. The senior advisor to the Ministry of Health was responsible for reopening clinics, supplying medicines, surgical supplies, and pharmaceuticals to Iraqi dispensaries, and paying doctors, nurses, and other medical personnel. Most of these activities were conducted by Iraqi employees of the ministries, but because CPA controlled funds and also held contracting authority, the senior advisor and his support staff had to ensure that

---

[19] The exception was the senior advisor for national security, who oversaw the creation of a new Ministry of Defense and, at times, also oversaw overall Iraqi security policy. This included interaction with CJTF-7 and some oversight over the Ministry of Interior as well.

[20] In addition, the Ministry of Public Works was renamed the Ministry of Municipalities and Public Works.

funds were released, contracts signed, and supplies delivered where they were needed. This meant that the small ministerial staffs at CPA, often just a handful of people, tended to concentrate on day-to-day tasks and had little time for longer-term strategic thinking. The institution building mission tended to suffer as a result. Exceptions were the staffs responsible for creating new ministries, such as the Ministry of Defense, which had the tasks of planning, structuring, and recruiting, giving them the opportunity for a broader strategic role.

The Iraqi ministries had difficulty in working effectively under their CPA advisors. Civil servants had been discouraged from taking initiative during the Saddam Hussein regime, and this legacy persisted. Many CPA staff members were surprised at the unwillingness of competent Iraqi staff to make recommendations. One Army Corps of Engineers officer recounted his experience with an Iraqi counterpart. In the course of their working together, the Iraqi engineer provided a complete solution to fix a transformer. However, when the Corps of Engineers employee asked that he present it in the form of a recommendation, the Iraqi was terrified and refused to present his solution in an official form. Problems were exacerbated by the lack of facilities. Most ministries were gutted in the looting that occurred during and following the capture of Baghdad. In a number of instances, staff had nowhere to go and no way to work immediately following the end of the conflict.

CPA staff generally concurred that most Iraqi civil servants were inefficient. They put in truncated hours, frequently failed to fulfill assignments, and, in a number of instances, did not have clear tasks and responsibilities. There were exceptions, however. The advisory staff to the Ministry of Foreign Affairs generally reported that they were impressed by the professionalism and competence of their Iraqi colleagues.

Increases in government salaries provided by CPA made government employment, outside of some security jobs, more attractive than employment in the private sector. As a consequence, applicants queued for government jobs. Civil servants and ministers, once appointed, frequently rewarded friends and relatives by providing jobs. For example, the number of director general positions (equivalent to the assistant secretary level) in the Ministry of Electricity rose from 12 to 80 between August 2003 and February 2004. Many of these positions were awarded to individuals tied to members of the IGC or to ministers. CPA staff did what they could to put a lid on such practices, but their limited capacity to oversee all aspects of operations, combined with the need to put an Iraqi face on hiring and other efforts, limited their ability to be effective.

CPA recognized early that it had to not merely get the Iraqi government machinery operational again; to enable this government to operate in a transparent, accountable, and efficient manner, CPA also had to institute a major overhaul of the civil service. This began on an ad hoc basis as CPA advisors to individual ministries sought to apply best practices from their home departments, such as establishing merit-based personnel structures, devolution of authority, and modern financial management practices. It became evident, however, that a more comprehensive program of reform was required. The UK's Department for International Development (DFID) issued a comprehensive plan for emergency public administration reform. As of the transfer of authority on June 28, 2004, implementation of this plan was only in its very early stages. The program focused initially on key cross-cutting ministries (finance, public works, and defense) but also provided initial support to the machinery of central government (cabinet, prime minister's and president's offices), envisioning a strong cabinet committee role. The program laid out a multiyear effort to turn the top-heavy, hierarchical, bloated, and inefficient civil service into something resembling an efficient bureaucracy centered around providing cost-effective citizen services rather than serving the ruling elite.

This was obviously a tall order, and the first indications in early summer 2004 were that many of the ministers and senior officials in the Iraqi Interim Government were unhappy with a program that threatened their ability to use government appointments as part of their patronage, their "spoils of office." Thus, the full implementation of DFID's program by the Iraqi government was significantly in doubt after the transfer of authority on June 28, 2004.

Organization of CPA

CPA was structured and run in a way that reflected its dual missions of governing Iraq and building Iraqi government institutions. Figure 8.1 shows a July 2003 organization chart for CPA. As can be seen from the chart, CPA had a somewhat unwieldy organizational structure at that time. It consisted of "staff" and "operations" components. The "staff" included the General Counsel's office, Intelligence, an Operations Support Group that included the Facilities Management office and others, the Executive Secretariat, the Strategic Policy Office, Financial Oversight, and a Requirements Coordination Office. Strategic Communications, CPA's public affairs office, was eventually added to this group.

The "operations" components consisted of the senior advisors, described above, and their staffs. As noted, those staffs in most cases focused on immediate issues of maintaining the ministries' core operations. Policy planning and formulation were to be conducted by even more senior "directors" who often covered several ministries. In a number of instances, senior advisors reported to these individuals, who reported to Bremer. Other senior advisors reported directly to Bremer.

Bremer sat at the pinnacle of CPA. He had responsibility for ruling Iraq. Power was fully concentrated in his office, with minimal delegation. On occasion, Bremer's management style was to focus on detail to the point of micromanagement. At times, this may have hampered his ability to set and supervise the implementation of clear

Figure 8.1
Original Organization of CPA, July 2003

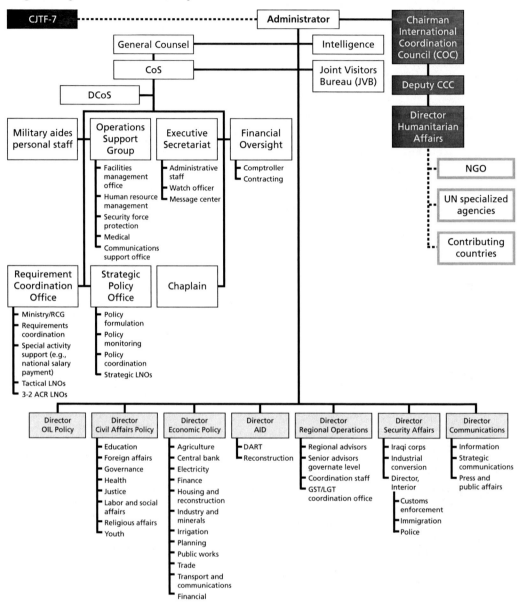

SOURCE: Coalition Provisional Authority.
RAND MG642-8.1

priorities for CPA and Iraq. At the same time, a substantial amount of his time was spent interacting with Iraqis, giving speeches and press conferences to communicate CPA goals, decisions, and thinking to Iraqis, to the citizens of coalition partners, and to the international community at large. He also spent a substantial amount of time communicating with his superiors, Secretary of Defense Rumsfeld and President Bush.

Bremer took a hands-on approach to managing CPA. He held short (10-minute) daily staff meetings and longer strategy meetings every Friday. In addition, his day was filled with meetings with senior administrators at CPA. Functional and policy decisions were made at these meetings.

The organization shown in Figure 8.1 served Bremer poorly. Directors and senior advisors who participated in meetings with him did not always follow through on the directives they received. This failure stemmed partly from the organizational structure. No one was responsible for following up to ensure that action was taken. While Bremer did not have time to write down decisions and ensure follow up, others in the organization lacked the authority to enforce discipline. Ambassador Clay McManaway, senior counselor to Bremer, partially filled this role, but he spent much of his time filling in for Bremer or pursuing special projects, such as finding a site for training Iraqi police. The follow-up function was not built into the organization.

Coordination within CPA was also somewhat haphazard. The urgency, real and perceived, of the work at hand created strong disincentives for approaching other offices for approval of plans and projects. As a result, decisions were generally made by small subsets of the CPA staff, often in isolation from other offices in the organization.

In addition to these internal organizational problems, the Department of Defense lacked the organization and staffing resources appropriate for a nation-building effort. Although the State Department and USAID have had their share of problems with these types of operations, they do have staff with significant experience in post-conflict situations who can be detailed to such assignments. More important, these officials know that their performance on these assignments will have career consequences. In contrast, DoD does not have a civilian staff whose job description and training prepare them specifically for nation-building efforts. DoD had to recruit personnel from inside and outside the government to staff CPA. A number of these individuals were drawn from the State Department, but the majority of them were not.

Stays in Baghdad were short, from three to six months, so turnover was much more rapid than in other post-conflict situations. Because most staff did not return to departments that continued to be engaged in Iraq, much of what they learned was lost. Institutional memory was short.

To compound matters, while CPA in Baghdad was undergoing a revolving door of staff changes, CPA "West" in the Pentagon was undergoing a very rapid buildup, drawing on individuals throughout the government. Whereas State Department–led operations in other post-conflict situations had an institutional anchor already in place when the operation began, DoD had to construct its own office to oversee CPA, almost from scratch. The role of CPA "West" was to handle recruiting, vetting, and the logistics of deploying civilian employees. It also ostensibly served as a liaison between CPA and the interagency community in Washington. In practice, however, staff at CPA Baghdad generally communicated directly with senior officials at the White House, the Pentagon, and other agencies in Washington, often leaving CPA "West" out of the loop on actions and plans.

Senior officials in Washington also had no compunctions about picking up the telephone to get in touch with staff in Baghdad. In principle, CPA "West" had a coordinating role, and Secretary of Defense Rumsfeld oversaw the entire effort. In practice, various CPA offices at times seemed to be reporting to their home organizations in Washington—that is, those from which staff were drawn and that were their substantive counterparts. Each of CPA's core "foundations" tended to have its own chain of command: In general, Economics reported to Treasury; Governance reported to State; Security reported to both the NSC and DoD (creating additional complications); and Essential Services interacted with the Army Corps of Engineers. Finally, while Bremer and the security offices reported to DoD, all of CPA also reported to Bremer himself, and the organization as a whole often operated on a shorter time horizon than one permitting extensive coordination with home agencies. Although some CPA staff believed that they continued to report to agencies back home, others saw their role as that of a detailee to CPA, which meant that they reported to Ambassador Bremer and not their "home" agencies. These multiple command chains and reporting channels were confusing, and they resulted in substantial duplication of reporting efforts, misunderstanding, and occasional rancor.

Not surprisingly, the agencies and organizations in Washington often thought they had insufficient information about activities in Baghdad. This classic embassy-capital problem was greatly magnified by the lack of established channels, the absence of trained reporting officers, and the chaos and conflict all around. Some offices had better records of coordination than others. CPA's Office of General Counsel held regular, often daily, discussions with the Defense Department's Office of General Counsel (as well as other organizations) to discuss relevant legal issues and maintain communications. The Office of National Security Affairs remained in close contact with both the Defense Department and the White House, although this did not prevent miscommunication and confusion about plans and actions. However, the Governance team was broadly seen in Washington as not making sufficient effort to keep the capital apprised of its activities, particularly in the first months of CPA's existence.[21]

The presence of a significant number of non-U.S. staff further complicated matters. Like the U.S. staff, these personnel felt a dual loyalty to CPA and to their home agencies. Unlike the U.S. staff, their home agencies were not in the U.S.

---

[21] Interviews with CPA officials, July and August 2004.

government. Many of them represented the defense ministries of their home countries. Some represented their foreign offices and other agencies. The United Kingdom also had a senior ambassador in place in Baghdad. The first of these was John Sawyers, who was followed by Sir Jeremy Greenstock, who was in turn replaced by David Richmond. Ostensibly, the UK ambassador was to be Bremer's counterpart, but with Bremer as the head of CPA and the Iraqi government, the precise outlines of this U.S.-UK relationship were never fully clear. Lack of clarity created some tension between the two positions and the individuals who filled them, and, at times, between the U.S. and British contingents.

In mid-November 2003, CPA revamped its structure to address some of the shortcomings. Figure 8.2 shows a bare-bones organizational chart. Under the new organization, Ambassador Richard Jones became the Deputy Administrator and the Director of Policy. General Joseph (Keith) Kellogg (ret.) became the Director of Operations. His primary responsibility was the reconstruction effort, and the senior advisors to the "hard" infrastructure ministries reported to him. Other senior advisors reported to Jones in principle, although they too often had direct access to Bremer as well.

Ambassador Jones was simultaneously the U.S. ambassador to Kuwait. A senior Foreign Service officer, he brought the experience as well as the seniority and prestige to chair policy meetings without Bremer. He could also perform the "nag" function, insisting that senior administrators follow through on their commitments.

This delegation of responsibility in November 2003 improved the operations of CPA. The structure was cleaner, and it enabled a somewhat more cohesive policy-formulation process, although reach-back continued to be largely to home agencies in Washington and other capitals. Kellogg brought in additional management staff who were able to address some of the logistical and programmatic blockages hindering such ministries as interior and oil. For most people, this meant multiple reporting channels, both within CPA and back home, which were relatively effective in keeping various agencies informed of developments. It was problematic when different guidance was provided by the different agencies, and it imposed a tremendous time burden on ministerial advisors.

Problems of coordination within CPA also continued, although improved personal relations and concerted efforts by some senior staff members to improve coordination on specific issues (security institution building, for example) did help to increase transparency. Recognition of the benefits of coordinating with staff from other offices also grew over time, as a result of both the efforts of various CPA staff to remain apprised of events relevant to their work and a push by the Executive Secretariat to ensure the appropriate coordination of memoranda and position papers provided to Ambassador Bremer. Still, the process remained deeply imperfect throughout the existence of CPA.

Figure 8.2
Revised Organization as of November 2003

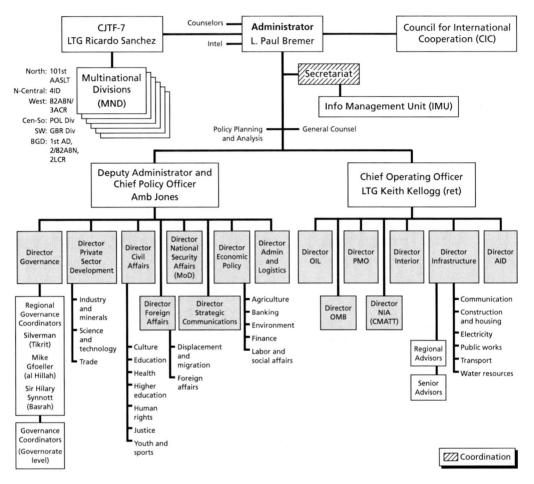

SOURCE: Coalition Provisional Authority.
RAND *MG642-8.2*

The change in CPA structure coincided with a crucial policy shift. On November 15, 2003, after consultations with President Bush, CPA and the IGC agreed that sovereignty would be returned to an interim Iraqi government by July 1, 2004, as will be discussed in Chapter Ten. After this November 15 agreement, CPA's capacity to fulfill its dual tasks of governance and institution building was further strained by an immediate need to meet the increasingly close deadline for transferring authority. While many individuals in CPA remained committed to building effective institutions, the time constraint left them with little choice but to find short-term answers that would enable the transfer of sovereignty, rather than long-term approaches that might in the end have been more effective. In a number of policy areas, particularly economics, governance, and security, this policy decision resulted in CPA's abandoning important policy changes or accelerating the standing up of new Iraqi institutions at such a pace that the results were counterproductive.

Location and Staffing

CPA was headquartered at the former Republican Palace next to the Tigris River in a secure area of several square miles called the "Green Zone," shown in Figure 8.3.[22] Most CPA staff worked in the palace, traveling to ministries and other destinations with security guards or secure vehicles or by military convoy. Over time some staff

Figure 8.3
Location of the Palace in Baghdad

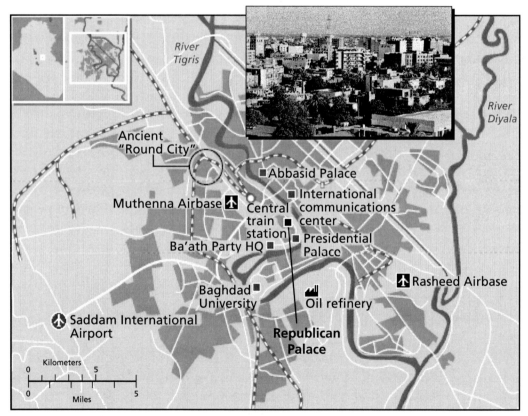

SOURCE: Coalition Provisional Authority.
RAND *MG642-8.3*

---

[22] In addition, several CPA staff members were assigned to governance teams in the provinces. The issues with regard to those staffs, and relations between Baghdad and the regions, are discussed in Chapter Ten.

were provided with their own vehicles, and they traveled to their destinations without escorts. Nonetheless, providing security for CPA personnel traveling to and from downtown Baghdad and to points outside the city became a major task for CJTF-7. There were always substantially more requests than available convoys. In the fall of 2003, trips had to be requested at least one week ahead of time. When the security situation was perceived as relatively benign, CPA staff felt more comfortable moving around Baghdad. However, two sharp turning points—the New Year's Eve 2003 attack on a popular Baghdad restaurant and the early April 2004 flare-up of violence in Baghdad and throughout the country—significantly constricted CPA staff movement outside of the secure area. Staff who had security detachments provided by private security firms, including those from USAID, the UK, and Australia, were able to travel around Baghdad more easily.

The food to feed the staff was subject to U.S. Department of Agriculture regulations regarding food fed to U.S. Army forces and was for the most part shipped over from the United States, resulting in occasional shortages, particularly in April 2004 when attacks on coalition vehicles slowed supplies into CPA. Until the October 2003 attack on the Al Rasheed Hotel on the edge of the Green Zone, substantial numbers of CPA staff were housed in that facility. It was evacuated after the attack and not reopened for housing while CPA remained in existence. Throughout that time, the majority of CPA and CJTF-7 staff members assigned to the palace were housed in the numerous trailer parks built and administered by Kellogg, Brown and Root (KBR). These were scattered throughout the palace grounds and further out into the Green Zone, though there were exceptions: USAID had its own compound, for example, and the British moved into a separate compound. However, there were invariably more staff members than space in trailers, which were shared by four people, two to a room (a situation exacerbated by a growing underground market in trailer beds). Personnel awaiting a trailer assignment slept in bunk beds in the palace ballroom or chapel, often for weeks on end. When space became short in that facility, two large tents were set up to accommodate more personnel. Other CPA staff slept in their offices.

CPA had two major branch offices: CPA North in Irbil, the capital of the Kurdish governorates (administrative regions), and CPA South, covering Basra, Muthanna, Maysan, and Dhi Qar. By the end of 2003, CPA offices had been set up in every one of Iraq's 18 governorates. These offices were staffed by a combination of government detailees (from the State Department and the British Foreign and Commonwealth Office [FCO]), contractors, and military civil affairs personnel. CPA South was very much a British enclave, with the majority of its staffing comprising FCO and DFID. Given that the south was also run by a division under British command, CPA South at times operated semi-autonomously from Baghdad.

Total employment at CPA probably numbered about 1,500 people, but there was never any clear count of CPA staff. The combination of detailees, direct hires, political appointees, and so forth made it difficult to keep track. About half the people in the palace at any one time were military officers assigned to CJTF-7. Particularly in the early stages of its work, CPA benefited from large numbers of civil affairs officers, almost exclusively reservists, who worked with the senior advisors.[23] Some USAID contractors were also located in the palace. A large support staff was provided by contractors to run telecommunications systems, handle the motor pool, and provide laundry services, haircuts, security, and other functions.

As noted above, CPA staff were an eclectic crew. A number of retirees from the private sector and the State Department agreed to work for CPA, often at the request of senior administration officials. At the other end of the age spectrum, the Defense Department hired a substantial number of people in their twenties to staff the ministry advisory teams. After some unfavorable news articles printed interviews with former CPA employees, new employees were closely vetted on the basis of party affiliation. Internships in conservative public policy institutes or recommendations from Republican Party leaders were a common route into CPA for such young people.[24] Much of the rest of CPA was staffed with U.S. government civil servants and some military personnel.

All government departments were asked to encourage staff strongly to volunteer for duty in Iraq. The quality of the staff deployed was mixed, however. While there were a great many supremely competent personnel at CPA, representing the broad range of groups and organizations that contributed staff, there were also personnel who did not perform adequately. Various agencies and organizations had different attitudes toward sacrificing their best people to this mission, and several CPA staff members reported having had to fight their leadership to be allowed to deploy to Baghdad. Moreover, the broad mix of personnel had advantages and disadvantages. For example, the variety of perspectives was likely beneficial, while the lack of government experience of some staff created difficulties when interacting with Washington. The overall lack of experience among the junior staff also led to some difficulties in defining and implementing programs effectively.

Staff background varied. There were a number of functional experts, particularly among retirees and civil servants, but few regional specialists deployed. Of those who were in theater, very few also had subject matter expertise. This created a requirement for communication and information sharing that was not often met. Moreover, even the subject matter experts were often quite out of date in their understanding of Iraq; maintaining a clear sense of developments there had become difficult after the 1991 Gulf War. The result was that policies were often developed in ignorance of their potential impact on Iraqi society. The IGC thus came to be in-

---

[23] This number declined when the initial cohort of CJTF-7 officers rotated out in the spring of 2004.

[24] Ariana Eunjung Cha, "In Iraq, the Job Opportunity of a Lifetime," *Washington Post,* May 23, 2004; Rajiv Chandrasekaran, "Iraq's Barbed Realities," *Washington Post,* October 17, 2004.

creasingly used as a source of local expertise that would guide acceptance or rejection of policies on the basis of how they would work in Iraq.

As noted above, other coalition countries' civil servants and military personnel were represented. While the British contingent was the largest, there were also significant numbers of Australians, Italians, and Spaniards, as well as some Czechs, Ukrainians, Poles, Romanians, and others. The palace also housed the headquarters of the Iraq Reconstruction and Development Council (IRDC). This organization of about 150 Iraqi expatriates, primarily U.S. citizens, was a contract entity run by SAIC for the Defense Department. Its members provided links to Iraqi society, and some worked as advisors at ministries and other government institutions.

Relations with CJTF-7

The first order signed by Bremer defined the powers of CPA, including its relations with the coalition forces. CPA Regulation 1 states that "1) The CPA shall exercise powers of government temporarily in order to provide the effective administration of Iraq."[25] The order goes on to state, "2) The CPA is vested with all executive, legislative, and judicial authority necessary to achieve its objectives." However, CPA was strictly a civilian entity. "3) As the Commander of the Coalition Forces, the Commander of U.S. Central Command shall directly support CPA by deterring hostilities; maintaining Iraq's territorial integrity and security; searching for, securing, and destroying weapons of mass destruction; and assisting in carrying out Coalition policy generally."[26] In other words, CENTCOM through CJTF-7 was to support CPA, but CPA definitely did not have the authority to command CJTF-7.[27]

The degree of coordination between CPA and CJTF-7, at both senior and staff levels, varied over time. CJTF-7 had a large office at the palace where CPA was located, although its headquarters was at Camp Victory, near Baghdad International Airport. Lieutenant General Ricardo Sanchez, the CJTF-7 commander, had daily meetings with Ambassador Bremer. Operations were generally discussed in detail before a decision was made to proceed. CPA and CJTF-7 attempted to coordinate political strategy, reconstruction efforts, security policies, and strategic communications with military operations, but this was done in an ad hoc way. The conventional warfare approach of CJTF-7 and the failure of CPA to deploy sufficient staff and re-

---

[25] Coalition Provisional Authority Regulation Number 1, Section 1. All of the CPA's regulations are available at a web site, no longer updated, that has been left open for historical purposes. As of October 2007: http://www.cpa-iraq.org/regulations/#Regulations

[26] Coalition Provisional Authority Regulation Number 1.

[27] CJTF-7 was not the only coalition structure not under CPA control. Intelligence agency representatives in Baghdad did not report to Ambassador Bremer, although they were represented at the palace, and neither did the Iraq Survey Group (ISG), whose mission was to seek evidence of WMD in Iraq.

sources to fulfill many of its mission areas created tensions. With no requirement to coordinate and no single official in charge in theater, it was often difficult to convince busy people of the need to share information—even if the repercussions of a failure to do so could be quite serious.

Although relations were relatively cordial at the highest level, relations between civil affairs officers assigned to the regional commands (and lower echelons) and CPA staff were sometimes fractious, particularly early in the CPA governance teams' deployments. The civil affairs officers were the interface between Iraqis living within the military commands' areas of responsibility and various areas within each command assigned to units under the commander's control. These officers heard complaints from Iraqis, worked with the local councils and government agencies, and also operated as liaisons between local state-owned enterprises, the private sector, and ministries in Baghdad. In short, their days were spent solving problems, many of which needed to be resolved quickly. They were eventually replaced by individual teams in each of Iraq's 18 governorates. Both groups felt they had insufficient input into the policy and funding decisions taken by CPA in Baghdad. Civil affairs officers would constantly come across needs within their territories for funds to repair schools and other government buildings, pay Iraqi government employees, or invest in local enterprises. For their part, CPA employees focused on establishing transparent budgeting and payment systems in the ministries and in tracking funds. Requests from civil affairs officers sometimes received short shrift.

CPA also had to contend with ad hoc policy decisions taken within the commands. For example, despite a nationally announced policy of no tariffs on imported goods, Major General David Petraeus, who commanded operations in the north, levied small charges in dollars on trucks entering Iraq through his sector and spent those funds on reconstruction activities in his area of operations.[28] A bigger policy headache arose from wages paid to local security guards, whose wages were initially funded from Commander's Emergency Response Funds (CERF), money allocated to commanders for immediate expenses related to reconstruction and assistance efforts. Facilities protection service employees in some parts of the country received higher wages than those in other parts of the country. When wage payments for these employees were transferred to the Iraqi government in 2004, CPA and the Ministry of Finance had to contend with a host of unhappy employees whose salaries had to be reduced to fit into the new national pay scale.

---

[28] Interview with U.S. civil affairs officer, October 2003.

Concluding Observations

The CPA faced the daunting task of simultaneously governing Iraq and building new institutions to manage the political transition and reconstruction effort. It was at once a U.S. government organization and an international one. It was one of two representatives of the occupation on Iraqi soil, alongside CJTF-7. It was understaffed and faced significant logistical and structural problems. Its staff represented various U.S. and foreign government organizations and structures, staff members had varying levels of experience, and individuals were generally in country for only short periods. As time went on, CPA was increasingly driven by near-term deadlines.

It is not surprising, therefore, that certain tensions were inherent in its work. The difficulty of governing and responding to short-term needs, while working to build lasting institutions during a time of conflict (when the short-term needs are particularly imperative), cannot be overstated. The need to balance the interests of the home countries and home agencies of CPA staff with the interests of CPA itself also created a range of challenges. Coordination, within CPA, with CJTF-7, and with home governments, in a context requiring decisions to be made quickly and actions taken immediately, often seemed a burden rather than a useful mechanism for facilitating effective work.

CPA's role was to administer Iraq for a short period and to make it possible for Iraqis to govern themselves after it left. That was a tall order. While it will be some decades before historians judge whether the overall endeavor was a success or a failure, now at least we can draw lessons from CPA's experience, good and bad, to improve postwar administration and nation-building efforts in the future.

CHAPTER NINE

Building New Iraqi Security Forces

Defining and Building Iraqi Security Forces

Certain assumptions lay at the core of prewar and pre-occupation planning for the creation of new Iraqi security forces.[1] One of the most significant of these was that not very much would have to be done. It was expected that police forces, being largely non-Ba'ath Party and professional, would remain cohesive and be able to maintain law and order in occupied Iraq with limited involvement from the coalition. It was further expected that certain units of the Iraqi armed forces would stand aside from the fight and could later be used to assist in reconstruction and security efforts. CENTCOM PSYOP campaigns instructed army units to capitulate, i.e., to stay in formation and surrender. The processing of these surrendered units was to be coordinated through a sizeable disarmament, demobilization, and reintegration (DDR) plan developed by a contractor. Plans were also developed by U.S. government personnel for reforming justice and police functions. These plans assumed functioning bureaucracies and an advisory role.[2]

These assumptions proved faulty. Coalition forces found that rather than standing aside from the fight as intact units, the Iraqi army was nowhere to be seen. Some of its soldiers may have changed into civilian clothes and gone home, as suggested by piles of abandoned military footwear. Many, seeing their senior leadership flee, no doubt saw no reason to stick around themselves. They took weapons and other equipment that seemed valuable with them, and coalition staff found Iraqi military facilities quite thoroughly looted. The police, for their part, had scattered, and there was no one to respond to widespread looting—not just of military facilities but of government and office buildings, stores, and entire cities. Coalition forces, which had

---

[1] This chapter was co-authored by RAND analysts Heather Gregg and Olga Oliker, the latter of whom worked for CPA in Baghdad. All information in this chapter, including that not otherwise attributed, is informed by that personal experience.

[2] Mark Magnier and Sonni Efron, "Arrested Development on Iraqi Police Force," *Los Angeles Times*, March 31, 2004; interviews with DoD, ORHA, and CPA officials, November 2003, July 2004, and August 2004.

not seen the "post-conflict" phase as a stability operations mission, did not see policing Iraq as one of their core missions.[3]

The decision to disband the Iraqi armed forces was apparently made in Washington, D.C., in early May 2003, before the deployment of Ambassador L. Paul Bremer to Baghdad. The view of decisionmakers at the highest levels of the Defense Department, in consultation with Bremer, was that the Iraqi military had already "self-demobilized" by leaving their units and going home, and that it would be impossible to bring them back into barracks (which no longer existed) to implement the DDR plan that had been developed by the contractor in expectation of a significant effort. Instead, on May 23, 2003, shortly after Bremer's arrival in Baghdad, the second CPA order issued declared the Iraqi army, Ministry of Defense, Intelligence Service, and other related organizations and agencies to be disbanded and dissolved.[4]

Thus, CPA was charged with building from scratch a full complement of security services. On the one hand, this presented a remarkable opportunity to establish democratic and effective security structures for a new Iraq. On the other hand, it was to prove a formidable task, as CPA struggled to define Iraqi needs even as it faced the requirement to train forces adequately at a time when actual security requirements were increasing. The tradeoffs between immediate requirements and long-term institutional goals were perhaps more stark in this sphere than in any other of the occupation.

It was clear from the beginning that CPA staff would have to oversee the process of building effective security forces and structures. In March 2003, Walter B. Slocombe, former Under Secretary of Defense for Policy in the Clinton administration, agreed in principle to head up the effort to restructure the Iraqi Defense Ministry. In May, the Defense Department asked Bernard Kerik, former Chief of the New York Police Department, to be the senior advisor to the Ministry of Interior. The Ministry of Interior oversaw a broad range of internal security functions, including police, border police, customs, fire, and emergency personnel. The Defense Ministry function was more narrowly defined, but Slocombe's mission was actually broader—he was charged not only with overseeing the Defense Ministry, but also with the effort to build Interior Ministry forces and any Iraqi intelligence functions to be developed. He and his office were also responsible for building broad national security strategy and coordination functions for the Iraqi government.[5]

---

[3] Interviews with OSD, ORHA, CPA, and CJTF-7 officials and personnel, November 2003, June 2004, July 2004, and August 2004.

[4] Interviews with ORHA and CPA officials, November 2003, December 2003, July 2004, and August 2004. CPA Order Number 2, along with all of the CPA's orders, can be found at a web site, no longer updated, that has been left open for historical purposes. As of October 2007:
http://www.cpa-iraq.org/regulations/#Orders

[5] Interviews with CPA officials, December 2003, July 2004, and August 2004.

The structure that eventually emerged comprised several sets of forces. Because of Iraq's historical experience with the domestic use of the Iraqi armed forces, CPA and IGC agreed that the primary role for domestic security should lie with the Ministry of Interior (MoI). Iraq's army (including its air and coast guard components) would have the responsibility for defense from external threats, although after significant discussion between CPA and the IGC it was agreed that in situations of dire need, the new Iraqi army could be called upon to support Interior Ministry forces within the country. Thus, the forces under MoI control would be the police, the border police, the facilities protection service, and two specialized forces: the civil intervention protection service and the emergency response unit. All of these would be in various stages of development when CPA went out of existence in June 2004, although none of them would be complete at that time.

Because the Defense Ministry had been abolished, the forces that would eventually be placed under it were built and trained by coalition forces without an Iraqi ministry involved. First among these were the Iraqi armed forces, which consisted of ground, air, and coastal defense forces to start. In time the Iraqi Counterterrorism Force, developed by coalition special operations units, was also to be assigned to the Defense Ministry, as was the Iraqi Civil Defense Force, which became the Iraqi National Guard after the Iraqi Interim Government took power. It was initially built and developed by individual CJTF-7 commanders. All of these structures are discussed at length below, along with the development of Iraqi intelligence structures (including those in the Defense and Interior ministries); the creation of the Ministerial Committee for National Security, whose role is to coordinate national security police among the ministries and agencies of the Iraqi government; and the question of militias.

Building the MoI

If the Ministry of Defense (MoD) and the Iraqi armed forces were to be built from scratch, the Ministry of Interior continued as a functioning organization, stripped of senior Ba'athists to the best of CPA's ability but otherwise retaining its own staff, as did most of Iraq's ministries. Kerik had limited preparation for his role as senior advisor to the MoI, leaving for Kuwait just days after agreeing to serve in that role. His previous experience was entirely at the municipal level. He was unfamiliar with Iraq, with the U.S. federal government, and with the international institutions that had supported police training and security sector reform in previous post-conflict environments. Beyond being informed that the ministry had oversight of a number of forces and structures beyond the police force itself (including customs, immigration, border patrols, fire department, and emergency services),[6] Kerik had little information about the organization he was to take responsibility for. A team of U.S. State Department and Justice Department personnel were already in place as advisors to the ministry on his arrival, and they had begun to carry out assessments of needs and requirements. The most obvious and fundamental task, however, was to build up a police force.[7] Kerik hoped not only to reform these services but also to change the public's perceptions of law enforcement; under Saddam, it was seen as a tool of oppression and not as a service to the population.[8]

The CPA team worked closely with Iraqi MoI staff to attempt to rid the ministry of senior Ba'athists and to develop a strategy to get police on the streets quickly. They were hampered in this effort by limited information about the backgrounds of their interlocutors and others. Throughout CPA's tenure, advisors to the MoI would be torn between the need to define Iraq's long-term domestic security requirements and build forces to address them and the requirement to respond to the immediate imperatives of crime in Baghdad and throughout the country. Many of the MoI advisors had police experience in their home countries and were able to apply it to problems in Baghdad. However, this meant that they were often out on the streets of the city, carrying out policing functions themselves. That left them little time to build the institutions Iraq would need to enable Iraqis to take on these roles.[9]

Kerik left Baghdad after four months, replaced by former Drug Enforcement Agency Intelligence Chief Steven Casteel. Over time, CPA's bureaucratic structure for its MoI-advisory role grew to encompass the many forces under MoI's control and to include a policy office. But the team remained consistently understaffed, for policy development, for training functions, for policing, and for mundane tasks of keeping the ministry functioning, finding a new ministry building, and so forth.

Creating the Iraqi Police Service (IPS)

Unlike other Iraqi security forces under Saddam's regime—such as the army and the Mukhabarat (secret police)—the estimated 20,000-strong prewar Iraqi police force was not a primary instrument of dictatorial repression. This led coalition authorities to expect that the police could be relied on to maintain law and order after Saddam was removed from power, with perhaps some additional training. However, the failure of police to report for duty after the Iraqi government was ousted, as well as the

---

[6] Amy Waldman, "U.S. Struggles to Transform a Tainted Iraqi Police Force," *New York Times*, June 30, 2003.

[7] Interviews with CPA officials, November 2003 and December 2003.

[8] Waldman, "U.S. Struggles to Transform a Tainted Iraqi Police Force."

[9] Interviews with CPA officials, November 2003 and December 2003.

widespread descent into looting and lawlessness, proved this assumption to be faulty. The fact was that police under Saddam had little authority, little competence, and little commitment to service. They were notoriously corrupt, and they were the least trusted of the security services under Saddam. After the war, they also lacked infrastructure: the widespread looting had included Iraq's police stations, jails, and courthouses. In Baghdad alone, 60 of the city's 61 police stations were destroyed and their office equipment stolen.[10]

Military forces responsible for the various regions took stopgap measures. In early May 2003, coalition forces distributed emergency stipends of $20 to police and firemen as incentives for them to return to their jobs.[11] Following this initiative, Iraqi police began to return to the streets in Baghdad, wearing their old uniforms with new insignia.[12] Also at this time, newly deployed MPs of the 709th Military Police Battalion began training police in Baghdad and prioritizing equipment needed for the force, including radios, patrol cars, and side arms.[13] By mid-May, roughly 7,000 of Baghdad's 20,000 police had responded to coalition efforts to recall the force; that number reached 10,000 by the end of the month.[14] Despite this turnout, looting, kidnappings, and armed robbery continued to plague the city.[15] The situation was not dissimilar elsewhere in the country.

British forces in southern Iraq moved somewhat more quickly to establish police forces, working with local religious leaders through mosques to build schools to provide emergency police training, which addressed issues such as detention, interrogation, and torture. In this way, they got some of the first "trained" police onto the streets.[16]

Fewer problems arose in the Kurdish-controlled areas in the north, where police forces continued to function. There were problems in Mosul and Kirkuk, however. Major General David H. Petraeus, commander of the 101st Airborne Division, was

---

[10] Kevin Whitelaw, "Law and Disorder," *U.S. News and World Report,* May 26, 2003; interview with CPA officials, December 2003.

[11] Rubin, "U.S. Struggles in Quicksand of Iraq."

[12] Mark Fineman, "U.S. Forces Eager to Relinquish Baghdad Beat," *Los Angeles Times,* May 3, 2003.

[13] David Josar, "MPs Start Training Police in Baghdad," *European Stars and Stripes,* May 5, 2003; John Daniszewski, "Baghdad Is Asking, Where Are the Police?" *Los Angeles Times,* May 27, 2003; Sandra Jontz, "U.S. Gives Iraqi Police 3,000 Radios," *European Stars and Stripes,* June 5, 2003; Vincent J. Schodolski, "In Baghdad, Army MPs Reshaping Wary Police Force," *Chicago Tribune,* June 6, 2003.

[14] Betsy Pisik, "U.S. Sees Organized Crime in Baghdad," *Washington Times,* May 15, 2003; John Daniszewski, "Policing Isn't Black and White in Baghdad," *Los Angeles Times,* May 20, 2003; Marni McEntee, "Joining Forces: Iraqis, U.S. on Patrol," *European Stars and Stripes,* May 20, 2003.

[15] Daniszewski, "Baghdad Is Asking, Where Are the Police?"

[16] Patrick Healy, "Policing Iraq's Police," *Boston Globe,* May 24, 2003; Matthew Campbell, "Quiet Britons Outpace U.S. in Taming Iraq," *Sunday Times* (London), December 28, 2003.

responsible for the north as a whole, and he took some controversial steps, such as reportedly hiring 1,000 former Iraqi army officers to help restore order in Mosul.[17]

On his arrival in Baghdad, Kerik saw the police as the first priority, and saw his most critical task as getting officers out on the streets, planning to increase the Baghdad force, for example, from 8,000 to 18,000 officers, roughly the prewar force size.[18] By the end of Kerik's four-month tenure, his office had overseen the opening of 35 police stations housing 40,000 police.[19] But the challenge was not simply one of numbers. Determining which of the former police officers willing to return to work were not former Ba'athists, in accordance with CPA's de-Ba'athification program, meant that MoI advisors had to rely on reports from their former colleagues and often make gut decisions on whom to trust.

Kerik would eventually fire a reported 7,000 officers for their connections with the former regime.[20] His control of this process outside of Baghdad was limited, however, and there were reports that de-Ba'athification was less stringently applied in recruiting and restaffing police forces in Tikrit, Kirkuk, Fallujah, and elsewhere.[21]

Moreover, vetting was more than a question of Ba'ath Party ties. The poor training and capacity of Saddam's police force and its rampant corruption meant that CPA's broad plans for reform would require lengthy, expensive, and complex measures. It was one thing to remove the military trappings of the police force formally and to abolish the structures of the police that had been termed "administrative" under Saddam's regime: intelligence, security police, and internal government investigations. It was quite another to transform the Iraqi police into the community policing structure that Kerik and his staff envisioned.[22]

On June 29, 2003, Kerik's office initiated the first Transitional Integration Program (TIP) in Baghdad, a three-week course designed to retrain Iraqi police officers.[23] In mid-July, 4,600 Iraqis began training and another 445 applications were under review.[24]

---

[17] Mary Beth Sheridan, "For Help in Rebuilding Mosul, U.S. Turns to Its Former Foes," *Washington Post*, April 25, 2003.

[18] Jontz, "U.S. Gives Iraqi Police 3,000 Radios"; Daniel Williams, "Lack of Security Hampers Efforts to Aid Baghdad Police," *Washington Post*, June 9, 2003.

[19] Interview with former U.S. government official, December 2003.

[20] Edmund Sanders, "A Delicate Duet of Policing," *Los Angeles Times*, August 26, 2003.

[21] Interviews with CPA officials, December 2003.

[22] Interviews with CPA officials, December 2003. It is worth noting that the intelligence function would eventually be determined sufficiently critical to bring back.

[23] TIP training retrained approximately 14,000 Iraqi police officers by the middle of April 2004. See "Police Training Fact Sheet," Iraqi Ministry of Interior, April 15, 2003.

[24] Higgins.

Formal recruitment and training for the Iraqi police services did not get under way until the end of August, following the appointment of Ahmed Ibrahim as Iraq's new Deputy Minister of Interior and national chief of police. The Pentagon awarded DynCorp the contract to train the police force.[25] The CPA's initial goal was to have 71,000 Iraqi police deployed throughout the country by September 2004; this goal was later accelerated to include 75,000 Iraqi police by June 30, 2004.[26] To reach these numbers, CPA set the salary for police at $120 a month, prompting thousands of Iraqis to apply for the job.[27]

Recruitment remained generally successful. By the end of October, the Pentagon reported that the Iraqi police force numbered 40,000, comprised primarily of police and soldiers from Saddam's regime.[28] A report in *Jane's Defence Weekly* noted that coalition forces had recruited around 65,000 Iraqis to train for the police force, including former police officers, which would form the backbone of Iraq's new security forces.[29] In November, CPA announced that the number of Iraqi security forces deployed throughout the country totaled 85,000, including 50,000 police.[30] By June 2004, there were some 90,000 police on the books.[31]

But recruiting was only part of the equation. The tension between the need to get police out on the streets and the need for vetted and trained officers would continue to be a problem well beyond CPA's existence. The plan that emerged, for phased training and vetting of police already hired, even as new police were brought into the force, meant that untrained and unvetted personnel remained in their jobs. Many of those who returned to work in response to the call from coalition forces would prove problematic for reasons well beyond ties to the Ba'ath Party.

Initially the Pentagon considered training 28,000 Iraqi police in eight-week intensive courses in Hungary, where it had trained exiled Iraqi militia before the war.[32] However, the Hungarian government did not approve the initiative.[33] On September 29, 2003, the Jordanian government announced that it would train 30,000 Iraqi po-

---

[25] Borzou Daragahi, "Use of Private Security Firms in Iraq Draws Concern," *Washington Times*, October 6, 2003; Higgins.

[26] Bradley Graham and Rajiv Chandrasekaran, "Iraqi Security Crews Getting Less Training," *Washington Post*, November 7, 2003.

[27] Tim Ripley, "Mean Streets," *Jane's Defence Weekly*, October 9, 2003; David Filipov, "Harsh Legacy Slows Training of Iraq Police," *Boston Globe*, October 19, 2003.

[28] Mike Dorning, "Policing Baghdad Not of the Meek," *Chicago Tribune*, November 2, 2003.

[29] Ripley, "Mean Streets."

[30] Glenn Kessler, "Rumsfeld: No Need for More U.S. Troops," *Washington Post*, November 3, 2003.

[31] "Iraq Police Training a Flop," *Associated Press*, June 10, 2004.

[32] Dexter Filkins, "U.S. to Send Iraqis to Site in Hungary for Police Course," *New York Times*, August 25, 2003.

[33] Higgins; Thomas Fuller, "Hungary Is Cool to U.S. Idea to Train Iraqis," *International Herald Tribune*, September 3, 2003.

lice and other security forces, roughly a third of the projected forces slated for train-ing.[34] Alongside Jordan's initiative, France and Germany expressed a willingness to provide officers to help train the new Iraqi police force, although this never bore fruit.[35] Efforts were further supported by an additional police-training center in the United Arab Emirates.[36] Training for new recruits was broken down into eight-week courses that graduated 1,500 Iraqi officers each.[37] For a summary of police force training programs, see Table 9.1.

Implementation of these training efforts was complicated by the difficulty of at-tracting qualified civilian trainers. Although a number of countries had expressed willingness to help, the actual trainers remained in short supply. In late May 2003, a U.S. Justice Department team developed a training plan that called for 6,500 civilian police advisors from the United States and other coalition countries.[38] Subsequent plans spoke of numbers between 500 and 1,000, but even those numbers proved dif-ficult to reach, with the actual count never rising above the dozens during CPA's ten-ure.[39] Civilian police were deterred by the security situation in Iraq, and little in the way of a consistent effort was made by coalition nations to attract qualified people. Governments were also concerned about the prospect of sending large numbers

Table 9.1
Iraqi Police Service Training

Subdivision of Training	Purpose	Training Facility
TIP training	Three-week course to retrain former police	Al Anbar, Baghdad, Baquba, Basra, Diwaniyah, Hillah, Karbala, Kirkuk, Mosul, Muthanna, Najaf, Al Qaim, Tal Afar, Tikrit, Wasit
New recruit training	Eight-week course to train new recruits	Muwaqqar (Jordan), United Arab Emirates, Baghdad, Mosul
Field training	24-week training and mentoring on the job (by international police)	Throughout the country
Specialized training	Select officer and special unit training in investigation and counterterrorism	Adnon Palace (except for counterterrorism)

---

[34] Jack Fairweather, "Jordan to Train Iraqi Police Force," *Daily Telegraph* (London), September 30, 2003; "40,000 Police Recruits to Train in Jordan," *USA Today*, October 14, 2003.

[35] Barbara Slavin, "France Willing to Help Train Police," *USA Today*, September 19, 2003.

[36] Judy Dempsey, "Several European States Ready to Help Train 25,000 Iraqi Police," *Financial Times* (London), November 18, 2003.

[37] "Police Training Fact Sheet"; Ariana Eunjung Cha, "Crash Course in Law Enforcement Lifts Hopes for Stabil-ity in Iraq," *Washington Post*, December 9, 2003.

[38] Higgins.

[39] Email correspondence with CPA officials, May 2004.

of civilians into harm's way. Training was further hampered by difficulties with funding streams and other bureaucratic hassles. For example, continuing delays in the release of $800 million in U.S. State Department funds to support civilian police trainers throughout Iraq created a significant roadblock.[40]

Several steps were taken to change this situation. In November 2003, as part of the November 15 timetable to hand over sovereignty to Iraqi leaders, the Pentagon announced plans to accelerate the training and deployment of Iraqi security forces with the aim of reducing the size of U.S. forces in the country.[41] This, however, raised questions of whether the quality of training was being compromised in the interest of pushing people through the programs faster. Then, in March 2004, responsibility for training the police was shifted from the CPA MoI team to CJTF-7, the military command (in the process putting it under the same umbrella as military training, also shifted from CPA to CJTF-7). While CJTF-7 was able to employ military personnel as trainers to make up for the absence of civilian trainers, the quality of training remained problematic. Major General Paul Eaton, who took over the police training mission with its shift to CJTF-7, stated in an interview prior to his departure from Iraq in June, "We've had the wrong training focus—on individual cops rather than their leaders."[42]

By June 2004, with slightly over 90,000 police on the books, some two-thirds had not received training.[43] For about 21,000 of those who had, the training in question was TIP training, which varied significantly—trainers had tremendous discretion in designing and implementing the programs. The result was that some personnel received a full three-week complement, some only three days; some received classroom and field training, others did not; and so forth.[44] Vetting also continued to be a problem, as procedures for confirming the acceptability of recruits continued to rely on word of mouth and limited centralized data.[45]

Training and vetting were not the only problems faced by the Iraqi police. CPA's MoI (until March 2004) and CJTF-7 advisors (after that date) also had responsibility for equipping the police force and renovating police stations. As previously mentioned, 60 of the 61 police stations in Baghdad alone had been gutted and burned in the weeks following Saddam's fall; all of these buildings needed to be re-

---

[40] Magnier and Efron; Higgins; email correspondence with CPA officials, May, July, and August 2004.

[41] Craig Gordon, "Iraq Security Plan Carries Own Risks," *Long Island Newsday*, November 4, 2003; Andrew Koch, "Training More Locals Is Key to U.S. Strategy for Iraq," *Jane's Defence Weekly*, November 5, 2003.

[42] "Iraq Police Training a Flop."

[43] See "Iraqi Security Forces Personnel & Training Summary," U.S. Department of Defense Working Papers, June 8, 2004.

[44] U.S. General Accounting Office, *Rebuilding Iraq: Resource, Security, Governance, Essential Services, and Oversight Issues*, GAO-04-902R, Washington, D.C., June 2004, p. 57.

[45] Thanassis Cambanis, "Stability of Iraqi Police Eroding," *Boston Globe*, April 18, 2004.

built and refurbished for the police force to function effectively.[46] Moreover, MoI oversaw the awarding of contracts to firms that would supply weapons and other necessary equipment to the new police force. In January 2004, the contract for equipping Iraq's forces was awarded to Nour USA Ltd. of Vienna, Virginia—$327 million for "operational equipment," including guns, ammunition, armor, vehicles, and communications.[47] Various problems would plague the equipping effort throughout the coming months, including a contract dispute that held up the distribution of materiel.[48] Supplies were slow in reaching police on the ground, stemming in large part from a contract dispute that left equipment sitting in warehouses in the United States and Europe.[49] Regional military commanders tried to combat this shortage by purchasing necessary equipment with discretionary funds; however, the Iraqi police, and other security forces, continued to suffer from the lack of necessary equipment through the end of CPA's tenure.[50]

As of March 28, 2004, CPA reported that the Iraqi police had only 41 percent of their required patrol vehicles, 63 percent of uniforms, 43 percent of pistols, 21 percent of hand radios, 7 percent of vehicle radios, and 9 percent of the protective vests deemed necessary.[51]

The new Iraqi police services were also hindered by the absence of legal codes and due process throughout the country. Iraq had an antiquated civil and penal code that needed overhauling. Moreover, most citizens were not familiar with the law after 30 years of oppressive, lawless rule under Saddam Hussein. The country lacked public defenders and trained judges, there were no proper jails in which to hold suspects, and only four functioning courthouses were in existence to hear cases.[52] All of these issues slowed and hampered the process of restoring (or, perhaps more accurately, creating) law and order in Iraq.

---

[46] Whitelaw, "Law and Disorder."

[47] This contract included equipment for the police, the ICDC, and the border police. See Thom Shanker and Eric Schmitt, "Delivery Delays Hurt U.S. Efforts to Equip Iraqis," *New York Times*, March 22, 2004. The Iraqi Ministry of Interior names the contract amount for the police in particular as $190 million. See "Police Training Fact Sheet."

[48] Ken Dilanian, "Iraqi Police Force Taking Shape, but It Still Lacks Some Basics," *Philadelphia Inquirer*, September 24, 2003; Ripley, "Mean Streets"; "'Nobody Said This Was Going to Be Easy,'" *National Journal*, October 25, 2003.

[49] Shanker, "Delivery Delays Hurt U.S. Efforts to Equip Iraqis."

[50] Shanker, "Delivery Delays Hurt U.S. Efforts to Equip Iraqis."

[51] U.S. General Accounting Office, *Rebuilding Iraq: Resource, Security, Governance, Essential Services, and Oversight Issues*, p. 58.

[52] Maureen Fan, "Trying To Restore Order in Iraq, With Little Help from Law," *Philadelphia Inquirer*, June 8, 2003; Rick Scavetta, "Fair Treatment of Prisoners Crucial to Policing of Baghdad," *European Stars and Stripes*, June 15, 2003.

Although public opinion polling suggested that of all Iraq's security services, the police were the most trusted; insurgents viewed them as lackeys of the occupation and, perhaps more important, as soft targets that clearly and symbolically demonstrated the coalition's incapacity to provide security. Police stations and recruitment sites increasingly became the targets of insurgent terrorist attacks.[53] For example, on October 27, 2003, suicide bombers attacked three police stations in Baghdad, killing 15 and wounding over 65 officers.[54] Again on October 30, a suicide bomber disguised as a repairman detonated a car full of explosives at a police station in the capital.[55] In November, suicide bombers targeted the police station in Baquba, killing nearly 20 people.[56] By the end of 2003, attacks on Iraqi police stations had become a weekly occurrence, and an estimated 208 Iraqi police had been killed since the country's liberation, 60 in Baghdad alone.[57] By the end of March 2004, this number had soared to an estimated 350 police officers killed as a result of attacks on police stations and individual officers.[58] This trend persisted throughout 2004, including a massive attack in Basra on April 21 that killed more than 60 people, including children on a nearby school bus.[59]

Under these circumstance, it is perhaps not surprising that many police stations failed to stand up to insurgent attacks that began in April 2004. Moqtada al-Sadr's forces found little resistance in the Sadr City section of Baghdad, in Najaf, or in other predominantly Shi'ite areas when they moved to take over police stations starting in that time frame. Similar problems occurred in Fallujah. Some police officers even reportedly joined the militants.[60]

These continued problems created significant questions regarding the capacity of Iraqi police to take on the core duties of providing for security and law and order in Iraq—and fighting the many internal threats to the country, still deemed to be their primary role. Prior to the April attacks, the coalition had taken steps to transfer certain authorities to Iraqi police. In February 2004, the Pentagon announced that it

---

[53] "Public Opinion in Iraq: First Poll Following Abu Ghraib Revelations: Baghdad, Basra, Mosul, Hillah, Diwaniyah, Baqubah," briefing slides, Coalition Provisional Authority and the Iraq Center for Research and Strategic Studies, 14–23 May 2004. As of February 2008:
http://www.globalpolicy.org/security/issues/iraq/poll/2004/06iiacss.pdf

[54] Tyler Marshall, "Iraqi Police Struggle to Hold Back a Crime Wave," *Los Angeles Times*, October 28, 2003.

[55] Mike Marshall, "Training of Iraqi Police Seen as Key to U.S. Success," *Newhouse.com*, December 2, 2003.

[56] Nicholas Blanford, "Iraqi Police Walk Most Perilous Beat," *Christian Science Monitor*, November 25, 2003.

[57] Scott Peterson, "Iraqi Police Walk Perilous Beat," *Christian Science Monitor*, January 23, 2004; Higgins, "As It Wields Power Abroad, U.S. Outsources Law and Order Work."

[58] Kevin Johnson, "Attacks on Iraqi Police Increase," *USA Today*, March 31, 2004.

[59] Jeffrey Fleishman, "Several Blasts Kill 30 in Basra," *Los Angeles Times*, April 21, 2004; Pamela Constable and Khalid Saffar, "Blasts at Iraqi Police Facilities Kill 68," *Washington Post*, April 22, 2004.

[60] Hamza Nedawi, "Iraqi Police Desert Stations," *Washington Times*, April 9, 2004; Bradley Graham, "Iraqi Security Forces Fall Short, Generals Say," *Washington Post*, April 13, 2004.

would reduce its force size in Baghdad, cutting the number of coalition posts in the capital from nearly 60 to 26 and further reducing the number to eight by mid-April; in their place, newly trained Iraqi police forces would be deployed.[61] U.S. Major General Martin Dempsey, the head of the 1st Armored Division in Baghdad, told reporters that he believed Iraqi forces were ready to handle the city's security: "My personal opinion is I think the . . . police are a powerful team for this kind of urban warfare and urban security challenge."[62] Coalition forces experimented with pulling back from certain cities and allowing Iraqis to manage security, such as the Sunni city of Ramadi, but they were quick to assert that not all cities were ready to be patrolled by only Iraqis.[63]

The Facilities Protection Services

To combat chronic looting and theft in Iraq following the fall of Baghdad in April 2003, coalition authorities created a force to guard infrastructure, warehouses, government buildings, schools, hospitals, and other important sites. This force, named the Facilities Protection Services (FPS), was designed to work alongside an estimated 15,000 private security forces—such as Custer Battles, tasked with protecting the Baghdad airport and Erinys, which was hired to guard Iraq's oil fields—and replace coalition forces charged with guarding Iraq's infrastructure and public buildings.[64]

Coalition authorities and private contractors began training the FPS in June 2003, making it one of the first forces to receive formal training in Iraq.[65] The FPS was based in part on a Saddam-era force, the Vital Facilities Protective Service.[66] FPS recruits were drawn primarily from ex-Iraqi soldiers and guards.[67] Candidates were given a physical examination and asked to sign two forms denouncing the former Ba'athist regime and agreeing to a code of ethics.[68] Brigade commanders used CERP funds (allocated to them by CPA) to hire individual Iraqis, train them, and put them

---

[61] Thom Shanker, "G.I.'s to Pull Back in Baghdad, Leaving Its Policing to Iraqis," *New York Times*, February 2, 2004.

[62] Rowan Scarborough, "General Optimistic over Iraqi Policing," *Washington Times*, February 3, 2004.

[63] Dexter Filkins, "A U.S. General Speeds the Shift in an Iraqi City," *New York Times*, November 1, 2003.

[64] Daragahi, "Use of Private Security Firms in Iraq Draws Concern"; Nicholas Pelham, "Rival Former Exile Groups Clash over Security in Iraq," *Financial Times* (London), December 12, 2003; interview with CPA official, August 2004.

[65] Kent Harris, "Army Training Iraqis for Guard Duty," *European Stars and Stripes*, June 25, 2003.

[66] Interview with CPA official, August 2004.

[67] Scott Peterson, "U.S. Shifting Guard Duty to Former Iraqi Soldiers," *Christian Science Monitor*, July 15, 2003.

[68] Douglas Jehl, "U.S. Considers Private Iraqi Force to Guard Sites," *New York Times*, July 18, 2003; Harris, "Army Training Iraqis for Guard Duty."

to work in their sectors. CPA policy was that pay for FPS troops was $50 per month, but the individual basis of the recruiting meant that in the early days of the program, some FPS personnel were paid as much as twice that.[69] Training was conducted by coalition forces in coordination with Kroll Inc., a New York-based private security firm.[70] Most forces were trained at the Baghdad Public Academy, which was also the site of training for Iraqi police and border police.[71]

The idea behind FPS was that the guards would belong to the specific ministries whose facilities they were guarding, which would also be responsible for their pay and other costs, while overall policy, training, and standards would be set and coordinated by the MoI. After initial concerns and confusion about resourcing, by fall 2003 this range of responsibilities was formally transferred from coalition military commanders to the Iraqi ministries (and, during CPA's tenure, their advisors).[72] CPA Order Number 27, which formally created the FPS, was issued on September 4, 2003; it spells out the details of this arrangement.[73] The majority of FPS personnel were trained to guard warehouses, schools, hospitals, and other public buildings. Some of them were also trained to guard sites surrounding the oil industry, the country's electrical grid, and its port at Umm Qasr. As the June 2004 handover date drew closer, more and more FPS personnel were tasked with guarding these sites.[74] In addition, a reported 2,200 FPS personnel were trained for Diplomatic Protection Services (DPS), which were tasked specifically with guarding embassies in Iraq.[75] For a summary of FPS training and tasks, see Table 9.2.

Coalition authorities set the projected number of FPS personnel at 30,000 by the time of the June 28, 2004 handover; this number was later amended to 40,000.[76] At the beginning of August 2003, more than 4,000 FPS had been hired to guard Iraqi infrastructure and sites. By the end of August, total personnel had grown to 17,000.[77] This number climbed to a reported 20,000 guards protecting more than

---

[69] Interview with CPA officials, August 2004.

[70] Jehl, "U.S. Considers Private Iraqi Force to Guard Sites"; Harris, "Army Training Iraqis For Guard Duty."

[71] "Police Training Fact Sheet."

[72] Interview with CPA officials, August 2004.

[73] Text for CPA Order Number 27 can be found at the web site below. As of October 2007: http://www.cpa-iraq.org/regulations/#Orders

[74] Globalsecurity.org, "Facility Protection Service." As of February 2008: http://www.globalsecurity.org/intell/world/iraq/fps.htm

[75] CPA, "Ministry of Interior—Security Forces Information Packet." As of February 2008: http://cpa-iraq.org/security/MOI_Info_Packet.html

[76] Ravi Rikhye, "Iraqi Police and Paramilitary Forces," *Orbat.com*, January 25, 2004.

[77] Globalsecurity.org, "Iraqi Military Reconstruction." As of February 2008: http://www.globalsecurity.org/military/world/iraq/iraq-corps3.htm

Table 9.2
Facilities Protection Services

Subdivisions	Purpose	Training
Basic security guards	Guard warehouses, museums, schools, hospitals, and other public buildings	Three-day training by coalition forces and Kroll, Inc.
Oil infrastructure guards	Guard oil infrastructure from saboteurs	Some specialized training
Electricity guards	Guard electrical power-plants from saboteurs	Some specialized training
Diplomatic Protection Services (DPS)	Guard embassies, diplomats, and politicians	Specialized training; may also be hired and trained by diplomats and politicians independently

240 sites by mid-October 2003.[78] In 2004, the number of FPS continued to climb, reaching 40,000 by mid-April.[79] At the time of the handover, the reported number of FPS was at 74,000, including private contract forces.[80]

The FPS had a mixed record in its first months of operation. As previously mentioned, the service's tasks included protecting a wide array of infrastructure: power plants, oil fields, convoys, schools, ministries, warehouses, etc.[81] However, the initial 9,000-man force proved to be unreliable. By the end of 2003, a *Washington Post* article reported that roughly 2,500 FPS personnel were not showing up for work.[82] Furthermore, FPS personnel were reported to have abandoned their posts after receiving threats or coming under attack.[83] Iraq's oil and electric infrastructure continued to come under attack from saboteurs throughout 2004.

As the June 2004 transfer of authority grew closer, the 40,000-strong FPS force appeared to be working well with private security teams and coalition troops to mitigate threats to Iraq's infrastructure and to reduce looting of public buildings. As with other newly emerging Iraqi security forces, the FPS continued to work alongside international forces following the transfer of political authority, providing continuity and further training for these new forces.

---

[78] Globalsecurity.org, "Facility Protection Service."

[79] CPA, "Ministry of Interior—Security Forces Information Packet."

[80] Jeffrey Gettleman, "U.S. Training a New Iraqi Military Force to Battle Guerrillas at Their Own Game," *New York Times,* June 4, 2004.

[81] Keith Johnson and Alexei Barrionuevo, "U.S. Declares Local Force Will Police," *Wall Street Journal,* June 26, 2003.

[82] Ariana Eunjung Cha, "Flaws Showing in New Iraqi Forces," *Washington Post,* December 30, 2003.

[83] Cha, "Flaws Showing in New Iraqi Forces."

The Border Police

Iraq's borders represent one of its most crucial security concerns. Iraq shares frontiers with six nations, making for thousands of miles of land borders. Since the end of Saddam's reign, these borders have been exploited by a flow of illegal migrants, weapons, goods, and terrorists.[84] This situation is further complicated by the several Shi'ite Muslim holy sites in Iraq, particularly in Karbala and Najaf, to which Shi'ite pilgrims from Iran flock for religious events and festivals, something they could not do when Saddam ruled.[85]

CPA chose to retain MoI responsibility for the borders, but it also sought to revamp the border policing and guarding service and make it more effective and merit-based than Saddam's force had been. As with other services, an additional challenge was to integrate border control efforts in the Kurdish-controlled north with those in the rest of the country. The Department of Border Enforcement, under the MoI, comprised four separate forces: the Border Police, the Bureau of Civil Customs Inspection, the Bureau of Immigration Inspection, and the Bureau of Nationality and Civil Affairs.[86]

In January 2004, coalition forces began to train Iraqi police for the specific task of guarding Iraq's borders. The border police received the same eight-week training course as the regular police force. Also similar to the IPS, the border police earned the same salary of $60 per month, plus hazardous-duty pay. Following basic training, the border police also underwent a two-week course that taught specialized skills for border patrol.[87] The border police were trained at the Baghdad Public Safety Academy, along with the police and other Iraqi security forces. However, in 2004, CPA began construction of a training facility specifically for border enforcement, slated to open at the end of the calendar year. This new training facility, located near Sulamaniya in the predominantly Kurdish north, was designed to train up to 1,000 recruits at a time.[88] For a summation of the Department of Border Enforcement's training, see Table 9.3.

The CPA set the projected number of the Department of Border Enforcement at 25,000 by the time of the June 2004 handover.[89] By mid-October 2003, from

---

[84] Anne Barnard, "U.S. Trains 'New Breed' of Iraqi Border Guards," *Boston Globe,* January 13, 2004.

[85] Barnard, "U.S. Trains 'New Breed' of Iraqi Border Guards."

[86] CPA, "Ministry of Interior—Security Forces Information Packet."

[87] CPA, "Ministry of Interior—Security Forces Information Packet."

[88] CPA, "Ministry of Interior—Security Forces Information Packet."

[89] Rikhye, "Iraqi Police and Paramilitary Forces."

Table 9.3
Department of Border Enforcement

Subdivisions	Purpose	Training
Border Police	Maintain territorial integrity of Iraq, man fixed points of entry into country, monitor and defend Iraq's borders	Same eight-week course as regular police plus two-week specialized training
Bureau of Civil Customs Inspection	Regulate flow of goods across Iraq's borders	Specialized training under development
Bureau of Immigration Inspection	Monitor flow of people across Iraq's borders	Specialized training under development
Bureau of Nationality and Civil Affairs	Monitor presence of foreign nationals in Iraq	Specialized training under development

4,000 to 5,000 border police were reportedly deployed and defending Iraq's borders alongside coalition forces.[90] Coupled with other officers of the Department of Border Enforcement, the total number of personnel was estimated to reach 12,000 by the end of November 2003.[91] This number climbed to a reported 23,000 by February 2004,[92] of which an estimated 9,000 were border police.[93] At the time of the June 2004 handover, the number of border police trained and deployed was reported to have reached 17,000.[94]

These numbers, however, belie continuing problems with border controls. Coalition planners recognized that controlling Iraq's borders would be an issue but lacked the manpower and resources to address it adequately.[95] As a result, the borders were left unattended for a number of months. Later, as efforts by Iraqi and coalition forces to patrol borders were launched, it remained a difficult task, with insufficient personnel and reports of corruption and local "taxes" being levied at the borders. Despite the training plans detailed above, fewer than 300 of the personnel hired had been trained when the Iraqi Interim Government took power.[96]

Efforts to better the border regime for crucial Shi'ite holy days, when large numbers of Iranian pilgrims crossed into Iraq, sparked some improvement, but prob-

---

[90] Globalsecurity.org, "Iraqi Military Reconstruction."

[91] Globalsecurity.org, "Iraqi Border Police." As of February 2008: http://www.globalsecurity.org/intell/world/iraq/ibp.htm

[92] Globalsecurity.org, "Iraqi Border Police."

[93] CPA, "Ministry of Interior—Security Forces Information Packet."

[94] Gettleman, "U.S. Training a New Iraqi Military Force to Battle Guerrillas at Their Own Game."

[95] Interview with DoD officials, November 2003.

[96] Department of Defense, "Working Papers—Iraqi Security Forces Personnel & Training," June 2004, and "Iraq Index: Tracking Variables of Reconstruction & Security in Post-Saddam Iraq," the Brookings Institution, updated July 7, 2004, p. 15.

lems remained, some arising from the nature of these territories (mountains, marshland, desert), which makes them particularly difficult to patrol effectively. Tribal and family links often cross national boundaries, and border guards themselves may overlook smuggling and other activities by those with whom they have tribal or blood links.[97] Nomads cross the border to Saudi Arabia frequently. Thus, Iraq's borders remained highly porous. According to Iraqi officials, between June 2003 and July 2004, up to 10,000 foreign agents and spies infiltrated the country.[98]

Moreover, it proved no less difficult to recruit effective border police trainers from abroad than to recruit police trainers. Even the U.S. Office of Homeland Security reportedly declined to support this effort.[99] Border units remained undermanned and lacked the equipment that could have enabled them to perform surveillance on border areas effectively without high numbers of personnel. Border police had less effective equipment and vehicles (two-wheel drive versus four-wheel drive) than those they were tasked to track and stop.[100]

High-End MoI Forces

Recognition of the need for Iraqi counterinsurgency and counterterrorism forces surfaced early in the occupation. Whether such forces would be housed within the Ministry of Interior or the Ministry of Defense created some debate within CPA and the IGC. Past experience with oppression—both on the part of the domestic intelligence service, the Mukhabarat, and by the armed forces, especially as they were used against the Kurds—raised significant concerns. However, with primary responsibility for domestic security, the MoI unquestionably needed the capacity to respond to high-end security threats. The exact structures of such forces took some time to define, however, and their development and creation took even longer.

The structure that eventually emerged consisted of some high-end units within the police structure (locally based public order and SWAT units) and two additional, centrally deployed high-end units within the MoI's force structure. This Civil Intervention Force was envisioned as having primarily a riot control and counter-civil-disturbance mission. Its Emergency Response Unit was designed to be more of a comprehensive counterterror/counterinsurgency force. In addition, CPA advisors and Iraqi personnel agreed in spring 2004 on the importance of developing a structure similar to the UK Special Branch, which would work closely with the national intel-

---

[97] Ann Scott Tyson, "Iraq Battles Its Leaking Borders," *Christian Science Monitor*, July 6, 2004.

[98] Darien McElroy, "New Iraq Government Accuses Iran and Syria of Backing Insurgents," *Sunday Telegraph* (London), July 4, 2004.

[99] Email exchange with CPA officials, July 2004.

[100] Tyson, "Iraq Battles Its Leaking Borders."

ligence service but would, as an MoI entity, have arrest and detention authority that the intelligence service would not have.

Training and development of these forces was held up by a number of factors, including lack of funding, lack of personnel, and lack of high-level attention to the implementation of plans. At the time that CPA dissolved, the Civil Intervention Force, with a planned force size of 4,800, had still not begun training and had no personnel on the books. The Emergency Response Unit, planned for a total force size of 280, had a few dozen personnel on active duty or in training. No progress had been made on developing a Special Branch–type structure.

Ministry of Defense

The Ministry of Defense had been abolished along with the Iraqi armed forces by CPA Order Number 2. The creation of a new Defense Ministry was coordinated out of the Office of National Security Affairs. The process of designing and structuring the ministry was carried out in large part by that office's staff of advisors, which included military and civilian defense specialists from the United States, United Kingdom, Spain, Italy, Australia, the Czech Republic, and Estonia. This team also oversaw general defense and national security policy issues, the creation of national security institutions, intelligence issues, and, until March 2003, the creation and training of the Iraqi armed forces.

The design of the new ministry reflected in large part the advisors involved in this effort, in that in many ways it resembled U.S., British, and to some extent Australian models of defense policy, acquisition, and personnel. The ministry's role was to provide policy oversight of the Iraqi armed forces (including the Iraqi Counterterrorism Force and the Iraqi Civil Defense Corps) and to formulate overall defense policy. It was to be responsible for military policy; budget and financial matters; human resources, including recruitment and training of civilian and military personnel; acquisition, sustainment, and logistics; infrastructure; and defense intelligence analysis and requirements. The ministry was to be civilian-controlled, transparent, professional, merit-based, and broadly representative of the Iraqi people.[101]

The effort to build the Ministry of Defense covered a broad range of tasks, ranging from the very mundane to high policy. CPA staff had to identify qualified personnel to form the core of the MoD staff and provide for their training (an effort carried out through the National Defense University and the U.S. Institute of Peace, which provided training modules for newly hired MoD staff). The CPA Office of National Security Affairs worked with the IGC Security Committee, headed by Iyad Allawi, to identify senior personnel for the MoD. A budget for the ministry had to be

---

[101] CPA Order Number 67; CPA working papers, April 2004.

developed in concert with the Ministry of Finance (although the assumption that U.S. budgetary allocations would end after 2004 led to a significant shortfall in budget plans for 2005 and beyond). The legal framework that created the ministry, eventually expressed in CPA Order Number 67, had to be defined and framed. Property for the ministry's offices had to be identified, acquired, and refurbished (with a short-term location also identified for use until the ministry building was ready). A public information campaign had to be designed. In the absence of an Iraqi Defense Ministry, all of these tasks fell to CPA.

In fact, by the time the ministry was officially created by Order Number 67, signed on March 21, 2004, it did indeed have a number of key staff members in place. Until the opening of the Ministry of Defense Annex building, a former school converted to house the fledgling ministry until its permanent building was complete, the staff initially reported for work to the Office of National Security Affairs in the palace. On April 4, an interim Defense Minister, Ali Allawi, formerly interim Minister of Trade, was named. He was subsequently replaced by Hazem Sha'alan when the interim Iraqi government was appointed at the end of May 2004.

The Iraqi Armed Forces

Dissolution of the Iraqi Armed Forces: Aftermath

The decision to dissolve the Iraqi armed forces in May 2003 was to prove a lightning rod for criticism of CPA and Administrator Bremer. A December 2003 report published by the International Crisis Group (ICG) exemplified the widespread criticism of this decision by arguing, "Disbanding the former army was almost certainly the most controversial and arguably the most ill-advised CPA decision."[102]

The critique of this decision had two key components. First, critics argued, the dissolution of the military put roughly 350,000 Iraqi men out of work. Following the fall of Baghdad on April 9, thousands of Iraqi troops demonstrated in several cities and demanded stipends from the coalition authorities, arguing that these payments were necessary for the survival of their families.[103] Military personnel had not been paid by Saddam since February of 2003, and prices in postwar Iraq were rising. Many felt they were owed several months of back pay, and some had begun to register for emergency payments (coalition personnel were working to establish the structures for such payments) when CPA decided to formally dissolve the armed forces.[104]

---

[102] International Crisis Group, "Iraq: Building a New Security Structure," ICG Middle East Report No. 20, December 23, 2003, p. i.

[103] Marc Lacey, "Their Jobs in Jeopardy, Iraqi Troops Demand Pay," *New York Times*, May 25, 2003; Azadeh Moaveni, "Thousands of Ex-Soldiers in Iraq Demand to Be Paid," *Los Angeles Times*, June 3, 2003.

[104] Interviews with ORHA and CPA officials, November and December 2003, March, July, and August 2004.

Furthermore, many of the soldiers expressed feelings of betrayal by coalition powers, claiming that they had followed prewar instructions not to fight in good faith and with the expectation that they would be rewarded after the fall of Saddam.[105] One Iraqi soldier claimed that "we did what Bush told us to do. He told all the soldiers to go home, and we could have been executed if we had been found out."[106]

It is important to reiterate that there had never been any intent to leave the Iraqi armed forces exactly as they had been during Saddam's reign. CENTCOM planning included an assumption that members of the Iraq military would undergo a DDR process after major combat operations ended. The Ronco Corporation presented an outline of a comprehensive and costly DDR program to ORHA in March 2003.[107] In April and early May 2003, ORHA considered more detailed plans for a pilot program that would set up three DDR centers for former military personnel.[108] However, by the time that Bremer and Slocombe arrived in Iraq to establish CPA, the effective self-demobilization of the Iraqi armed forces made a formal DDR process unworkable.

In June, CPA announced that it would pay former officers and NCOs a monthly stipend that reflected their military rank and prior pay. The program was formally established in July. Eligible personnel included professional (nonconscript volunteer) soldiers and officers, Defense Ministry civilians, cadets enrolled in Iraq's military colleges, and the families of POWs and missing in action from the Iran-Iraq War. With the exception of the latter, all had to have been on active duty or the equivalent as of April 16, 2003, and could not be or have been senior Ba'ath Party officials, members of the Special Republican Guard, or Fedayeen Saddam. Stipends ranged from $50 per month for enlisted personnel, NCOs, MoD civilians, and cadets to $150 per month for three-star generals. In addition, conscripts, who did not receive monthly payments, received a lump-sum payment of $40. Finally, the two main Kurdish parties, the KDP and PUK, received bulk payments for distribution to their fighters, which were calculated on the basis of those groups' estimates of their numbers and a similar stipend structure (these payments were stopped after two months because of a failure by the KDP and PUK to provide promised data on the personnel to whom they were distributing the money).[109]

Stipends were initially paid at nine payment centers, but this proved to be too manpower-intensive, so payment was shifted to 159 bank locations. Payments were

---

[105] Lacey, "Their Jobs in Jeopardy, Iraqi Troops Demand Pay."

[106] Patrick E. Tyler, "U.S.-British Project: To Build a Postwar Iraqi Armed Force of 40,000 Soldiers in 3 Years," *New York Times*, June 24, 2003.

[107] Ronco Corporation, "Decision Brief to Department of Defense Office of Reconstruction and Humanitarian Assistance on the Disarmament, Demobilization, and Reintegration of the Iraqi Armed Forces," March 2003.

[108] Interviews with ORHA and CPA officials, November 2003 and August 2004.

[109] Interviews with CPA officials, January and August 2004.

also shifted from monthly to quarterly, with the Ministry of Finance authorizing the release of funds. Personnel arriving for payment were checked against a database of military personnel acquired by CPA. About 280,000 people were receiving regular payments by January 2004. Although the protests halted when the program was announced, some problems continued, as individuals complained of not receiving their pay and numerous efforts were made to cheat the system and receive multiple and/or unsanctioned payments. The program was criticized both in Iraq and abroad. Some Iraqi veterans complained about the payment system and about their lack of work or a means of providing for their families.[110] The December 2003 ICG report claimed that "The humiliating treatment meted out to former soldiers and the absence of a plan to get them back to work on reconstruction and humanitarian tasks alienated a significant part of the population."[111] A *Washington Times* article even argued that the number of soldiers put out of work coupled with their disappointment over stipend payments helped to fuel the growing insurgency against coalition forces.[112]

The second criticism of the dissolution decision was that it underestimated the army's importance as a symbol of Iraqi nationalism. Iraq's military, created in 1921 and the fourth largest in the world prior to the 1991 Gulf War, stood as a testimony to the country's strength and perseverance.[113] Following Bremer's order to disband the army, protesting soldiers carried banners that read: "Dissolving the Iraqi army is a humiliation to the dignity of the nation."[114] IGC interviews with former soldiers and officers of the Iraqi military found that "a powerful mix of nationalism, humiliated pride and nostalgia (chiefly among the senior corps) for the old institutional benefits [of the army] is fueling anger against the decision to dismantle the army."[115] Disbanding the military, therefore, was interpreted by many as an attack on Iraqi identity, not as a means by which to purge the country of Saddam's influences.[116]

Designing a New Force

The creation of a new structure for Iraq's new armed forces was never in doubt, however. As early as April 2003, CENTCOM planners reportedly were discussing the recruitment and training of some former Saddam military personnel, brought on as individuals rather than units, to assist with internal security functions. The plan that

---

[110] Borzou Daragahi, "Disbanded Army Draws Praise, Condemnation," *Washington Times*, January 6, 2004.

[111] International Crisis Group, "Iraq: Building a New Security Structure," p. i.

[112] Daragahi, "Disbanded Army Draws Praise, Condemnation."

[113] For a brief history of Iraq's armed forces, see International Crisis Group, "Iraq: Building a New Security Structure," especially pp. 1–2.

[114] Mark Fineman, Warren Veith, and Robin Wright, "Dissolving Iraqi Army Seen by Many as a Costly Move," *Los Angeles Times*, August 24, 2003.

[115] International Crisis Group, "Iraq: Building a New Security Structure," p. 10.

[116] Daragahi, "Disbanded Army Draws Praise, Condemnation."

emerged for Iraq's new force envisaged a three-division structure of about 12,000 soldiers per division and a total force size of some 40,000 personnel. Such a force could, of course, change to reflect Iraq's future needs, but this basic structure would constitute its initial core. The force would be defensive in nature and configured so as not to be perceived as aggressive by neighbors. Thus, it would have limited capacity to project or sustain power away from bases. Its primary mission would be defensive against external threats, although it, like any other nation's military force, could be called upon to assist with internal security in a situation of extreme need. CPA staff initially planned to call this light, wheeled force without tanks or significant artillery the New Iraqi Corps (until it was determined that the acronym was unacceptable in Iraqi Arabic). After some debate about whether to build the force from the top down (that is, by recruiting senior leadership for all three divisions and the over structure, then building the remaining force), or bottom up (one unit at a time), the latter approach largely prevailed, driven in part by considering the potential dangers inherent in appointing senior Iraqi military leaders too early.[117]

A concept for the force was developed in May and June of 2003, drawing somewhat on the experience of building the Afghan National Army, and presented to the Pentagon for approval in early June. The Coalition Military Assistance Training Team (CMATT) was created in early June, and Major General Paul Eaton was selected to head this structure, which was to have responsibility for recruiting and training the military force. Eaton's deputy was UK Brigadier General Jonathan Riley (who was succeeded by Nigel Aylwin-Foster). The staff would eventually grow to some 200 people, and included personnel from the United States, United Kingdom, Italy, Spain, Australia, and other coalition countries. A $48-million, one-year contract to assist CMATT in the provision of trainers and other support was awarded to the Vinnell Corporation, a subsidiary of Northrop Grumman. Vinnell subcontracted the job of training the force to MPRI—a firm based in Alexandria, Virginia, that had helped build forces in Bosnia and Croatia.[118]

Coalition leaders were adamant that they must forge a corps that balanced Iraq's ethnic, religious, and regional makeup. Saddam's regime had greatly favored Arab Sunnis in the officer corps while Shi'ites were compelled to serve as foot soldiers; coalition personnel strongly believed that this bias had to be corrected in the new army.[119]

---

[117] Interviews with CPA officials, November and December 2003, July and August 2004; Rajiv Chandrasekaran, "U.S. to Form New Iraqi Army," *Washington Post*, June 24, 2003.

[118] Interviews with CPA officials, November and December 2003, July and August 2004; Nathan Hodge, "Pentagon Agency May Train Iraqi War-Crimes Prosecutors," *Jane's Defence Weekly*, September 15, 2003; "Northrop Grumman to Train New Iraqi Army," *Jane's Defence Weekly*, July 7, 2003; Mark Fineman, "Arms Plan for Iraqi Forces Is Questioned," *Los Angeles Times*, August 8, 2003; Daragahi, "Use of Private Security Firms in Iraq Draws Concern."

[119] Daragahi, "Disbanded Army Draws Praise, Condemnation."

Building the New Force

Recruiting began in July, with the establishment of a handful of recruiting stations in Mosul, Baghdad, and Basra, to start. In the Kurdish-dominated north, CPA and CMATT staff relied on local officials to identify personnel; throughout the rest of the country they relied on the recruiting centers and an advertisement campaign featuring recruiting posters to bring in young men. By August 2, 1,000 people were undergoing training.

Initial plans for recruiting were based on assumptions that proved faulty. For example, CPA staff believed that military rank and Ba'ath Party membership were correlated and that people above the rank of colonel in the old Iraqi military were likely to have been high-level party members, disqualifying them from positions in public service, including in the new Iraqi armed forces. Later, when Iraqi personnel provided records of the old Iraqi military (records that had disappeared in the looting of government and military facilities but were used to develop the stipend payment lists), they learned that some lower-level military personnel were senior Ba'ath Party members and some high-level personnel, including general officers, were not party members at all.[120] This realization resulted in a revision of some policies.[121]

Nonetheless, the effort to build a new Iraqi military had little need of the large number of former generals in Iraq, a result of the top-heavy structure of the old army, or even of most former colonels, even if they had no Ba'ath affiliation. The new army was envisioned as free of the detritus of the old, with its thousands of generals and heavy rank inflation. Former colonels did not take well to being told that they might be given the rank of major in the new force. This was true not only of

---

[120] This database was used to vet personnel and has a somewhat interesting history. Soon after CPA was established, CPA staff were approached by former Iraqi Ministry of Defense staff who reported that they had in their possession several hard drives full of personnel records of the old Defense Ministry. These personnel were hired by CPA to create a new database for CPA on the basis of this one, amending it to make it searchable for such information as Ba'ath Party membership, criminal records, and so forth. Once the stipend program was begun, the database was further updated on the basis of the registration forms provided by stipend recipients. This made it possible to validate and update the information in the original database, as well as to acquire better information where there were significant gaps, such as in the records for noncommissioned officers. Iraqi personnel developed the database, entered information into it, and ran searches for CPA staff, all with minimal supervision. In the spring of 2004, CPA staff learned that the personnel responsible for creating the database had written the software such that certain information, specifically Ba'ath Party membership and criminal records, was "firewalled" and readily accessible only to the original coders. Thus, a search carried out by somebody else would not, for example, reveal Ba'ath Party membership. In fact, searches were at times carried out by other Iraqi staff and, as a result, some information provided to CPA and CJTF-7 staff reflected the full record and some did not. When questioned, the Iraqi personnel reported that this firewalling, or password protection, was only put in place for when they were not available, but there is no easy way to confirm this. CPA staff interviewed for this study who doubted this claim speculated that these Iraqis might have been motivated by a concern that individuals not be denied pensions, although some also voiced concerns about possible infiltration, and reported that there had been accusations by other Iraqi personnel to that effect. Regardless, the end result was clearly problematic. It is not clear whether the original MoD database was retained. Interviews with CPA officials, December 2004 and January 2005.

[121] Interviews with CPA officials, December 2003, July and August 2004.

officers in the old Saddam military, but also of personnel recruited from the militias that had fought Saddam, such as in the Kurdish areas. Generally speaking, CPA staff sought to recruit younger people, who, if they had military experience as officers, were not above the ranks of captain or major.[122]

The initial salary for Iraqi recruits was $60 per month. Over time, salaries would be adjusted to rank and position.[123] Coalition authorities worked to enlist prospective soldiers from across Iraq's ethnic and religious groups, seeking physically fit personnel between the ages of 18 to 40.[124] In July 2003, Major General Eaton announced plans to have a 1,000-troop light mechanized infantry battalion formed and operational by October 2003, with an additional eight battalions ready by the end of 2004.[125]

The new Iraqi army was officially called into being on August 7, 2003, via CPA Order Number 22.[126] The first recruits reflected coalition aims of creating a new army that represented Iraq's ethnic and religious diversity: 60 percent were Shi'ites, 25 percent Sunnis, 10 percent Kurds, and 5 percent from other minority groups; 75 percent of the soldiers came from Saddam's former army.[127] An effort was made to ensure a similar ethnic mix down to the smallest unit levels, insofar as possible.[128] The first battalion, which graduated in the fall of 2003, comprised 700 recruits: 65 officers, 230 noncommissioned officers, and 405 soldiers.[129] The second battalion graduated on January 6, 2004, Iraq's national holiday to celebrate the army.[130] In Jordan, the first batch of 560 Iraqi army officers began their three-month training course in 2004.[131] In April 2004, ground was broken for a new Iraqi army training base outside of Kirkuk.[132]

---

[122] Interviews with CPA officials, December 2003, July and August 2004.

[123] Interviews with CPA officials, November 2003 and January 2004.

[124] John Daniszewski, "Hundreds Line Up to Join New Iraqi Army," *Los Angeles Times*, July 22, 2003.

[125] "U.S. to Begin Recruiting for New Army," *USA Today*, July 10, 2003; Andrew Koch, "Iraq Occupation: Questions Remain," *Jane's Defence Weekly*, July 16, 2003.

[126] As of October 2007:
http://www.cpa-iraq.org/regulations/#Orders

[127] Theola Labbe, "Iraq's New Military Taking Shape," *Washington Post,* September 16, 2003; John Daniszewski, "New Iraqi Army Makes Its Debut," *Los Angeles Times*, September 16, 2003; "First Battalion of 700 Troops Completes Its U.S. Training," *USA Today,* October 6, 2003.

[128] Interview with CJTF-7 official, April 2004.

[129] Ian Fisher, "Iraqi Army Takes Shape as Recruits End Training," *New York Times*, October 5, 2003.

[130] John F. Burns, "In Hussein's Shadow, New Iraqi Army Strives to Be Both New and Iraqi," *New York Times*, January 7, 2004; Nicholas Riccardi, "In Iraq, an Army Day for No Army," *Los Angeles Times*, January 7, 2004.

[131] Riad Kawaji, "Forging a New Iraqi Army—in Jordan," *Defense News*, February 9, 2004.

[132] "Iraqi Military Reconstruction."

Changing Goals and Parameters

Planning for the Iraqi armed forces would continue to evolve throughout CPA's tenure. Following a series of bomb attacks in August—including attacks on the Jordanian embassy, the UN headquarters, and the main Shi'ite mosque in Najaf—and growing attacks against coalition forces, the Pentagon announced plans to speed up the deployment of Iraqi forces by cutting training time by two and a half weeks.[133] Slocombe announced that the Pentagon's new goal was to have 27 battalions ready within 12 months, which was twice as fast as the initial deployment plan envisaged.[134] Moreover, an air force and a coastal defense force contingent were added to the planned-for Iraqi force structure.[135] The latter two services were included to ensure that Iraq would have a full complement of capabilities, albeit on a small scale for air and naval components.

In addition, the Iraqi Counterterrorism Force, a small (envisioned at 200 personnel), rapidly deployable special operations structure, was assigned to the Iraqi armed forces per CPA Order Number 67 (which created the Ministry of Defense).[136] The force had initially been designed with this capability included in the new structure, but in spring 2004 there had been some discussion within CPA and CJTF-7 regarding the appropriateness of a requirement to include counterterrorism forces within the externally focused Iraqi armed forces.

Debates over the future of Iraq's new army intensified when the first battalion of new recruits graduated on October 4, 2003, and the recruits began to desert in large numbers. By the beginning of December, soldiers reporting for duty had dropped from 694 to 455.[137] The desertion rate was blamed on the low salaries paid, particularly in comparison to other security forces; army soldiers earned between $60 and $180 a month relative to other forces such as the police and ICDC, which earned $10 to $40 a month more.[138] Another factor may have been the lack of a clear

[133] Dana Priest, "Iraqi Role Grows in Security Forces," *Washington Post*, September 5, 2003; Christopher Cooper, "Troop Training in Afghanistan Outpaces Similar Efforts in Iraq," *Wall Street Journal*, September 8, 2003.

[134] Thom Shanker, "U.S. Is Speeding Up Plans for Creating a New Iraqi Army," *New York Times*, September 18, 2003.

[135] Anthony H. Cordesman, "One Year On: Nation Building in Iraq," *CSIS Working Paper*, April 6, 2004, available at http://www.csis.org, accessed May 2004.

[136] As of Octber 2007:
http://www.cpa-iraq.org/regulations/#Orders

[137] Christine Spolar, "Iraqi Soldiers Deserting New Army," *Chicago Tribune*, December 9, 2003.

[138] Spolar, "Iraqi Soldiers Deserting New Army"; "Iraqi Battalion Loses a Third of Its Recruits," *Los Angeles Times*, December 11, 2003.

mission for these forces.[139] To combat desertion, coalition forces began offering the newly trained forces an additional $72 a month for hazardous duty.[140]

The effort to build a new Iraqi military continued. In March 2004, a Pentagon decision transferred the CMATT operation from the CPA chain of command to CJTF-7, at the same time assigning to it the police training that had previously also been under CPA (as discussed above). This changed little in the overall approach to training and policy for the Iraqi military, although it did create some confusion over reporting chains, because the Iraqi military forces were to be placed under the civilian control of the Defense Ministry, which was still the responsibility of CPA.

As the end of CPA's existence neared, the appointment of senior-level military personnel took on new importance. CPA and CMATT staff interviewed large numbers of former military, civilian, and other Iraqi personnel, seeking to identify who could serve in the highest positions of the new Iraqi armed forces, as was done in identifying senior leadership for the MoD. Personnel were identified through a wide variety of methods, including recommendation by IGC members and other Iraqis, through CPA and CMATT contacts, and so forth. Personnel were vetted against available information in Iraq and in coalition capitals, including through the IGC Security Committee, although the process of collecting this information was never perfected. In mid-April 2004, General Amer Bakr al-Hashimi was named chief of staff of the Iraqi armed forces and Kurdish General Babaker Zebari was named senior military advisor to the Defense Minister. Other senior military personnel, such as division commanders and the air force's commanding general, were also named prior to the transfer of power.

Success or Failure?

Amid concerns about the growing number of attacks against coalition forces and Iraqis in the fall of 2003, several U.S. military officials and members of Congress proposed calling up the former army as a means of providing better security and intelligence, particularly logistical support units and mid-level officers that were not in Saddam's inner circle.[141] Some in the IGC also reportedly began to question the complete dissolution of the former military and called for the reconstitution of certain elements of the Saddam-era army.[142] CPA rejected this proposal, with Slocombe arguing that "Given our objective of replacing Hussein's regime and not just its

---

[139] Interview with CPA officials, December 2003.

[140] Burns, "In Hussein's Shadow, New Iraqi Army Strives to Be Both New and Iraqi."

[141] Thom Shanker and Eric Schmitt, "U.S. Considers Recalling Units of Old Iraq Army," *New York Times*, November 2, 2003; Susan Sachs, "An Angry Former Iraqi Officer Says U.S. Is Wasting His Talent," *New York Times*, November 2, 2003; Tom Bowman, "U.S. Examining Plan to Reconstitute Iraqi Army," *Baltimore Sun*, November 10, 2003.

[142] Peter Slevin, "U.S. Scrambles to Rebuild Iraqi Army," *Washington Post*, November 20, 2003.

leader, it would have been a mistake, I think, to try and convert an army that was a principal tool of his oppressive system into the armed guardian of a new democracy."[143] Nonetheless, officials in Washington and Baghdad continued to press for the recall of former Iraqi troops.[144]

Although the first two battalions to graduate from training were deployed in early 2004, the new Iraqi army was not asked to fight until the first coalition-led Fallujah offensive in April that followed the brutal murder and mutilation of four U.S. private contractors in late March. In an effort to ensure a visible Iraqi role in the military operation, the Iraqi army's Second Battalion was to be deployed to Fallujah to help U.S. Marines conduct the offensive (ICDC and police units also participated). However, the battalion came under fire while transiting a Shi'ite neighborhood in Baghdad and refused to continue on. A later effort to airlift them to Fallujah was aborted because of continued refusal by battalion personnel to proceed.[145] The Iraqi police and military that fought during the Fallujah offensive were not as reliable as the United States had hoped. Major General Martin Dempsey estimated that "about 50 percent of the security forces that we've built over the past year stood tall and firm . . . About 40 percent walked off the job because they were intimidated. And about 10 percent actually worked against us."[146]

The experience of using Iraqi armed forces units in Fallujah highlighted the critical issue of employing Iraqi armed forces internally. As discussed above, this was a difficult historical issue in Iraq, and it also raised questions about the appropriate role of military forces in a democracy. The training for Iraqi military forces had also, prior to April, been geared primarily to an external threat. In spring 2004, more counterinsurgency-type training was added to the program.[147]

Iraqi Civil Defense Corps (ICDC)

The ICDC developed from the need for coalition commanders to identify Iraqis who could perform a range of functions, including static defense and human intelligence collection, and to have Iraqis alongside them on patrols. It is not entirely clear whether the ICDC evolved from the rather spontaneous initiatives of individual commanders to hire and train Iraqis to help them, or whether it was a top-down ini-

---

[143] Walter B. Slocombe, "To Build an Army," *Washington Post,* November 5, 2003.

[144] Anthony C. Zinni, "Restore Regular Iraqi Army to Assist with Reconstruction," *Atlanta Journal-Constitution,* February 5, 2004.

[145] Thomas E. Ricks, "Iraqi Battalion Refuses to 'Fight Iraqis,'" *Washington Post,* April 11, 2004.

[146] Connie Cass, "General: 10 Percent of Iraqi Forces Turned on U.S. During Attacks," *USA Today,* April 22, 2004.

[147] Interviews and discussions with CPA officials, January–June 2004.

tiative directed by CENTCOM. Regardless, the concept was fully developed by the commanders working through CJTF-7 and CENTCOM, and it was presented to CPA and Pentagon officials in the summer of 2003. The CPA's national security leadership found the concept acceptable as long as it was not viewed as an alternative to the Iraqi armed forces as a whole. The ICDC, after all, would be minimally trained and developed fundamentally as a support force, unable to operate independently. In many ways, that made it possible to begin to employ a sizeable number of former Iraqi military personnel (as well as some members of former anti-Saddam militia structures) while carrying out the slower, more deliberate process of building the new Iraqi armed forces.[148]

The initial concept for the ICDC envisaged very small, platoon-size units equipped only with uniforms, boots, and a weapon. They would be locally based and thus needed no housing (they could live in their homes) or other support. Commanders funded the ICDC out of CERP funds. The level of training for ICDC forces, as well as their duties, varied widely, depending on the coalition forces they were attached to. Over time, however, both the responsibilities and the requirements of the ICDC as a whole grew. Although it remained a support force for the coalition, it grew to company-sized units that provided linguists, intelligence, fixed-site security, drivers, disaster relief, humanitarian assistance, patrols, convoys, cordons, check points, and security for coalition forces. In making it possible to put an Iraqi face on operations, through joint patrols and so forth, the ICDC provided a valuable asset for coalition forces. Funding allocated to the force by Congress enabled the provision of helmets, armor, and vehicles, transforming these units into something resembling a wheeled infantry.[149] The ICDC was formalized by CPA Order Number 28, signed on September 3, 2003.[150]

The ICDC structure grew as well. Initially envisioned as perhaps one battalion for each subordinate command, for a total of five, it was later decided to build 18 battalions, one for each governorate, and finally 36.[151] The 36th Battalion itself was a unique structure, made up of personnel representing various anti-Saddam militias. As such, they were more experienced, better trained, and far more capable than other ICDC forces (or, perhaps, many other Iraqi security forces). Initially a response to the political representatives of the groups that had sponsored those militias, the 36th Battalion of the ICDC would prove a crucial unit for the Iraqi armed forces.

---

[148] Interviews with CPA officials, November and December 2003, August 2004.

[149] Interview with CPA official, August 2004.

[150] As of October 2007:
http://www.cpa-iraq.org/regulations/#Orders

[151] Interviews with CPA officials, August 2004.

The numbers grew alongside the structure: at the time of its creation, the Pentagon reported that the ICDC had reached 2,300 troops with plans for a rapid increase in its numbers over the coming months.[152] By the end of September, DoD announced that the number would likely increase to 9,000.[153] By mid-February 2004, more than 25,000 troops had been hired and trained, with another 3,600 in training.[154] The projected number of ICDC members at the time of the June 2004 handover was set at 40,000.[155]

The future of the ICDC was a policy concern almost from its inception. Created as a temporary force assigned to coalition units, it had no Iraqi government structure to which it was responsible, and CPA, Pentagon, CENTCOM, and IGC officials debated how long the force would last, what it would transition into (a national guard, a gendarmerie, or a military reserve were among the options discussed), and which ministry (Interior or Defense) it would report to. Eventually, consensus was reached that the force should fall under the Ministry of Defense, and this was reflected in CPA Order Number 67, creating the ministry, and CPA Order Number 73, transferring the ICDC to it.[156] The decision about the ICDC's final role was left to the Iraqi Interim Government that would take over upon the dissolution of CPA.

Unlike the Army forces, the ICDC faced military action to varying extents from its inception. Although quality and training varied, the *Washington Post* reported that these forces became the most respected troops in certain areas, particularly in the Shi'ite neighborhoods of Baghdad.[157] In certain regions, however, ICDC troops expressed fears of becoming too closely associated with occupation forces and thus a target of insurgents; this was particularly true in the Sunni Triangle.[158] In the early April offensive in Fallujah, as well as Moqtada al-Sadr's Mahdi Army attacks throughout the country at the same time, some ICDC units fought bravely, while others deserted or disappeared. Some ICDC checkpoints posted photographs of Grand Ayatollah Mohammed Sadiq Sadr, Moqtada al-Sadr's father, although it was not clear whether this was done in an effort to ward off attacks by the Mahdi Army

---

[152] Douglas Jehl and Dexter Filkins, "Rumsfeld Eager for More Iraqis to Keep Peace," *New York Times*, September 5, 2003.

[153] Tom Squitieri and Glen C. Carey, "U.S. Pushing to Transfer Security Duties," *USA Today*, September 19, 2003; Carol J. Williams, "U.S. to Expand Iraqi Corps," *Los Angeles Times*, December 8, 2003.

[154] Globalsecurity.org, "Iraqi Civil Defense Corps." As of February 2008: http://www.globalsecurity.org/military/world/iraq/icdc.htm

[155] Cordesman, "One Year On."

[156] As of October 2007: http://www.cpa-iraq.org/regulations/#Orders

[157] Gary Anderson, "On Patrol with Iraq's 'Homeboys,'" *Washington Post,* January 28, 2004.

[158] Jason Keysor, "U.S. Trained Civil Defense Corps Has Started to Show Some Cracks," *St. Louis Post-Dispatch*, January 4, 2004; Anthony Shadid, "Iraqi Security Forces Torn Between Loyalties," *Washington Post*, November 25, 2003.

or in support of it. The effort to use the 36th Battalion in Fallujah led to accusations that the coalition forces had sent Kurds to fight Arabs (an accusation strengthened by desertions from the unit at that time, increasing the Kurdish component of its ethnic mix).[159] In all, the ICDC performance in this crisis was mixed and, as with the armed forces as a whole, it raised questions about the effectiveness and reliability of the new Iraqi forces.

Intelligence

At the inception of the CPA, U.S. officials assigned to it believed that the question of an Iraqi intelligence structure and agency was best left to a future Iraqi government. Iraq's history of government oppression, in which the intelligence agency played a key role, made CPA staff think that Iraqis would establish an intelligence agency when they were ready to do so and that it was not CPA's role to create one for them. It was clear that an intelligence agency was being built, largely with the assistance of the CIA, and if any CPA office was to oversee that effort, it would have to be the Office of National Security Affairs. Moreover, as MoD and MoI structures developed and the scope of the insurgency and counterterror challenge to Iraq became clear, it also became necessary to consider which intelligence structures should be embedded in those ministries as well. Finally, because the ICDC also played a role in intelligence collection, and because the 1st Armored Division had created an Iraqi intelligence cell, called the Collection, Management, and Analysis Directorate (CMAD) to support CJTF-7 intelligence efforts,[160] the question of how these structures would function in Iraq's new government had to be considered.[161]

The concept for an Iraqi National Intelligence Service (INIS) was first briefed at CPA in November 2003. Efforts at that time to create a coordinating structure within CPA to oversee the development of this service were stillborn. Although Walter Slocombe did successfully argue that his office should have a role in the process, the nature of that role was unclear. The arrival of David Gompert in December as senior advisor for National Security Affairs energized the effort to ensure that CPA played a role in the development of policy for the INIS.

---

[159] Tony Perry, "At Least One Iraqi Battalion Is Ready to Help U.S.," *Los Angeles Times*, April 13, 2004; Deborah Horan, "U.S. Scales Back Hopes for Iraqi Security Forces," *Chicago Tribune*, April 22, 2004; Conversations with CJTF-7 personnel, April 2004.

[160] CMAD was made up of personnel representing the same parties that had contributed to the 36th Battalion of the ICDC. Major General Martin E. Dempsey, Commander, 1st Armored Division, "Coalition Provisional Authority Briefing," March 18, 2004. As of February 2008:
http://globalsecurity.org/military/library/news/2004/03/mil-040318-dod01.htm

[161] Interviews with CPA officials, August 2004.

The INIS structure that emerged was that of a collection and analysis agency with a domestic focus, without a foreign collection capacity, and with no authority to arrest or detain. Although some former members of Iraqi intelligence services would initially be hired as INIS staff, the IGC's Security Committee, which met to review the draft charter for the INIS, was assured by CPA that such hires would be minimized to the extent possible and that the involvement of former Iraqi intelligence personnel would decrease over time. On April 1, 2004, CPA Order Number 69 authorized the IGC to create the INIS by issuing a charter for that organization.[162] Mohammed Abdullah Mohammed al-Shehwani, an ethnic Turkoman who had lost two sons to Saddam's regime, was named interim director general of the INIS on April 4, 2004.

The INIS was also to be the "first among equals" as the coordinating body for Iraqi intelligence as a whole. The CPA order creating the Defense Ministry includes its defense intelligence analysis role.[163] MoD intelligence was created as part of a directorate of communications and intelligence; it was to have an external, rather than domestic, purview and an analytic role. This, of course, raised the question of whence such an analytical body would receive the information it was to analyze, because the INIS had no foreign collection role and the MoD had no collection role at all. This issue remained unresolved at the time of CPA's dissolution, as did the question of the status of Iraqis involved in intelligence collection for the coalition.[164]

MoI's intelligence functions as envisioned by CPA advisors were to focus on criminal intelligence and, insofar as needed to support MoI's counterterrorism and counterinsurgency roles, on those issues as well. Also, as discussed above, CPA staff advised that MoI develop a Special Branch-type structure to work with the INIS to ensure that INIS information could be acted on as necessary. At the time of CPA's dissolution, little or no progress had been made in establishing any of these entities, stemming primarily from resource constraints but also from bureaucratic incapacity and confusion about roles and missions by both the Iraqis and CPA personnel.[165]

National Security Decisionmaking Structures

Within CPA, the Office of National Security Affairs took it upon itself to ensure coordination of security policy between its own staff, MoI advisors, CJTF-7 personnel,

---

[162] CPA Order Number 69 was the delegation of authority and Annex A to Order Number 69 the charter. http://www.cpa-iraq.org/regulations/#Orders

[163] CPA Order Number 67. http://www.cpa-iraq.org/regulations/#Orders

[164] See footnote 162.

[165] See footnote 162.

and others at CPA with an interest or involvement in these issues. Regular meetings at senior and working levels helped facilitate the sharing of information, although the structures for doing so remained imperfect (see Chapter Eight for a discussion of coordination within CPA and between CPA and CJTF-7). As the security structures of the Iraqi government developed, it was clear that mechanisms of coordination among them would be needed as well.

In March 2004, Ambassador Bremer convened the Ministerial Committee for National Security (MCNS). This committee included the ministers of Defense, Justice, Interior, Foreign Affairs, and Finance (in the case of the Defense Ministry, the minister had not yet been named, and its secretary general attended the initial meetings); it was chaired by Ambassador Bremer, as the head of government (it would be chaired by the prime minister once one was named). The group met several times before being formally established by CPA Order Number 68 on April 4, 2004.[166] In addition to the head of government and the ministers listed, MCNS meetings were also attended by the senior military advisor to the government of Iraq, the director general of the INIS, and Iraq's national security advisor. In accordance with the CPA order, the commander of the Multinational Force-Iraq could also be invited to participate, and the CJTF-7 commander, Lieutenant General Ricardo Sanchez, sat in on the meetings in the spring of 2004. Other ministers could be invited to participate as relevant, and this, too, was done in several of the initial meetings of this body.

The MCNS role, as defined by the CPA order, is "to facilitate and coordinate national security policy among the Ministries and agencies of the Iraqi government tasked with national security issues" and to serve as the "primary forum for Ministerial-level decisionmaking on these issues." To support the government and the MCNS, the Office of the National Security Advisor was established to serve as the primary advisor to the head of government and the MCNS on national security issues and to manage and supervise the National Security Advisory Staff. As such, the national security advisor has two roles: to provide balanced, impartial advice to the head of government and the MCNS, and to facilitate interagency coordination both in preparation for meetings of the MCNS and more broadly. IGC member Mowfaq al-Rubaie was named interim national security advisor on April 9, 2004.

The establishment of these structures in the spring of 2004 had an almost immediate effect in one critical sense. Regular meetings between Ambassador Bremer, General Sanchez, and the men who had been named to senior ministerial and agency posts resulted in freer, more direct communications than were possible through the IGC, and a number of frank exchanges of views took place, particularly as the security situation turned dire in early April. Through the MCNS, officials of the Iraqi

---

[166] As of October 2007:
http://www.cpa-iraq.org/regulations/#Orders

Interim Government could participate in decisions made about the security of their country in an immediate way and to affect the choices made by CPA and CJTF-7.

Iraqi Armed Forces and the Handover of Power

Coalition plans for military actions following the handover of power to an interim Iraqi government, scheduled for the end of June 2004, had to address the question of how Iraqi forces would be used. Critical questions in this regard pertained to which forces would be assigned to the MNF-I, what authorities the Iraqi government would have to direct those forces, and how those forces would interact with coalition troops.

CJTF-7 planning expected that Iraqi units would operate as other members of the coalition did. They would have their own rules of engagement, and they would be free not to participate in certain operations deemed outside the scope of those rules. However, this left open the question of internal use of the Iraqi army, which had been highlighted by efforts to use Iraqi army personnel in Fallujah; it raised questions about the relationship between the MNF-I and the Ministry of Interior troops, which ostensibly had responsibility for internal security; and it left open the issue of intelligence sharing with Iraqi forces and personnel.

The structure that evolved reflected the leading role of MoI in Iraqi security. MNF-I would operate in support of MoI forces; it would work and coordinate with them to the fullest extent possible. The small size and slow training schedule of Iraqi army forces also helped postpone the question of their use. ICDC forces could continue to operate alongside coalition forces much as they had before. Other MOD forces could be assigned as necessary over time. They would continue to belong to the Defense Ministry, but they would be assigned to the MNF-I just as other country's forces were. Thus, Iraq would be contributing units to the coalition force responsible for ensuring its security and sovereignty.

The intelligence question raised a knotty issue. Intelligence sharing with alliance partners already presented a variety of problems, including logistical difficulties in passing classified information, problems related to specific classifications used as a matter of course (such as NOFORN, which prevents U.S. personnel from providing the material to a national of any other nation), and technical problems created by two classified computer networks: one U.S. and one coalition. With Iraq as part of the coalition, the problems would be exacerbated. Iraq had yet to establish security clearance procedures or security classifications, a necessary first step to developing intelligence-sharing mechanisms. Moreover, it was likely that other coalition forces would have information that they wished to share with one another but not with the Iraqis. This was the case with other countries already, of course, but given the size and scope of Iraq's contribution to the coalition, it was likely to cause problems.

Integrating the Armed Forces and Militias Not Under Government Control

One of the most critical questions that faced coalition personnel was the question of how to handle the several militias that had fought against Saddam.[167] These included an estimated 70,000 peshmerga Kurdish forces in the North, several Shi'ite militias in the South, and a few secular, Sunni, and other groups throughout the country. These militias needed to be either disbanded or somehow brought under the control of the newly emerging government in order for Iraq to be a viable country.[168]

Some thought was given to this challenging task in the first nine months of CPA's existence, but concerted efforts to address it did not begin until February 2004. During the first nine months, arrangements had been made that gave various groups quasi-official security roles. The peshmerga of the KDP and PUK, for example, continued to play a role akin to a Kurdish national guard, as authorized under Kurdistan regional government law. This led other groups to resist initiatives to disband or transition their militias and to oppose or thwart reintegration processes. Furthermore, a distinction was made early on between those groups that were willing to work inside the political process and those that chose to fight the coalition and the new Iraq. This section focuses on those who were willing to work with the government.

The CPA effectively enfranchised several political parties whose sponsors prior to OIF (and in some cases afterward as well) covered a wide ideological spectrum. So, for example, the Iraqi Communist Party and various Shi'ite parties sponsored by Iran were included in the IGC; they came to see their futures as part of the political process rather than as resistance groups fighting the coalition. However, this perception did not apply to all of the political parties that had resisted Saddam and maintained militias or armed forces that would be needed. Others, such as Moqtada al-Sadr, could have been made part of the political process but were not. For those who did participate in the process, continuing to do so was more attractive than reverting to armed resistance. Such participation was instrumental in the success of efforts to disband the militias.

A critical step in facilitating the transition and reintegration of armed forces and militias was the adoption of the Transitional Administrative Law (TAL) on March 8,

---

[167] The distinction between "armed forces not under government control" and "militias" is both real and politically important. There was a spectrum of armed forces in Iraq that varied in terms of professionalism and capabilities. By March 2003, at the beginning of OIF, the KDP and PUK had developed what amounted to professional armies, of which they were justly proud. The Badr Corps, operating out of Iran, had also developed a professional organization, though not organized and operated as an army. Other groups possessed paramilitary organizations with various levels of training and professionalism, none of which were standing, full-time organizations.

[168] This point is echoed in International Crisis Group, "Iraq: Building a New Security Structure," p. i, and in Tyler, "U.S.-British Project: To Build a Postwar Iraqi Armed Force of 40,000 Soldiers in 3 Years."

2004. According to the TAL, once it entered into effect, all armed forces and militias not under federal control would be illegal, except as provided by law. CPA Order Number 91 put this provision into effect prior to the effective date of the TAL and defined in greater detail a transition and reintegration program to disband illegal organizations. It also stipulated penalties for those groups or individuals who failed to disband.

CPA Order Number 91 sought to strike a balance between recognizing the contributions and sacrifices of those who fought against Saddam and the need to eliminate the potential security problem created by armed organizations not under government control. It also laid the legal foundation for creating transition and reintegration plans for (and with) the various armed forces and militia leaders. These plans were created for three reasons: to facilitate bringing former fighters into the Iraqi security forces as individuals; to provide pensions for those qualified; and to provide education, job training, and job placement services that would enable them to pursue other employment. The plans also established timelines for this to happen and for each of the armed groups to draw down as its members moved into one of the three tracks. Militia personnel undergoing this process would be legally registered and, thus, covered by federal law during the reintegration process. Those that were not would be illegal and were subject to legal sanction.

The process of negotiating these agreements and finding mechanisms to integrate militia members into Iraqi security services proved challenging. Senior members of several militias assumed senior positions in the government or the Iraqi security forces. Some militia members joined the ICDC and the Iraqi armed forces. The ICDC's 36th Battalion and MoD's Iraqi Counter Terrorism Forces (ICTF) also included representatives from former militia members, but a great many remained to be integrated at the time CPA was dissolved.

A flurry of negotiating activity in the spring of 2004 resulted in agreement with nine key parties to disband their militias, including the KDP, the PUK, and the SCIRI Party's Badr Corps (which was renamed the Badr Organization after Saddam was overthrown). The others that agreed to transition and reintegration programs were the Iraqi Communist Party, one of the oldest and largest groups that remained in Iraq during Saddam's reign; the Iraqi National Accord, the party and militia of interim Prime Minister Iyad Allawi; the Da'wa Party of Ibrahim al-Jaafari;[169] the Iraqi National Congress of Ahmed Chalabi; Iraqi Hezballah, led by Abu Hatem;[170]

---

[169] There are several Da'wa parties. Jaafari's is one of the largest and most influential. A second Da'wa party approached CPA and the Iraqi Interim Government to express interest in disbanding its militia as well.

[170] Abu Hatem's given name is Abdul Karim Muhamadawai. The Iraqi Hezballah party is not related to the Lebanese terror organization that uses the same name. Hezballah, perhaps more properly written as Hezb Allah, means "Party of God" and so is a common name for political movements in the Middle East.

and the Iraqi Islamic Party, the only religious Sunni group to have reached a transition agreement.

It is worth noting in this context that even as CPA and CJTF-7 worked to eliminate existing militias, short-term requirements led to the creation of at least one new one.[171] The decision by military commanders on the ground to create a local "Fallujah Brigade" composed of local fighters trying to keep order in a city that coalition forces and their Iraqi comrades had continuously failed to pacify, set a strange precedent. This brigade represented a structure that stood outside of the Iraqi armed forces, composed in part of personnel who had fought the coalition. Coalition military commanders gave the brigade control of the city and withdrew U.S. Marines to the city's outskirts.[172] After the brigade had spent nearly two months managing the city's security, a U.S. Marine officer questioned the success of the strategy, noting that those responsible for killing and mutilating U.S. contractors in March had still not been apprehended and the insurgents' heavy weapons had not yet been confiscated.[173] Certainly, this brigade sent a conflicting signal to those Iraqis who had joined or were urged to join the formal structures of Iraq's security forces, including those representing the anti-Saddam militias. By the time of the June 2004 handover of power, coalition forces had not disbanded the Fallujah Brigade.

Concluding Observations

CPA and CJTF-7 faced a tremendous challenge in their efforts to simultaneously govern Iraq, improve its security situation, and build security forces and institutions for it. Assisting in the development of a country's military and police in a period of peace is difficult enough. In Iraq, there was no such luxury: there was no peace. As a result, tradeoffs had to be made between hiring police to patrol the streets and ensuring that the police on the streets were well trained; between ensuring an appropriate leading role for the MoI in internal security and recognizing the low level of training and readiness of MoI forces; between finding people to fill high-level positions prior to the transition of power and making sure that the most qualified and most acceptable people took on those posts.

CPA and CJTF-7 also had to do all of this without much preparation. Planning assumptions had not anticipated either the deterioration of Iraq's infrastructure or a difficult security environment. The decision to dismantle the Defense Ministry and

---

[171] The Fallujah Brigade is a different type of militia than those for which the transition and reintegration program was created. As such, its dissolution would not fall under the legal structures set up to take care of those militias that fought against Saddam.

[172] Gregg Zoroya, "Falluja Brigade Tries U.S. Patience," *USA Today*, June 14, 2004.

[173] Zoroya, "Falluja Brigade Tries U.S. Patience."

the Iraqi armed forces may or may not have been the only one possible, but the absence of any structures in those areas created a variety of additional tasks and challenges for those who had to build new institutions to replace the old. They had to take action in an environment of limited information, and with limited resources.

That the development of security forces and structures lagged behind the initial (and evolving) plans for them is therefore not surprising. That training programs were not fully effective, and that forces proved less than reliable when called on to fight also reflects a variety of understandable factors. That said, some responsibility must be laid at the feet of those whose job it was to plan and develop these structures, both in the coalition capitals and in Baghdad. The failure to recognize the critical importance of police forces from the very beginning, as well as to resource adequately the effort to build police forces, was a key reason for the many problems faced by MoI throughout the occupation and since.

Other oversights include the failure to recognize the need for effective intelligence structures, as well as a means to integrate them, from an early date. The hope that intelligence could be avoided was remarkably optimistic given Iraq's security environment even before the scope of the insurgency became clear. As internal unrest developed, the need for intelligence collection and analysis capacity within Iraqi security structures, particularly the MoI, grew. The failure to recognize this need and to act upon it early and effectively is an indictment of the planners, funders, and executors of this effort.

New Iraqi military forces initially were built in a more benevolent context. With little expectation that these troops would be called on to serve immediately (as was the case with police), there was time to recruit and train, it seemed, as well as to ensure an adequate ethnic mix. Table 9.4 summarizes the various Iraqi security forces that were established during the occupation period. However, insufficient attention was paid to the potential problems inherent in sending Iraqi military forces, built and trained for an external defense role, to fight in their own cities. Moreover, while coalition officials sought to ensure that both sectarian and ethnic groups were represented in the armed forces, little attention was paid centrally to ensuring that such personnel got along sufficiently well to fight together. The effort to promote reconciliation was left to the discretion of individual trainers.

The tradeoffs between short-term goals and long-term goals were everywhere. The ICDC responded to an immediate need, but raised the question of what to do with a huge force incapable of independent action and dependent on coalition personnel. Excellent plans for much-needed structures, such as high-end MoI capabilities, were drafted but never implemented, as Iraqi and U.S. personnel struggled to respond to the emergency of the day.

Table 9.4
New Iraqi Security Forces

Force	Target Number	June 2004 Number	Tasks	Training	Ministry
Iraqi Police Services	85,000	92,000	Maintain domestic order, fight crime	Eight weeks for new recruits, three weeks for former police, mentoring	Interior
Facilities Protection Services	74,000	74,000	Guard public buildings, warehouses, embassies and infrastructure	Three-day basic training, specialized training for oil, electric, and diplomat guarding	Interior
Border Police	25,000	17,000	Maintain flow of people and goods over borders	IPS eight-week training plus two weeks specialized	Interior
Emergency Response Unit[a]	270	51	Counterterrorism, counterinsurgency	Seven-week training program plus variety of additional training based on specialty	Interior
Civil Intervention Force[a]	4,800	0	Public order	Police training plus specialized training	Interior
ICDC	40,000	25,000	Domestic disturbances, natural disaster relief	Three weeks	Defense
Armed forces	40,000	7,000	Defend Iraqi sovereignty	Six to nine weeks	Defense

[a] Data from Anthony H. Cordesman, "Inexcusable Failure: Progress in Training the Iraqi Army and Security Forces as of Mid-July 2004," Center for Strategic and International Studies, July 20, 2004. Dates for data not clear.

It would be easy to say that the lesson of security sector institution building in Iraq is that security forces cannot be built in an environment where security is not present. In any post-conflict nation-building situation, however, security will, to varying extents, be a problem. So perhaps the lesson is rather that we must learn how to build such structures and institutions under abysmal conditions—and how to improve the security situation, relying more on our own forces and assets when necessary, as we do so. The United States and the broader international community have accumulated a good deal of experience on these issues over the years, much of which was ignored during the early months in Iraq.

In the short term, Iraq's success or failure is a matter of just holding on. In the long term, it is a question of being able to stand on its own to defend itself against threats from within and without. The occupation provided the opportunity to lay the groundwork for this capacity. To some extent, this was done, albeit imperfectly. How well it holds up, and what is built on this foundation, will be the measure of success—both of coalition efforts, and of the new Iraq.

CHAPTER TEN

Governance and Political Reconstruction

This chapter addresses governance issues in Iraq before, during, and after major combat operations. It starts by examining prewar planning for postwar governance. It then assesses each of the major national developments during the occupation period: the formation of the Iraqi Governing Council (IGC), the November 15 agreement and the transfer of authority, and the adoption of the Transitional Administrative Law (TAL). The chapter also examines governance efforts at the provincial and city levels, and concludes with several lessons learned about establishing governance structures in the aftermath of major combat.

Prewar Planning for Postwar Governance

Prior to the beginning of combat operations on March 19, 2003, U.S. government planning for political transition relied primarily on Iraqi exiles to run the country after the fall of Saddam Hussein. Media reports suggest that several Iraqi exiles lobbied Congress and the White House years before the commencement of combat operations in Iraq and that certain individuals had developed influential relationships with different branches and agencies in the U.S. government.[1]

In the months leading up to war, the White House took measures aimed at including Iraqi exiles in both the political and military aspects of the war with Saddam Hussein. In October 2002, President Bush allocated $92 million to train Iraqi exile militias to aid in toppling Saddam Hussein's regime. This initiative was an addition to the 1998 Iraqi Liberation Act, which allocated $98 million for training Iraqi exiles but was restricted to nonlethal means.[2] The *Washington Post* reported that even

---

[1] For example, the *Washington Post* reports that Ahmed Chalabi, head of the Iraqi National Congress, lobbied "the Bush administration to go to war to topple Hussein." See Rajiv Chandrasekaran, "Exile Finds Ties to U.S. a Boon and a Barrier," *Washington Post,* April 27, 2003.

[2] See DeYoung and Williams, "Training of Iraqi Exiles Authorized"; Daniel Williams, "U.S. Army to Train 1,000 Iraqi Exiles," *Washington Post,* December 18, 2002; and Daniel Williams, "Iraqi Exiles Mass for Training," *Washington Post,* February 1, 2003.

though six Iraqi exile groups were named as recipients of U.S. funding and training, Ahmed Chalabi's INC essentially drafted the list of volunteers from within his organization and received the lion's share of training and equipment; a reported 5,000 exiles were trained in all.[3]

Inside Iraq, the U.S. government began working with several Kurdish leaders and their organizations prior to combat operations. *Jane's Defence Weekly* reported in March 2003 that CIA operatives were active in the U.S. and UK air-patrolled northern section of Iraq, training fighters and "buying loyalty" of Kurdish politicians.[4] Two Kurdish organizations in particular, the Patriotic Union of Kurdistan (PUK) and the Kurdistan Democratic Party (KDP), received U.S. aid and military training under the October 2002 presidential directive. The U.S. government also attempted to work with Shi'ites in the southern portion of the country, mainly by encouraging members of the Supreme Council for the Islamic Revolution in Iraq (SCIRI) to rally volunteers for its prewar U.S. Army training program. SCIRI, however, boycotted the program on the grounds that the U.S. government favored Chalabi and the INC.[5]

Thus prewar planning for postwar governance following Saddam Hussein's fall concentrated on the role of key Iraqi exiles and Kurds in the north to assume control of the government's infrastructure and pave the way for democratic transition. This strategy rested on the assumption that the population would accept exiles as legitimate leaders, Iraq's governmental infrastructure would be easily transferred to new leadership, and overall political transformation would be rapid and relatively easy.

Postwar Governance: The Iraqi Governing Council

During major combat operations, the U.S. government continued to rely primarily on Iraqi exiles to fill the political vacuum in a post-Saddam government. As retired Lieutenant General Jay Garner's Office of Reconstruction and Humanitarian Assistance (ORHA) staff prepared to move from Kuwait to Baghdad, Iraqi exiles also made plans to move into the capital. Media sources reported that the U.S. government stepped up its efforts to recruit Iraqi exiles after the commencement of combat operations and that, as before the war, Chalabi and the INC were receiving the lion's share of attention and resources.[6] The prominence of Chalabi and the INC is reported to have exacerbated tensions between the U.S. Defense and State depart-

---

[3] DeYoung and Williams, "Training of Iraqi Exiles Authorized"; Williams, "Iraqi Exiles Mass for Training."

[4] Andrew Koch, "U.S. Central Intelligence Agency Forces: Covert Warriors," *Jane's Defence Weekly*, March 12, 2003.

[5] Williams, "U.S. Army to Train 1,000 Iraqi Exiles."

[6] Peter Slevin, "U.S. Wary in Choosing Iraq's New Leadership," *Washington Post*, March 23, 2003.

ments, the latter of which feared that Chalabi would attempt to step into Iraq's power vacuum independently of U.S. governance.[7] Nevertheless, in early April—days before the fall of Baghdad—U.S. forces airlifted Chalabi and a reported 600 of his Free Iraqi Fighters into the southern city of Nasiriya.[8] Other exile groups were not given this assistance, a point of contention that surfaced as U.S. forces began to work with Iraqis in the postwar phase.

As U.S. troops moved into Baghdad, the city quickly fell into a state of anarchy. Angry citizens stormed through Baghdad's neighborhoods, looting and destroying buildings associated with Saddam's rule. The destruction caused by weeks of lawlessness in the city created several problems for political transition. First, looters took property that was essential infrastructure for running a government, including computers, phones, copy machines, and even desks and chairs. These items needed to be replaced before ministries could function again. Second, many government buildings were burned or demolished, destroying important documents necessary for running the government. Third, the sense of lawlessness had a negative psychological impact on the city, which made coalition forces' efforts to gain the public's trust more difficult.[9] All these factors hindered the development of national governance structures.

Prior to major combat operations, several key leaders and political parties existed both in the diaspora and within Iraq. Within weeks of Saddam's fall, many political groups sprang up with various designs for a postwar Iraqi government. One media source estimated that there were 100 political parties during the first few months of liberation.[10] Alongside these political parties and organizations, independent candidates came forward with hopes of participating in Iraq's political future.[11] Finally, tribal leaders also became active in Iraq's local and central political development.

From the explosion of political groups in Iraq following the fall of Saddam, six key parties emerged as the major political organizations in Iraq. They included two main exile groups, the INC and the INA (Iraqi National Accord); two main Kurdish groups, the PUK and KDP; and two Shi'ite groups, SCIRI and the Da'wa Party. Within these parties, Arab Sunnis—the group most strongly represented in Saddam's

---

[7] Slevin, "U.S. Wary in Choosing Iraq's New Leadership."

[8] Alan Sipress and Carol Morello, "U.S.-Led Gathering to Begin Remaking Iraq," *Washington Post*, April 15, 2003.

[9] John Kifner and John F. Burns, "As Tanks Move in, Young Iraqis Trek Out and Take Anything Not Fastened Down," *New York Times*, April 9, 2003; Daniel Williams, "Rampant Looting Sweeps Iraq," *Washington Post*, April 12, 2003.

[10] Monica Basu, "Iraqi Political Parties Vie to Fill Postwar Void," *Atlanta Journal-Constitution*, June 8, 2003; Peter Ford, "Democracy Begins to Sprout in Iraq," *Christian Science Monitor*, April 23, 2003.

[11] These included the diplomat Aqila al-Hashimi and former guerilla fighter Abdul Karim "Prince of the Marshes" Muhammadaw. See "The Iraqi Council," *Washington Post*, July 14, 2003.

Ba'ath Party—were greatly underrepresented.[12] This presented a growing problem for U.S. officials as Iraq continued to develop its own government. It was not until December 2003 that Sunni Arabs began to organize their own political group called the State Council for the Sunnis—which united three subdivisions within the Sunnis, the mystical Sufis, the orthodox Salafis, and the leftist Brotherhood—and demanded political representation in the new Iraq.[13]

The rapid collapse of the Iraqi government during major combat operations forced U.S. officials to quickly devise a strategy for setting up an interim governing authority. At the beginning of April, the Bush administration announced plans to hold a meeting of potential Iraqi leaders to discuss future political development for the country.[14] On April 11, 2003, Deputy Secretary of Defense Paul Wolfowitz testified before the Senate Armed Services Committee that U.S. and British forces would lead Iraq's political transformation and that he imagined it would be an interim authority that would be divided into "a representative council and a smaller executive committee"; the UN would play a small but important role.[15]

On April 12, three days after the fall of Baghdad, ORHA chief Garner announced plans for a "big tent" meeting that would bring together future Iraqi leaders of all backgrounds. The invitations were issued by General Tommy Franks, the commander of U.S. Central Command, and hosted by Department of State representative Ryan Crocker, Larry Di Rita from the DoD, and Zalmay Khalilzad, representing the White House National Security Council staff.[16]

The first meeting was held in the southern city of Nasiriya on April 15. An estimated 75 Iraqis participated, consisting of a mixture of tribal leaders, clerics, and political party officials.[17] "Local" Iraqis made up around two-thirds of participants and exiles the remaining third.[18] Despite the "big tent," certain groups—including SCIRI and the Iraqi Communist Party—protested the meeting, claiming that it was

---

[12] The prominent Sunnis in the INC and INA were exiles, and the Shiite and Kurdish groups did not have known Sunni Arab members.

[13] Edward Wong, "Sunnis in Iraq Form Own Political Council," *New York Times*, December 26, 2003; Alan Sipress, "Once-Dominant Minority Forms Council to Counter Shiite and Negotiate Future," *Washington Post*, January 6, 2004; Edward Wong, "Once-Ruling Sunnis Unite to Regain a Piece of the Pie," *New York Times*, January 12, 2004.

[14] "U.S. Prepares Meeting of Iraqi Exiles and Contenders for Leadership," *Washington Post*, April 9, 2003.

[15] Eric Schmitt and Steven R. Weisman, "U.S. to Recruit Iraqi Civilians to Interim Posts," *New York Times*, April 11, 2003.

[16] Paul Martin, "General Promises a 'Big Tent' for Iraqis," *Washington Times*, April 13, 2003.

[17] Marc Santora with Patrick E. Tyler, "Pledge Made to Democracy by Exiles, Sheiks and Clerics," *New York Times*, April 16, 2003; Keith B. Richburg, "Iraqi Leaders Gather Under U.S. Tent," *Washington Post*, April 16, 2003.

[18] Martin, "General Promises a 'Big Tent' for Iraqis."

a ruse for occupation or that they had been shut out from the meetings.[19] Notwithstanding these protests, the meeting produced a 13-point memorandum on the creation of a new Iraq, which included the points that the government must be democratic and that the plurality of ethnicity, race, tribe, and religion must be respected.[20] Of particular importance for future discussions between an emerging Iraqi political voice and coalition authorities was point 7: "That Iraqis must choose their leaders [and] not have them imposed from the outside." As discussed below, this point became a serious issue of contention in following months.

One of the first issues the U.S. government faced in forging a new Iraqi political landscape was the tension between Iraqi exiles, favored by the United States, and leaders who had remained in Iraq under Saddam's rule. In particular, the U.S. government's decision to airlift Chalabi and his Free Iraqi Fighters into Iraq toward the end of major combat operations sparked animosity among other political groups, which interpreted this act as U.S. favoritism toward the INC.[21] Furthermore, Chalabi created tensions between the INC and ORHA by boycotting the April 15 Nasiriya meeting, choosing instead to drive into Baghdad and there issue his own 14-point plan for political progress.[22] By the end of April 2003, it was reported that DoD and ORHA were taking steps to distance themselves from Chalabi and the INC.[23]

The Bush administration announced plans to have an appointed interim authority in place by June 3, to coincide with the expiration of the UN Oil for Food mandate.[24] Alongside this announcement, administration officials pledged to continue meeting with emerging Iraqi leaders and political organizations and, in particular, to bring SCIRI and other Shi'ite groups into the coalition-run process. To this end, ORHA began calling weekly meetings of Iraqi leaders and coalition forces. On April 28, a meeting of around 300 Iraqis, including members of SCIRI, agreed to hold a conference within a month in order to select an interim government.[25] A smaller follow-up meeting was held on April 30 to further discuss the selection process for attendees at May's meeting.[26]

---

[19] Richburg, "Iraqi Leaders Gather Under U.S. Tent."

[20] For all 13 points, see "A 'Federal System' with Leaders Chosen by the People," *New York Times*, April 16, 2003.

[21] Richburg, "Iraqi Leaders Gather Under U.S. Tent."

[22] Molly Knight, "Exile Groups Frustrated by U.S. Role in Meeting," *Baltimore Sun*, April 17, 2003.

[23] Tracy Wilkinson, "Fresh Rulers to Emerge, Garner Says," *Los Angeles Times*, April 25, 2003.

[24] Karen DeYoung and Glenn Kessler, "U.S. to Seek Iraqi Interim Authority," *Washington Post*, April 24, 2003.

[25] Rajiv Chandrasekaran and Monte Reel, "Iraqis Set Timetable to Take Power," *Washington Post*, April 29, 2003.

[26] Sharon Behn, "Iraqi Leaders Meet to Plan Reconstruction," *Washington Times*, April 30, 2003.

In response to these meetings, the Bush administration reiterated its vision of political transition, which called for the creation of a "transitional government" by the end of May that would have power over "nonsensitive" government ministries such as education and health care. This would be followed by the formation of a "provisional government" within "six months to two years after the interim authority was created." That government would have greater powers than its predecessor and be tasked with writing a constitution. The political transition would end with national elections for a permanent government.[27]

On April 24, just three days after his arrival in Baghdad, Garner was informed by the White House that he would be succeeded by Ambassador L. Paul Bremer, a terrorism expert and former career diplomat from the State Department.[28] With this change in leadership, the White House and DoD also announced plans for the creation of the Coalition Provisional Authority (CPA), which officially replaced ORHA on May 16.

Prior to Bremer's arrival, ORHA continued to work toward the creation of an Iraqi transitional authority. On May 5, the United States announced that "eight or nine" Iraqi leaders would be selected to form the interim authority, specifically naming the five leaders most likely to be selected: Ahmed Chalabi of the INC, Massoud Barzani of the KDP, Jalal Talabani of the PUK, Abdel-Aziz al-Hakim of SCIRI, and Iyad Allawi of the INA.[29] These political parties, in turn, announced that they would help organize a national assembly of around 350 deputies—two-thirds of whom would be "local" (not exiled) Iraqis—who would select an executive committee or possibly an interim Iraqi leader. They planned to identify the deputies by going out into each of Iraq's 18 provinces and meeting with local leaders.[30] Both ORHA and the above-mentioned leaders aimed to have an implementable plan prior to Bremer's arrival.

Bremer arrived in Baghdad on May 12. Within weeks, he and his staff instituted several controversial policy changes. On May 23, in CPA Order Number 2, Bremer announced that the Iraqi armed forces would be officially disbanded. As discussed in Chapter Nine, this policy came to be one of the most widely criticized decisions Bremer implemented while overseeing Iraq's reconstruction.

Another policy shift that Bremer instituted involved plans to purge more than just the highest-ranking Ba'athists from the government. This policy was announced

---

[27] Douglas Jehl, with Eric Schmitt, "U.S. Reported to Push for Iraqi Government, with Pentagon Prevailing," *New York Times*, April 30, 2003.

[28] See Chapter Five for more on this decision.

[29] Carol Morello, "'Nucleus' of Iraqi Leaders Emerge," *Washington Post*, May 6, 2003.

[30] Patrick E. Tyler, "Opposition Groups to Help Create Assembly in Iraq," *New York Times*, May 6, 2003.

on May 16, 2003, in CPA Order Number 1.[31] The policy appeared to have followed the logic that those attached to the former regime should be punished for their complicity in Saddam's crimes. However, as with the military, the policy left an estimated 15,000 to 30,000 civil servants unemployed, in addition to creating a shortage of qualified workers to run the government.[32] The decision also hit Sunni Arabs the hardest, further alienating this group from the postwar reconstruction process.

The alienation of Sunni Arabs from coalition plans for post-Saddam Iraq became an increasingly difficult problem for coalition forces. Under the former regime, Sunni Arabs received the lion's share of resources and held the most prominent positions within the government and military.[33] Saddam's fall, therefore, compromised their position of privilege, inspiring some to call for violence in hopes of returning to the prewar status quo. In June, a month after Bremer's decision to disband the Iraqi armed forces and expunge Ba'ath Party members from the Iraqi government, an Arab Sunni cleric in the "Sunni Triangle"—the area west and north of Baghdad including Fallujah, Tikrit, Ramadi, and Samara—called for a jihad against the United States and for Arab Sunnis to rise up and overthrow their occupiers.[34] Throughout CPA's tenure, an increasingly well-organized insurgency against coalition forces grew in the Sunni Triangle.

At the end of May, Bremer announced that he would delay the selection of a transitional Iraqi authority—slated under Garner's leadership to occur before June 3—claiming that the decision could not be made for at least another seven weeks due to security concerns.[35] Furthermore, Bremer announced that CPA would select the leaders for the interim authority, reversing the proposal made by Garner's office to have a large assembly of around 350 delegates help in the selection process.[36] These changes exacerbated growing tensions within Iraq over the slowness with which normalcy was returning to the country and magnified perceptions that the United States

---

[31] CPA Order Number 1 states: "On April 16, 2003, the Coalition Provisional Authority disestablished the Ba'ath Party of Iraq . . . Full members of the Ba'ath Party . . . are hereby removed from the positions and banned from future employment in the public sector . . . Those suspected of criminal conduct shall be investigated and, if deemed a threat to security or a flight risk, detained or placed under house arrest." See "Coalition Provisional Authority Order Number 1: De-Ba'athification of Iraqi Society," May 16, 2003. As of October 2007: http://www.cpa-iraq.org/regulations/#Orders

[32] Interviews with CPA officials, July 2004; Pamela Constable, "New Policy Unveiled for Ex-Baathists," *Washington Post,* January 12, 2004.

[33] Susan Sachs, "Iraqi Tribes, Asked to Help G.I.'s, Say They Can't," *New York Times*, November 11, 2003; Kenneth M. Pollack, "After Saddam: Assessing the Reconstruction of Iraq," the Brookings Institution, January 7, 2004.

[34] Edmund L. Andrews and Patrick E. Tyler, "As Iraqis' Disaffection Grows, U.S. Offers Them a Greater Political Role," *New York Times*, June 7, 2003.

[35] Scott Wilson, "U.S. Delays Timeline for Iraqi Government," *Washington Post*, May 22, 2003.

[36] Rajiv Chandrasekaran, "U.S. to Appoint Council in Iraq," *Washington Post,* June 2, 2003.

had no intentions of ending its occupation of Iraq;[37] the United States was even accused of reneging on its previous promises.[38] In the following weeks, voices from within the key Iraqi political organizations threatened to boycott Bremer's plan and to hold a national assembly without CPA's blessing.[39]

On July 13, after several rounds of negotiations between CPA and key Iraqi leaders, Bremer announced the selection of an Iraqi interim authority, to be called the Iraqi Governing Council (IGC). The IGC consisted of six major political parties—the INC, the INA, the KDP, the PUK, SCIRI, and the Da'wa Party—in addition to minor parties and independent leaders.[40] As noted in Table 10.1, it had 25 members (including three women): 13 Shi'ites, 11 Sunnis (including five Kurds, five Arabs, and a Turkoman), and one Assyrian Christian. CPA staff tasked with selecting IGC members sought to balance Iraq's diverse ethnic, religious, and political groups in the new federal body. However, the lack of access to and information about Iraqis who had stayed in the country during Saddam's reign created challenges in identifying appropriate participants.[41] Sixteen members came from either outside the country or from the autonomous north.[42]

The IGC's role was not to govern, since CPA remained the highest governing authority in Iraq. It served instead as the Iraqi face of the occupation, tasked by CPA to carry out certain efforts. Over time, it increasingly became a real partner in governance, having a de facto veto over certain CPA actions. The IGC's first acts were to select a delegation to travel to UN headquarters in New York for a July 22 meeting and to establish three subcommittees tasked with drafting bylaws, a political statement, and an agenda for future meetings.[43] The CPA also requested that the IGC select an interim leader from among its ranks and that it begin to take steps to draft a constitution.[44]

The IGC faced domestic and international challenges to its standing and role within the first few months of operation. Domestically, most Iraqis had not heard of

---

[37] Wilson, "U.S. Delays Timeline for Iraqi Government."

[38] Rajiv Chandrasekaran, "Iraqis Assail U.S. Plans for Council," *Washington Post*, June 3, 2003.

[39] Rajiv Chandrasekaran, "Iraqis Vow to Hold National Assembly," *Washington Post*, June 4, 2003; Patrick E. Tyler, "Political Leaders Resisting U.S. Plan to Govern Iraq," *New York Times*, June 5, 2003.

[40] Patrick E. Tyler, "Iraqis Set to Form an Interim Council with Wide Power," *New York Times*, July 11, 2003.

[41] Interview with CPA officials, July 2004.

[42] "The Iraqi Council," *Washington Post*, July 14, 2003; "Iraqi Governing Council Members," *BBC News Online*, July 14, 2003; "A Look at New Iraqi Leaders," *Baltimore Sun*, July 16, 2003.

[43] Rajiv Chandrasekaran, "Appointed Iraqi Council Assumes Limited Role," *Washington Post*, July 14, 2003.

[44] Rajiv Chandrasekaran, "Iraqi Council Postpones Selection of a Leader," *Washington Post*, July 15, 2003.

Table 10.1
The Iraqi Governing Council, July 2003

Member	Party	Ethnicity/ Religion	Local/ Exile
Samir Shakir Mahmoud	Independent	Sunni	Local
Sondul Chapouk (woman)	Independent	Turkoman	Local
Ahmed Chalabi	Iraqi National Congress	Shi'ite	Exile
Naseer al-Chaderchi	National Democratic Party	Sunni	Local
Adnan Pachachi	Independent Democratic Movement	Sunni	Exile
Mohammed Bahr al-Ulloum	Independent	Shi'ite cleric	Exile
Massoud Barzani	Kurdistan Democratic Party	Kurd	Exile (north)
Jalal Talabani	Patriotic Union of Kurdistan	Kurd	Exile (north)
Abdel-Aziz al-Hakim	SCIRI	Shi'ite	Exile
Ahmed al-Barak	Independent	Shi'ite	Local
Ibrahim al-Jaafari	Da'wa Party	Shi'ite	Exile
Raja Habib al-Khuzaai (woman)	Independent	Shi'ite	Local
Aqila al-Hashimi (woman)	Independent	Shi'ite	Local
Younadem Kana	Assyrian Democratic Movement	Assyrian Christian	Local
Salaheddine Bahaaeddin	Kurdistan Islamic Union	Kurd	Exile (north)
Mahmoud Othman	Kurdish Socialist Party	Kurd	Exile
Hamid Majid Mousa	Communist Party	Shi'ite	Exile (north)
Ghazi Mashal Ajil al-Yawer	Independent	Sunni (trIbal leader)	Exile
Abdul Zahra Othman Muhammad (Izzedin Salim)	Da'wa Party	Shi'ite	Local
Mohsen Abdel Hamid	Iraqi Islamic Party (Ikwan)	Sunni	Local
Iyad Allawi	Iraqi National Accord	Shi'ite	Exile
Wael Abdul Latif	Independent	Shi'ite	Local
Mouwafak al-Rubai	Independent	Shi'ite (secular)	Exile
Dara Noor Alzin	Independent	Kurd	Local
Abdel-Karim Mahoud al-Mohammedawi	Hezballah	Shi'ite from the marsh lands	Local

the IGC, nor were they familiar with the council members. This continued to be a problem throughout the IGC's tenure.[45] The CPA attempted to spread the word on the IGC to outlying provinces, but the campaign failed to effectively engage Iraqi television and newspaper media, and information that reached the Iraqi public re-

---

[45] At the end of October, for example, a poll showed that the majority of Iraqis still did not know about the IGC; see Dan Murphy, "Baghdad's Tale of Two Councils," *Christian Science Monitor*, October 29, 2003.

mained scanty.[46] Moreover, the IGC also faced challenges to its legitimacy. Many criticized the selection of IGC members by CPA and called for a process that would allow Iraqis to choose their own leaders. In response, one of Iraq's most influential Shi'ite clerics, the Ayatollah Ali Sistani, issued a fatwa—a religious legal edict—on June 28 calling for direct elections of Iraq's leaders and those responsible for drafting a constitution.[47] As discussed below, Sistani's fatwa proved to be a major stumbling block to both the perceived legitimacy of the IGC and CPA's plans for political development in Iraq.

The growing insurgency also challenged the IGC's legitimacy. Insurgents accused members of collaborating with the United States, and IGC members became targets of violence.[48] On September 25, 2003, IGC member Aqila al-Hashimi was gunned down in her car while in Baghdad.[49] On May 17, 2004, the acting Iraqi president for the month of May, Izzedin Salim, was assassinated in Baghdad.[50] Other government officials were also targeted. On November 4, 2003, a judge was shot in Mosul after being abducted. A few weeks later, on November 18, an Iraqi education official involved in training the Iraqi police force was killed in the city of Latifiyah.[51] This troubling trend persisted throughout the transfer of power from coalition forces to Iraqis.

The IGC also faced challenges to its legitimacy in international circles. On July 22, 2003, three delegates from the IGC—Ahmed Chalabi, Adnan Pachachi, and Aqila al-Hashimi—traveled to New York to meet with UN Secretary General Kofi Annan and push for UN recognition of the IGC and its right to occupy Iraq's chair at the General Assembly. The UN did not grant this recognition immediately, citing the fact that the IGC was appointed and not elected as grounds for withholding full recognition. After prodding from the United States and the United Kingdom, the UN officially recognized the right of the IGC to represent Iraq on August 14 and agreed to establish a UN mission in Iraq, building on its preexisting presence in the country.[52] Tragically, UN headquarters in Baghdad were bombed just five days after

---

[46] Murphy, "Baghdad's Tale of Two Councils"; "Iraq's Interim Leaders Review Transition Plans in Face of Shiite Poll Demand," *Associated Foreign Press*, November 29, 2003.

[47] For a detailed account of the circumstances leading up to Sistani's fatwa and its effects on political development in Iraq, see Rajiv Chandrasekaran, "How Cleric Trumped U.S. Plan for Iraq," *Washington Post*, November 26, 2003.

[48] Yochi J. Dreazen, "Insurgents Turn Guns on Iraqis Backing Democracy," *Wall Street Journal*, December 10, 2003.

[49] Gregg Zoroya, "Danger Puts Distance Between Council, People," *USA Today*, October 21, 2003.

[50] Scott Wilson, "Iraqi Council's Leader Is Slain," *Washington Post*, May 18, 2004.

[51] Dreazen, "Insurgents Turn Guns on Iraqis Backing Democracy."

[52] Paul Richter, "Security Council Endorses Iraq's New Governing Body," *Los Angeles Times*, August 15, 2003.

this announcement, killing UN special envoy Sergio Vieira de Mello and over 20 others.[53] As a result, the UN withdrew its non-Iraqi officials from the country.

Regional organizations were also reluctant to recognize the IGC as the new representative of Iraq. The 22-member Arab League announced at the end of July 2003 that it would not recognize the IGC, citing fears that recognition would legitimize a U.S. presence in the country.[54] As with the UN, the Bush administration began rigorously lobbying the Arab League, attaining its recognition of the IGC on September 9, 2003.[55] The IGC also managed to gain recognition from the International Monetary Fund and World Bank in September and to occupy Iraq's seat in the Organization of the Petroleum Exporting Countries (OPEC).[56] In February 2004, the World Trade Organization granted the IGC observer status.[57]

Alongside domestic and international challenges to the IGC's legitimacy, the body also faced challenges in the tasks it was asked to perform. Some limited successes marked the IGC's first few months in office. It succeeded in declaring April 9, the day of Baghdad's liberation, a national holiday. It also made positive steps toward the creation of a tribunal, which will be discussed further below. However, the IGC was slow in implementing several tasks presented to it by CPA. First, CPA asked the IGC to select an interim leader from among its ranks. After several weeks of negotiations, the IGC announced that it would not have one leader but nine who would rotate, in order of the Arabic alphabet, on a monthly basis.[58] According to this logic, Ibrahim al-Jaafari, the leader of the Da'wa Party, would be the first president for the month of August.[59] The CPA greeted this decision with skepticism and raised concerns that it would prevent the council from making decisions.

The second task CPA asked the IGC to perform was to appoint cabinet members for its various ministries. After six weeks of negotiations, the IGC succeeded in selecting cabinet members, but only after considerable disagreement and a compromise that increased the number of portfolios from 21 to 25.[60] As Table 10.2 shows, the

---

[53] Dexter Filkins and Richard A. Oppel, Jr., "Truck Bombing; Huge Suicide Blast Demolishes U.N. Headquarters in Baghdad," *New York Times*, August 20, 2003. The number of dead was later raised to 25. See Colum Lynch, "Brahimi to Be U.N. Adviser on Iraq," *Washington Post*, January 12, 2004.

[54] Steven R. Hurst, "Shi'ite Picked to Be Iraq's First President," *Washington Times*, July 31, 2003.

[55] "Arab League Nations Agree to Grant Seat to Iraq's Council," *New York Times*, September 9, 2003; John Daniszewski and Jailan Zayan, "Iraqi Council's Foreign Minister Takes a Seat at the Arab League's Table," *Los Angeles Times*, September 10, 2003.

[56] Bruce Stanley, "Iraq to Attend Next Week's OPEC Meeting," *New York Times*, September 17, 2003.

[57] John Zarocostas, "WTO Expected to Grant Iraq Observer Status Despite Baghdad's Lack of Customs Control," *Washington Times*, February 8, 2004.

[58] Robyn Dixon, "Iraq Council Picks 9 Leaders Instead of 1," *Los Angeles Times*, July 30, 2003.

[59] Hurst, "Shi'ite Picked to be Iraq's First President."

[60] Carol J. Williams, "Iraqi Council's Most Pressing Task: Legitimacy," *Los Angeles Times*, August 28, 2003.

highest-profile ministries—Foreign, Finance, Interior, and Oil—went to Zebari (a Kurd), Kilani (a Sunni Arab), Badran (a secular Shi'ite), and Bahr al-Uloom (a religious Shi'ite), respectively.[61] The performance of the ministries was variable; some were excellent while others were less successful. Of the 25 ministers selected, eight were to stay on beyond the June 2004 political handover: Foreign Affairs, Electricity,

Table 10.2
Cabinet Appointments, September 2003

Ministry	Name	Ethnicity/Religion
Oil	Ibrahim Mohammad Bahr al-Uloom	Shi'ite
Interior	Nuri Badran (until April 8, 2004) Samir Shakir Mahmoud (April–June 2004)	Shi'ite Sunni
Trade	Ali Alawi	Shi'ite
Finance	Kamal al-Kilani	Sunni
Labor	Sami Azara al-Majun	Shi'ite
Culture	Mufid Mohammad Jawad al-Jazairi	Shi'ite
Education	Alla Abdessaheb al-Alwan	Shi'ite
Foreign	Hoshyar Zebari	Kurd
Planning	Mahdi al-Hafez	Shi'ite
Public Works	Nisrin Mustafa al-Barwari	Kurd
Transport	Behnam Ziya Bulis	Assyrian Christian
Electricity	Ayham al-Samarrai	Sunni
Health	Khodayyir Abbas	Shi'ite
Housing and Construction	Bayan Baqir Solagh	Shi'ite
Science and Technology	Rashad Umar Mandan	Turkoman
Human Rights	Abd al-Basit Turki	Sunni
Industry and Minerals	Muhammad Tawfiq Rahim	Kurd
Migration and Immigration	Muhammad Jasim Khudayyir	Shi'ite
Water	Abd al Latif Jamal Rashid	Kurd
Communications	Haidar Jawad al-Abadi	Shi'ite
Agriculture	Abd al-Amir Abbud Rahima	Shi'ite
Youth and Sports	Ali Faiq al-Ghabban	Shi'ite
Environment	Abd al-Rahman Saddiq Karim	Kurd
Defense	Iyad Alawi (assigned on April 4, 2004)	Shi'ite
Information	Not assigned	

---

[61] Dexter Filkins, "Iraqi Council Picks a Cabinet to Run Key State Affairs," *New York Times*, September 2, 2003.

Public Works, Culture, Planning, Water Resources, Communications, and Youth and Sports. The remaining positions were filled primarily by apolitical technocrats.[62]

The third task that CPA hoped the IGC would accomplish was to create a process whereby the constitution could be drafted, a document that CPA believed was necessary to create before holding elections. After considerable debate, the IGC succeeded in selecting a 25-member committee of lawyers, intellectuals, and former judges to propose the process whereby a constitution would be created, not actually drafting the document.[63] As will be discussed below, the committee quickly bogged down in debates over the structure of government and the role of religion and religious law in the new Iraq, preventing significant progress on this task.

In addition to performing tasks assigned by CPA, the IGC also autonomously asserted its power by initiating several controversial policies. Most notably, the IGC chose on several occasions to ban Arab media outlets that they accused of fostering violence in Iraq. On September 23, 2003, the IGC banned the Bahrain-based Al Jazeera news station along with Al Arabiya, the Qatari station, on grounds that they were encouraging violence and provoking sectarian strife.[64] On November 24, the IGC issued orders to have Al Arabiya's offices raided, claiming that it was encouraging the anti-coalition insurgency.[65] These decisions raised concerns with international groups committed to press freedoms.[66]

The November 15 Agreement and the Transfer of Authority

By November 2003, the lack of progress toward an elected government and improvements in the security situation in Iraq had generated domestic and international criticism, most of which was directed toward Bremer and CPA. In response, Bremer abruptly flew to Washington on November 10 to hold meetings with officials in the White House and the Pentagon. While Bremer was in the United States, a car bomb exploded in Nasiriya, killing 18 Italian servicemen. Also while Bremer was in Washington, a CIA report dated November 10 was leaked to the press. It claimed that Iraqi citizens were losing faith in the United States and reconstruction efforts, and

---

[62] Table 10.3, later in this chapter, includes a complete list of the members of the Iraqi Interim Government. The cabinet members from September 2003 through May 2006 can be found at "The Interim Government of Iraq from 2003–06," *Middle East Reference.* As of October 2007: http://middleeastreference.org.uk/iraqministers.html.

[63] Dexter Filkins, "Iraqis Name Team to Devise Way to Draft Constitution," *New York Times,* August 12, 2003.

[64] Rajiv Chandrasekaran, "Iraqi Council Denies Access to Two Arab Satellite Networks," *Washington Post,* September 24, 2003.

[65] Bassem Mroue, "Iraqis Raid Offices of Television Network," *Associated Foreign Press,* November 24, 2003.

[66] Chandrasekaran, "Iraqi Council Denies Access to Two Arab Satellite Networks."

that this loss of faith was providing a foundation for armed insurgency.[67] These developments further heightened the sense of urgency to create a new action plan in Iraq that would accelerate the transfer of political power to Iraqis.[68]

On November 15, CPA and the IGC released a new timeline for the transfer of power to an Iraqi governing body. It called for the construction of a "Basic Law"— the forerunner to a constitution—by February 28, 2004; the creation of a larger Iraqi council by May 2004; the dissolution of CPA by July 1, 2004; the election of an interim government by the transitional council; and full elections by the end of 2005.[69] This plan, while not spelling out the details of each stage, offered a timeline for political independence, something that CPA had not previously articulated.

The November 15 agreement infused new life into the process of political development in Iraq and brought into focus several key challenges to the creation of a new Iraqi government. One of the first issues to arise was the method of selecting the next central political body, the "transitional authority." The CPA envisioned the creation of caucuses within Iraq's 18 provinces that would, in turn, appoint members of the transitional authority.[70] However, Iraqis from most religious, regional, and ethnic constituencies called for national elections to select this governing body; the call included the IGC, which voted unanimously at the end of November 2003 for full national elections to select an interim authority.[71] Ayatollah Sistani's June 28 fatwa, which called for national elections to select both the country's political leaders and the body responsible for drafting Iraq's new constitution, continued to pressure CPA to open the political process up to Iraqi citizens immediately instead of in 2005, as CPA had planned.

The CPA objected to immediate direct elections, citing the need for a new national census (the last one had been taken in 1997) that would assess the total number of Iraqis and allow for voter registration; CPA argued that a proper census would take roughly a year to conduct. However, the newly formed Iraqi Census Bureau claimed that it could do the job in a matter of months, a proposal that was largely

---

[67] Douglas Jehl, "CIA Report Suggests Iraqis Are Losing Faith in U.S. Efforts," *New York Times*, November 13, 2003.

[68] William Douglas and John Walcott, "U.S. Focuses on Faster Handover to the Iraqis," *Philadelphia Inquirer*, November 13, 2003; Robin Wright and Daniel Williams, "U.S. to Back Re-Formed Iraq Body," *Washington Post*, November 13, 2003.

[69] Interviews with CPA officials, July 2004; "Transfer Does Not Affect Troops," *BBC News Online*, November 16, 2003; "Bremer: U.S. in Tough Fight in Iraq," *BBC News Online*, November 17, 2003.

[70] Rajiv Chandrasekaran, "Iraqis Say U.S. to Cede Power By Summer," *Washington Post*, November 15, 2003.

[71] This call for national elections most notably included Shi'ites, but it also included Sunni Arab leaders who argued that Kurds, also Sunnis, would prevent the Shi'ites from taking over the new government. See Alan Sipress, "Once-Dominant Minority Forms Council to Counter Shiites and Negotiate Future," *Washington Post*, January 6, 2004. For the IGC vote, see Joel Brinkley, "Iraqi Council Agrees on National Elections," *New York Times*, December 1, 2003.

ignored.[72] Proponents of direct elections further suggested that UN Oil for Food ration cards, which were well organized and pervasive throughout the country, could be used for voter registration. The CPA countered this proposal by pointing out that those imprisoned under Saddam's regime had had their UN identity cards revoked as punishment; this approach, therefore, could potentially exclude the groups most persecuted by the previous regime, specifically the Kurds and the Shi'ites.[73]

Increasingly, the debate over the method of selecting new leadership polarized into two camps: those who favored caucuses for selecting an interim authority, most notably CPA, and those who backed Sistani's fatwa calling for direct elections. The CPA was criticized for not having taken Sistani's fatwa seriously when it was first issued in June and for failing to realize that most Shi'ites in the country would not go against his edict. Sistani refused to meet directly with Bremer or other officials in CPA, corresponding with them only through letters and via SCIRI's director of the party's bureau, Adel Abdel-Mehdi.[74] This often resulted in confusion and miscommunication. For example, CPA staff reported that they had been under the impression that Sistani had approved the November 15 agreement, but this turned out not to be the case.[75] As a stalemate became evident, CPA proposed a compromise, suggesting that direct elections for a transitional authority could be held in certain provinces or districts and leaders could be chosen via caucus nominations in areas where elections were not feasible.[76]

Following several months of fruitless efforts to work around Sistani's fatwa, CPA and IGC turned to the UN as a possible mediating force. Calling on the UN, however, presented its own set of problems, because of lingering disagreements about the origins of the war and the fact that the UN had not been integrated into CPA's postwar reconstruction efforts. In January 2004, the IGC sent a delegation to the UN in hopes of securing UN aid in the electoral process. Its petition was backed by the U.S. government, which had begun to see that the plan to have caucuses select an interim government was too complicated.[77] Sistani and several members of the IGC asked, in particular, if the UN could help facilitate national elections before the July 1, 2004 departure of CPA. After dispatching a team of elections experts to Iraq at the end of January, the UN concluded that national elections could not be held before

---

[72] See Joel Brinkley, "U.S. Rejects Iraq Plan to Hold Census by Summer," *New York Times*, December 4, 2003.

[73] Brinkley, "Iraqi Council Agrees on National Elections."

[74] Chandrasekaran, "How Cleric Trumped U.S. Plan for Iraq."

[75] Interviews with CPA officials, July 2004.

[76] Chandrasekaran, "How Cleric Trumped U.S. Plan for Iraq."

[77] Steven R. Weisman and John H. Cushman, Jr., "U.S. Joins Iraqis to Seek U.N. Role in Interim Rule," *New York Times*, January 16, 2004. CPA officially dropped the caucus scheme on February 17, 2004. See Warren P. Strobel and Hannah Allam, "U.S. Alters Plans for Iraqi Handover," *Philadelphia Inquirer*, February 18, 2004.

July, but that Iraqis had other options for selecting an interim authority.[78] In February, Secretary General Annan dispatched to Iraq Lakhdar Brahimi—the former UN special envoy tasked with fostering political transition in Afghanistan. He concluded that Iraq could hold national elections at the end of 2004 or the beginning of 2005 at the very earliest. He cited the poor security environment as one key obstacle to holding elections.[79] Sistani accepted these findings but called for elections to be held at the end of 2004.[80]

In addition to debating the method of selecting an interim authority in Iraq, the IGC also clashed with CPA over the length of the IGC's tenure as a governing body and its role in the future Iraqi government. Amid outcries from local and international voices over the shortcomings of the IGC, CPA prescribed the dissolution of the IGC following the selection of an interim authority in May 2004.[81] Several members of the IGC, including Ahmad Chalabi, Iyad Allawi, and Adnan Pachachi, disagreed with this formula for the transfer of power and called for the maintenance of the IGC after July as some form of advisory committee or perhaps even as one chamber within a bicameral government.[82] These proposals raised fears that the IGC would not willingly relinquish power within the proposed timeframe put forward by CPA and the Bush administration.

Following the February UN report that discouraged national elections before the end of 2004, CPA and the UN began discussing possible alternatives to creating an interim government. In February, the UN proposed two plans for an interim government. The first called for convening "a national conference of tribal, political, and religious leaders that reflects Iraq's disparate population to select a provisional government"; such a conference would be similar to the loya jirga held in Afghanistan.[83] The second proposal suggested expanding the existing IGC to around 100 members to create an interim body that could minimally govern until full elections.[84] Rumors circulated that the U.S. government preferred the latter plan.[85]

---

[78] Betsy Pisik, "U.N. Sides with U.S. on Voting in Iraq," *Washington Times*, January 16, 2004; Betsy Pisik, "U.N. Iraq Advisor Issues Vote Warning," *Washington Times*, January 28, 2004.

[79] Rajiv Chandrasekaran, "U.N. Envoy Backs Iraqi Vote," *Washington Post*, February 13, 2004; Colum Lynch, "U.N. Plan for Iraq Transition Released," *Washington Post*, February 24, 2004; Nick Wadhams, "U.N. Says No to Iraq Elections Until at Least 2005," *Associated Press Worldstream*, February 23, 2004.

[80] Dexter Filkins, "Iraqi Ayatollah Insists on Vote by End of Year," *New York Times*, February 27, 2004.

[81] Tom Lasseter, "Top Cleric Spurns U.S. Plans," *Miami Herald*, January 26, 2004.

[82] Joel Brinkley, "Some Members Propose Keeping Iraqi Council After Transition," *New York Times*, November 25, 2003; Paul Richter, "Iraqi Officials Wage Political War in U.S.," *Los Angeles Times*, February 5, 2004.

[83] Colum Lynch and Robin Wright, "U.N. Chief to Urge Delaying Elections, Senior Officials Say," *Washington Post*, February 19, 2004.

[84] Lynch and Wright, "U.N. Chief to Urge Delaying Elections, Senior Officials Say"; Hamza Hendawi, "Expanded Council Likely When U.S. Cedes Power," *Miami Herald*, March 5, 2004.

[85] Barbara Slavin, "Power Could Transfer to an Expanded Council," *USA Today*, February 19, 2004.

On March 17, the IGC formally petitioned the UN for help in forming an interim government.[86] In response, Brahimi returned to Iraq in early April to determine how to form an interim government for the June transfer of power.[87] On April 14, Brahimi announced his recommendations: that the IGC should be completely dissolved; that the UN should appoint an interim president and two vice presidents consisting of a Sunni, a Shi'ite, and a Kurd; and that politically neutral technocrats should run the ministries. Brahimi further called for a national conference in July 2004 to elect a consultative assembly.[88] This proposal prompted outcries from several members of the IGC—most notably Ahmad Chalabi—who claimed that this approach was anti-democratic and would be politically destabilizing.[89] The Bush administration, however, agreed to the UN plan.[90] At the end of April, Brahimi proposed that the new interim body should be selected by the end of May in order to allow for a smooth transition in July.[91]

After several weeks of negotiations between Brahimi, Bremer, White House envoy Robert Blackwill, and the IGC, the new interim government, called the Iraqi Interim Government (IIG), was announced on June 1, 2004. Ghazi al-Yawer, a 45-year-old Sunni tribal chief, was named as the country's new interim president. It was reported that Yawar was neither Blackwill's nor Bremer's first choice—they had favored Adnan Pachachi, the 80-year-old former Iraqi foreign minister—but that the IGC had staunchly opposed that nomination and favored Yawar instead.[92] Similarly, the selection of the prime minister, the most influential position, was hotly debated among the different parties selecting the new government. Brahimi reportedly favored Hussain Shahristani, a Shi'ite nuclear scientist, who was also rejected by the IGC. Bremer and Blackwill pushed for Iyad Allawi, who eventually won the blessings of Brahimi and the IGC, despite concerns that he was too close to the U.S. government. Brahimi later complained that he was under "terrible pressure" to conform to Bremer's and Blackwill's preferences.[93] In addition to the new president and prime

---

[86] Rajiv Chandrasekaran, "Iraqi Council Agrees to Ask U.N. for Help," *Washington Post*, March 18, 2004.

[87] Maggie Farley and Sonni Efron, "U.N. Envoy May Provide the Key to a Transfer of Power in Iraq," *Los Angeles Times*, April 14, 2004.

[88] Rajiv Chandrasekaran, "Envoy Urges U.N.-Chosen Iraqi Government," *Washington Post*, April 15, 2004.

[89] Chandrasekaran, "Envoy Urges U.N.-Chosen Iraqi Government"; Sharon Behn, "Governing Council Huddles over Power Transfer," *Washington Times*, April 19, 2004; Sharon Behn, "Chalabi Says No to U.N. Oversight," *Washington Times*, April 29, 2004.

[90] Steven R. Weisman and David E. Sanger, "U.S. Open to a Proposal That Supplants Council in Iraq," *New York Times*, April 16, 2004.

[91] Warren Hoge, "U.N. Envoy Seeks New Iraqi Council by Close of May," *New York Times*, April 28, 2004.

[92] Rajiv Chandrasekaran, "Interim Leaders Named in Iraq," *Washington Post*, June 2, 2004; Steven Komarow, "Iraqis, U.S. Split on New Leaders," *USA Today*, June 1, 2004.

[93] Rajiv Chandrasekaran, "Envoy Bowed to Pressure in Choosing Leaders," *Washington Post*, June 3, 2004.

minister, two deputy presidents were also chosen, Ibrahim al-Jafaari of the Shi'ite Da'wa party and Rowsch Shaways of the Kurdish KDP.

Brahimi was also tasked with naming a new cabinet. As previously mentioned, Brahimi called for the interim government to be staffed primarily with apolitical technocrats, who would run the county until full national elections could be held. Brahimi succeeded in selecting a 32-member cabinet that included several women and had only six ministers affiliated with the major political parties: Foreign Affairs (KDP), Finance (SCIRI), Public Works (KDP), Communications (PUK), Youth and Sports (SCIRI), and Women (PUK); none of the cabinet ministers were from Chalabi's Iraqi National Congress party. Media sources reported that, unlike the selection of the president and prime minister, Brahimi was responsible for the selection of all but two of the cabinet members.[94] Table 10.3 lists all the members of the IIG. Unlike its predecessor, the IIG received validation from prominent individuals and organizations almost immediately. On June 3, the Ayatollah Sistani issued a statement that formally endorsed the IIG, although Sistani continued to call for full national elections by the end of 2004.[95] Likewise, on June 17, the 57-member Organization of Islamic Conference (OIC) officially acknowledged the new government and vowed to "actively assist" its leaders as it prepared to take political control of Iraq.[96] Perhaps most important to the new government, polling data taken at the end of June showed that the majority of Iraqis were familiar with the new government's various members: 68 percent had "confidence" in their leadership; 73 percent approved of Allawi as the prime minister; 84 percent approved of President Yawar.[97] Thus, the IIG received popular, religious, and organizational backing almost immediately following its creation.

The IIG faced several challenges leading up to the June 30, 2004 political handover. First, the new political body had to work with the UN Security Council to secure a new Security Council resolution officially ending political occupation of the country. Resolution 1546 was passed unanimously on June 8 after five versions of the proposal were debated among the Security Council's 15 members. The final version granted the IIG full political and economic control of Iraq, including its oil industry, with the caveat that contracts agreed to under CPA would be honored beyond the June handover. Furthermore, the resolution gave Prime Minister Allawi control of Iraqi forces and, in theory, the right to dismiss the multinational forces.

---

[94] Chandrasekaran, "Interim Leaders Named in Iraq."

[95] Edward Cody, "Influential Cleric Backs New Iraqi Government," *Washington Post*, June 4, 2004.

[96] "Islamic Group Endorses Iraqi Government," *Los Angeles Times*, June 17, 2004.

[97] Robin Wright, "Iraqis Back New Leaders, Poll Says," *Washington Post*, June 25, 2004.

Table 10.3
Iraqi Interim Government, June 2004

Position	Name	Party	Ethnicity/ Religion
President	Ghazi Mashal Ajil al-Yawer		Sunni
Deputy President	Ibrahim al-Ushayqir (al-Jafaari)	Da'wa	Shi'ite
Deputy President	Rowsch Shaways	KDP	Kurd
Prime Minister	Iyad Alawi	INA	Shi'ite
Deputy PM for National Security	Barham Salih	PUK	Kurd
Foreign Affairs	Hoshyar Mahmud Muhammad al-Zibari	KDP	Kurd
Oil	Thamir Abbas Ghadhban		Sunni
Finance	Adl Abd al-Mahdi	SCIRI	Shi'ite
Interior	Falah Hasan al-Naqib	Iraqi National Movement	Sunni
Justice	Malik Duhan al-Hasan		Sunni
Trade	Muhammad al-Jiburi		Sunni
Defense	Hazim Shalan		Shi'ite
Education	Sami al-Mudhaffar		Shi'ite
Electricity	Ayham al-Samarrai	Independent Democratic Movement	Sunni
Public Works	Nasrin Mustafa Sadiq Barwari (woman)	KDP	Kurd
Health	Ala al-Din Abd al-Sahib Alwan		Shi'ite
Higher Education	Tahir Khalaf Jabar al-Bakaa		Shi'ite
Housing and Construction	Umar al-Faruq Saim al-Damluji		Sunni
Science and Technology	Rashad Umar Mandan		Turkoman
Culture	Mufid Muhammad Jawad al-Jazairi	ICP	Shi'ite
Human Rights	Bakhtyar Amin		Kurd
Industry and Minerals	Hashim M. al-Hasani	Iraqi Islamic Party	Sunni
Labor and Social Affairs	Layla Abd al-Latif (woman)		Shi'ite
Migration and Immigration	Baskal Ishu Wardah (Pascale Isho Warda) (woman)	Assyrian Democratic Party	Assyrian Christian
Planning	Mahdi al-Hafidh		Shi'ite
Water	Abd al-Latif Jamal Rashid	PUK	Kurd
Communications	Mohammad Ali al-Hakim		Shi'ite
Agriculture	Sawsan Ali Majid al-Sharifi (woman)		Shi'ite
Transportation and Telecommunications	Luay Hatim Sultan al-Aras		Shi'ite
Environment	Mishkat Mumin		Sunni
Youth and Sports	Ali Faiq al-Ghabban	SCIRI	Shi'ite

Table 10.3 (continued)

Position	Name	Party	Ethnicity/Religion
Minister of State			
• No portfolio	• Qasim Daud		Shi'ite
• No portfolio	• Mamu Farham Utham Brali		Kurd
• No portfolio	• Adnan al-Janabi		Sunni
• Provinces	• Wail Abd al-Latif		Shi'ite
• Women	• Narmin Uthman (woman)	PUK	Kurd

However, the United States succeeded in preserving the right to conduct offensive military operations, with the promise that they would be closely coordinated with Iraqi authorities. Although unanimously agreed to, the new resolution did raise concerns as to how it would play out in practice. Overall, however, the resolution was warmly received by members of the Security Council and was heralded as a "major step forward."[98]

The second major issue that the IIG faced in its preparations for political handover was the security problem presented by the persistent insurgency, particularly in the Sunni Triangle and with al-Sadr's forces in Najaf and eastern Baghdad. On June 20, the *New York Times* reported that Prime Minister Allawi was considering imposing a state of emergency following the June handover of power in order to consolidate authority and fight the growing insurgency.[99] This suggestion received condemnation from Bremer and other U.S. diplomats, who maintained that only the United States, as head of the multinational forces, had the right to declare a state of emergency or martial law in Iraq. U.S. officials claimed that Resolution 1546 stipulated this by granting foreign forces "all necessary measures" to ensure security in Iraq including, they argued, martial law. IIG members backed away from the proposed state of emergency, despite continued insurgent activity.[100]

The IIG also faced time constraints in preparing for the political handover scheduled at the end of the month. On June 4, the UN handed over the management of preparations for Iraq's national elections to a seven-member team chosen at the end of May.[101] This new team was to prepare for the national conference, tentatively scheduled for July 2004, that was to elect Iraq's consultative assembly, in addition to preparing for full national elections in 2005. On June 9, CPA transferred authority over Iraq's oil industry to the newly appointed minister, Thamir Abbas

[98] Peter Sleven and Robin Wright, "UN Backs Plan to End Iraq Occupation," *Washington Post*, June 9, 2004.

[99] Dexter Filkins, "Iraq Government Considers Using Emergency Rule," *New York Times*, June 21, 2004.

[100] Nicolas Pelham, "Iraq Ministers Told Only U.S. Can Impose Martial Law," *Financial Times* (London), June 23, 2004.

[101] Peter Y. Hong, "UN Hands Task of Organizing Vote to Iraqi Panel," *Los Angeles Times*, June 5, 2004.

Ghadhban.[102] On June 25, just days before the political handover, CPA handed the final 11 ministries still under its authority over to the new Iraqi government, including some "sensitive" ministries such as Defense, Interior, and Trade.[103] On June 28, two days before the scheduled handover, with hopes of thwarting planned insurgency attacks, Ambassador Bremer officially handed the country's political reins over to the IIG, effectively dissolving CPA.[104]

The Transitional Administrative Law

One of the major obstacles to political development in Iraq involved the drafting of a full-fledged constitution. Initially, the U.S. government's plans for political development in Iraq called for the drafting of a constitution first, followed by national elections for an Iraqi government.[105] U.S. officials believed that a constitution would help mitigate Iraq's contentious issues—such as religious freedom, federalism, and the protection of minorities—and prevent radicalism and sectarianism from taking root. These issues were elucidated in the April 15, 2003 meeting in Nasiriya that produced the 13-point memorandum.[106]

Time and experience revealed, however, that in order to write a constitution, the divisive political issues facing Iraq had to be addressed first. As previously mentioned, one of CPA's first assignments for the newly formed IGC was to select a body that would be responsible for drafting a constitution, something that CPA hoped the new governing body would do quickly.[107] With difficulty, the IGC succeeded in selecting a 25-member Constitution Preparation Committee (CPC)—composed of judges, attorneys, and scholars—that would propose the process for drafting and ratifying a constitution, rather than actually writing the document.[108] As had happened with selecting leaders for the IGC, debates broke out over the qualifications and selection of the drafters. Some favored a nomination process, while others, most notably Ayatollah Sistani, called for elections.[109]

---

[102] "Iraqis Assume Control of Oil Industry," *Washington Times*, June 9, 2004.

[103] Jim Krane, "Iraq Takes Over Last of Ministries," *Philadelphia Inquirer*, June 25, 2004.

[104] "Handover Completed Early to Thwart Attacks, Officials Say," *New York Times on the Web*, June 28, 2004.

[105] Interviews with CPA officials, July 2004.

[106] "A 'Federal System' with Leaders Chosen by the People."

[107] Rajiv Chandrasekaran, "U.S. Presses New Iraqi Council to Begin Tackling Major Issues," *Washington Post*, July 16, 2003.

[108] International Crisis Group, "Iraq's Constitutional Challenge," Middle East Report No. 19, November 13, 2003, available at www.icg.org, accessed May 2004; Filkins, "Iraqis Name Team to Devise Way to Draft Constitution."

[109] Howard LaFranchi, "New Fast Track for Iraqi Sovereignty," *Christian Science Monitor*, November 17, 2003.

At the end of September, the committee presented its findings on how to proceed with drafting a constitution. It highlighted the key challenges it believed Iraq faced in creating a new political system, including how to treat the role of religion in the government, how to reconcile exiles and Iraqi locals, how to improve Iraq's security problems, and how to prevent the United States from dominating the political process.[110] The committee also announced that it was deadlocked over how to proceed with selecting drafters. Many believed that ignoring Sistani's fatwa for direct elections could result in the Shi'ites breaking away from the political process and that this issue had to be resolved before a constitution could be created. The committee concluded by claiming it would take at least a year for a constitution to be drafted; hence, political development should proceed by holding elections first.[111] Despite these findings and recommendations, CPA continued to pressure the IGC to begin drafting a constitution.[112]

The November 15 agreement between CPA and the IGC eased tensions over the imperative of drafting a constitution quickly. The establishment of an Iraqi governing body—first through the IGC, next through an interim authority, and then through national elections for a full-fledged Iraqi government—became the primary focus of political development. However, alongside the evolution of an Iraqi government, CPA set a February 28, 2004 deadline for the creation of a set of "Basic Laws," or governing principles that would effectively serve as an "interim constitution."[113] These Basic Laws, not unlike the 13-point memorandum developed at the April 15, 2003 Nasiriya meeting, would set the stage for the key challenges facing political development in Iraq: the role of religion in Iraqi law and politics, the structure of the new government, and the representation of religious, ethnic, and tribal groups.

In November, the IGC began to propose various drafts of an interim constitution, working closely with CPA staff to refine issues. Some of the key topics debated were the role of Islam in the new government, the creation of quotas for women in parliament, and the right to free speech and privacy. The question of regional auton-

---

[110] Alissa J. Rubin, "Select or Elect? Iraqis Split on Constitution Delegates," *Los Angeles Times*, September 29, 2003.

[111] Patrick E. Tyler, "Iraqi Groups Badly Divided over How to Draft a Charter," *New York Times*, September 20, 2003; Rajiv Chandrasekaran, "Iraqis Call U.S. Goal on Constitution Impossible," *Washington Post*, September 30, 2003.

[112] Guy Dinmore and James Harding, "Bremer in Talks on Timing of Iraq Pullout," *Financial Times* (London), October 29, 2003.

[113] Although the Transitional Administrative Law functioned as an interim constitution, coalition authorities avoided the use of this term, preferring to call it an administrative law. "Timetable Set for Iraq Transfer," *BBC Online*, November 15, 2003; "Handing Over the Keys in Iraq," *BBC Online*, November 13, 2003.

omy, specifically Kurdish autonomy, was sidelined for future discussion.[114] The CPA expressed particular concern over the role of Islam in the interim constitution, and in mid-February, Bremer threatened to block proposals that would make Islamic law the "backbone" of an interim constitution.[115]

On March 1—two days after the proposed deadline—the IGC agreed with "full consensus" to a Transitional Administrative Law (TAL), touted as the "most progressive" document of its kind in the Middle East. In addition to enshrining individual rights in a 13-article bill of rights, the document also succeeded in giving Islam a place as "a source" of law but stopped short of naming it as the primary source. Moreover, the document earmarked 25 percent of parliament's seats for women.[116] Administratively, the document stipulated that elections for a 275-member transitional assembly would be held in January 2005.[117] It also gave Kurds "broad autonomy," essentially preserving the status quo.[118]

Ratification of the document, however, was delayed following a series of attacks on Shi'ite shrines on March 2, 2004, during the religious holiday of Ashura. These attacks killed nearly 200 pilgrims in Karbala and Baghdad.[119] In the days following, Shi'ite leaders, backed by a statement from Ayatollah Sistani, retracted their approval of the document, claiming that one of its clauses could allow Kurds or Sunnis to veto a final constitution, even if it were ratified by popular vote. After some negotiation, however, religious Shi'ites in the IGC agreed to sign the document on March 8, bringing the Transitional Administrative Law to life.[120] Ayatollah Sistani issued a statement promising to amend the questionable clause in the future.[121]

Despite the IGC's success in drafting and ratifying an "interim constitution," the TAL contained several potential pitfalls. First, the document did not resolve Kurdish aspirations for independence and, in particular, left unanswered the fate of the ethnically diverse city of Kirkuk, which most Kurds claim as Kurdish by historic right. Second, some argued that the TAL set an unreasonable deadline for national elections of the drafters of Iraq's final constitution, calling for these to be held in January 2005. Finally, there were concerns that the Iraqi Interim Government might

---

[114] Interviews with CPA officials, July 2004; International Crisis Group, "Iraq's Constitutional Challenge"; Dexter Filkins, "Panel Starts to Draw Up Constitution for Short Run," *New York Times*, February 3, 2004.

[115] Robert H. Ried, "Bremer to Block Islamic Charter," *Washington Times*, February 17, 2004.

[116] Dexter Filkins, "Iraqi Leadership Gains Agreement on Constitution," *New York Times*, March 1, 2004.

[117] Rajiv Chandrasekaran, "Shiites to Seek Changes," *Washington Post*, March 9, 2004.

[118] Dexter Filkins, "Iraqis Receive U.S. Approval of Constitution," *New York Times*, March 2, 2004.

[119] Rajiv Chandrasekaran and Anthony Shadid, "Shiites Massacred in Iraq Blasts," *Washington Post*, March 3, 2004; "Iraq Constitution Delayed, Leaders Call for Calm," *New York Times on the Web*, March 3, 2004.

[120] Dexter Filkins, "Top Shiites Drop Their Resistance to Iraqi Charter," *New York Times*, March 8, 2004.

[121] Chandrasekaran, "Shiites to Seek Changes."

choose to ignore some—or perhaps even all—of the TAL's provisions after the transfer of authority.

Provincial and City Governance

Virtually independent of initial U.S. efforts to create a federal government with exiled Iraqis, coalition military efforts to create local Iraqi governing bodies developed alongside spontaneous "grass roots" movements to form city and regional governments as Saddam's government collapsed. Military personnel, as the coalition's representatives on the ground, sought to help structure local governing authorities both to serve as their interlocutors and to ensure functioning governance in the regions. The congruence of these efforts with indigenous Iraqi moves toward self-government varied. In the south, Shi'ite leaders organized groups to provide security, oversee the distribution of essential goods, and keep city governments running.[122] The ability of southern, Shi'ite-dominated cities to establish de facto governments and provide for their citizens raised fears within the U.S. government that they would impose an Iranian-style government in Iraq, either through force or—with 60 percent of the population—via majority rule in elections.

On April 11, 2003, the Pentagon awarded the Research Triangle Institute (RTI) in North Carolina a contract to help facilitate political development in Iraq.[123] RTI worked in coordination with CPA, the emerging Iraqi federal government, and USAID to build Iraq's political infrastructure "from the ground up."[124] Together, RTI and USAID launched three initiatives aimed at fostering local governance in Iraq: the Local Governance Program, which specifically worked to develop neighborhood and city councils; Community Action Programs, which focused on developing community groups to address critical needs in neighborhoods and supplying information through those groups; and the Iraq Transition Initiative, which concentrated on rebuilding critical infrastructure, such as municipal buildings, schools, and clinics, in addition to fostering civil society and the media.[125]

---

[122] Peter Ford, "Iraqi Holy Men Leap into Postwar Politics," *Christian Science Monitor*, May 14, 2003.

[123] Prepared Testimony of Andrew S. Natsios, Administrator, U.S. Agency for International Development, Senate Foreign Relations Committee, June 4, 2004. RTI's contract was renewed in April 2004. See "RTI Gets Contract Extension for Iraq Rebuilding," *Associated Press*, April 1, 2004.

[124] For more on these agencies' initiatives in Iraq, see "USAID Assistance for Iraq" and "Strengthening Local Governance in Iraq." As of October 2007:
http://www.usaid.gov/iraq/
http://www.rti.org/page.cfm?nav=356&objectid=91AD4ED9-F623-4C93-8CCF98BA9F7C82F5

[125] U.S. Agency for International Development, "Democracy in Iraq: Building Democracy from the Ground Up," May 2004. As of February 2008:
http://www.usaid.gov/iraq/pdf/iraq_demgov_0504.pdf

In May, just one month after Saddam's government fell, several cities began to show signs of political development by electing governing councils and mayors. On May 5, 2003, Mosul, with support of U.S. military personnel, chose a 24-member interim government, which in turn selected a mayor.[126] In the southern Shi'ite city of Amara, local resistance fighters overthrew Iraqi forces in the early days of the war and established a city council that included a body of 27 people picked to run local facilities.[127] On May 15, British forces handed over the southern port city of Umm Qasr to a 12-member council, followed by the transfer of the port to Iraqi leaders on May 22.[128] And on May 24, 300 Iraqi delegates elected a council in the northern city of Kirkuk. The U.S. military oversaw the election, which included all three of the city's major ethnic groups, Kurds, Arabs, and Turkomans.[129] The council, in turn, elected a mayor on May 28.[130]

However, coalition authorities also clamped down on the creation of local governments in certain areas. In June 2003, citizens of Basra protested British forces' selection of leaders and technocrats for the city, particularly criticizing the absence of clergy in the appointed government.[131] They demanded the right to run their own city.[132] CPA halted elections in the southern Shi'ite city of Najaf, claiming that conditions were not appropriate for elections. Despite U.S. Marines having overseen the process,[133] CPA appointed a former Iraqi military officer (who was later sacked on corruption charges) to the post of mayor.[134] Similarly, U.S. forces halted elections in Samara on June 28 and appointed a leader, angering citizens there. In total, CPA reportedly appointed mayors in a dozen cities.[135] These policies, coupled with censoring certain groups' political rhetoric, bred resentment within the population and led

---

[126] Sabrina Tavernise, "A Northern Iraqi City Chooses an Interim Government," *New York Times*, May 6, 2003.

[127] Nazila Fathi, "One Iraq City Tries on Democracy and Finds It Fits Well," *New York Times*, May 18, 2003.

[128] Interview with USAID official, July 2004; Tini Tran, "First Iraqi City Handed Over to Civilian Government," *Washington Times*, May 16, 2003; Marcy Lacey, "British Give Port Control to the Iraqis," *New York Times*, May 23, 2003.

[129] William Booth, "U.S.-Run Local Election Touted as 'First Step' for Iraqis," *Washington Post*, May 25, 2003.

[130] Danna Harman, "Baathists Need Not Apply: Kirkuk Elects a New Mayor," *Christian Science Monitor*, May 28, 2003; Sabrina Tavernise, "Kurds Celebrate Elections of Mayor in Kirkuk," *New York Times*, May 29, 2003.

[131] Marc Lacey, "Plans for a British-Appointed Ruling Council in Basra Go Awry," *New York Times*, June 2, 2003; William Booth, "Iraqis in Basra Chafe at Authority," *Washington Post*, May 30, 2003.

[132] "Iraqis Protest over Running of Iraq," *The Times* (London), June 16, 2003.

[133] David Rhode, "Iraqis Were Set to Vote, but U.S. Wielded a Veto," *New York Times*, June 19, 2003.

[134] William Booth and Rajiv Chandrasekaran, "Occupation Forces Halt Elections Throughout Iraq," *Washington Post*, June 29, 2003; Paul Martin, "Democracy Starts to Take Shape as Councilors Meet in Baghdad," *Washington Times*, July 8, 2003.

[135] Booth and Chandrasekaran, "Occupation Forces Halt Elections Throughout Iraq."

to accusations that the United States had no intention of bringing democracy to the country.[136]

This somewhat haphazard variety of means for establishing local governments—including elections held in some areas and appointments by CPA or the coalition in others—confused and frustrated the local populations.[137] So CPA began to construct several teams, each with two purposes: first, to coordinate different agencies working on political development in Iraq, including the military, civilians, government agencies such as USAID, and NGOs mandated to work on governance issues; and second, to foster local political development and government building at provincial, district, and city levels more consistently throughout Iraq.

Each of these teams had four subdivisions. One subdivision was made up of CPA representatives, primarily from the United States, United Kingdom, and Italy. It was tasked with overseeing developments at the provincial level and acting as a liaison between officials at the federal and local levels. A second subdivision was USAID and RTI workers, whose job was to help Iraqis facilitate town hall meetings, build civil society, and teach political culture. A third subdivision was the Iraqi Reconstruction and Development Council (IRDC), a body created in February 2003, consisting of exiles who could offer technical expertise in running the government.[138] The fourth subdivision was made up of coalition military forces, including civil affairs personnel, already on the ground and aiding in political development. Finally, the governorate coordinator, who acted as the shadow governor of a province, oversaw the four subdivisions.[139] The size of these teams varied from province to province. Basra's team, for example, had around 500 members, while others were much smaller.[140]

Beginning in November 2003, following creation of the governance teams, CPA initiated a "refreshment" process for local governance in which current leaders were evaluated on their progress. The goal of the refreshment process was to increase the legitimacy of the provincial and city governments with their constituents, particularly those governments appointed by CPA.[141] In each of Iraq's provinces, the process called for the creation of local caucuses made up of individuals from each of the province's social, religious, and ethnic groups. These caucuses, in turn, debated the performance of current provincial leadership and made recommendations for

---

[136] David Rhode, "America Brings Democracy: Censor Now, Vote Later," *New York Times*, June 22, 2003.

[137] Interview with USAID official, July 2004.

[138] For more on the IRDC, see "The Iraqi Reconstruction and Development Council," *Middle East Reference*. As of October 2007:
http://middleeastreference.org.uk/irdc.html

[139] Interviews with CPA and USAID officials, July 2004.

[140] Interview with CPA official, July 2004.

[141] Interview with CPA officials, July 2004.

new appointments. Overall, CPA viewed the refreshment process as successfully achieving its goal.

To further increase the legitimacy of "refreshed" provincial governments, CPA also initiated a policy that allowed provincial governments to spend discretionary funds independently of the federal government's budget. CPA gave $1 million to each provincial government that successfully refreshed its leaders, with the possibility for additional funds after the initial sum was spent. This money allowed local ministries to create proposals for projects, get bids from Iraqi contractors, hire laborers, and oversee projects. This initiative had a twofold effect. First, work was completed more quickly if the proposal did not have to go through Baghdad for approval. Second, it empowered local governments to respond directly to constituent needs, increasing their legitimacy.[142]

The CPA also worked to resolve structural pathologies created by Saddam's regime between the federal and provincial governments. Under Saddam's rule, legislative and fiscal policies were established almost exclusively in Baghdad. This meant that nearly all decisions at the provincial level had to go through Baghdad before action could be taken. Provincial and city governments, therefore, were weak and strongly tied to the federal government. The CPA aimed to build greater autonomy at the provincial level, particularly in the area of fiscal policy. It also gave local governments the authority to dismiss province-level ministerial authorities and to fire officials found guilty of corruption.[143]

In addition to CPA's efforts to develop provincial and city governments throughout Iraq, the creation of a government for the city of Baghdad presented its own challenges to coalition authorities and received special attention. Prior to Garner's arrival in the capital on April 21, Iraqi leaders had moved into Baghdad with independent plans to govern the city. Most notably, while U.S. officials convened the April 15 meeting near Ur, Chalabi and other members of the INC headed to the newly liberated capital, claiming that U.S. forces had granted them authority to govern the city.[144] On April 16, Mohammed Mohsen Zubeidi of the INC announced that the United States had selected him as Baghdad's mayor, an assertion that a CENTCOM officer later denied.[145] On April 22, the Pentagon announced it had selected a U.S. official to run the city, Barbara Bodine, and that it did not recognize Zubeidi's authority. In response to this announcement, the INC claimed that Zubeidi was only loosely connected to its organization and that he was "acting with-

---

[142] Interviews with CPA official, July 2004, and civil affairs personnel, August 2004.

[143] These rights were ratified in CPA Orders 71 and 100. See www.cpa.org, accessed August 2004.

[144] Sharon Behn and Paul Martin, "Chalabi's Forces Enter Baghdad," *Washington Times*, April 17, 2003.

[145] Behn and Martin, "Chalabi's Forces Enter Baghdad."

out orders" and "off the reservation."[146] Despite this disclaimer, Zubeidi continued to speak as the de facto mayor of the city until U.S. troops had him arrested on April 27.[147] Amid this political turmoil, the city continued to suffer from crime, looting, and general lawlessness. Bodine's tenure lasted only weeks before she was replaced in early May amid local and international criticism over the lack of progress in restoring order to the city, as well as interagency disagreements in Washington.[148]

A governing council for Baghdad slowly took shape during the month of June. Bremer's office announced the creation of the Interim City Advisory Council on July 7, consisting of 37 delegates—including six women—representing 88 neighborhoods. The council's powers were primarily advisory, with the hope of providing a link between Baghdad's five million citizens, its leaders, and CPA.[149]

Despite these successes in creating a Baghdad city council, the new governing body had some shortcomings. It was given no budget and therefore lacked the power to implement its own programs. This, in turn, undercut the council's credibility; citizens of Baghdad quickly learned that their complaints to council members would not result in any changes. The CPA took action to resolve this problem by giving the council a small discretionary fund, but only after months had passed and council members and citizens had complained of the body's inability to implement change.[150] However, despite these shortcomings, the creation of the Baghdad city council was heralded by a U.S. official as "a major step in turning over the running of Baghdad to Iraqis."[151]

Overall, efforts to develop local governance in Iraq had mixed results. Throughout CPA's tenure, RTI and USAID boasted that positive developments on local governance issues were considerable. The May 2004 USAID bulletin, *Democracy in Iraq*, reported that USAID's Local Governance Program had assisted "445 neighborhoods, 194 sub-districts, 90 districts, and 16 governorate councils" in their efforts to achieve greater participation in Iraq's civic and political life.[152] More than 660 community action groups had been established in 17 governorates, which helped to initiate 1,395 projects aimed at building civil society. And the Iraq Transition Initiative had issued "more than $48 million in 700 small grants for civic education, civil society and me-

---

[146] Rajiv Chandrasekaran, "Baghdad Figure Won't Vacate Post," *Washington Post*, April 23, 2003.

[147] Rajiv Chandrasekaran, "Iraqis Set Timetable to Take Power," *Washington Post*, April 29, 2003.

[148] Patrick E. Tyler and Edmund L. Andrews, "U.S. Overhauls Administration to Govern Iraq," *New York Times*, May 12, 2003.

[149] Martin, 2003.

[150] Interview with civil affairs personnel, August 2004.

[151] Under Secretary of Defense for Policy Douglas J. Feith, quoted in Vernon Loeb, "U.S. to Appoint Council to Represent Iraqis," *Washington Post*, July 9, 2003.

[152] U.S. Agency for International Development, "Democracy in Iraq: Building Democracy from the Ground Up."

dia, women's participation, conflict mitigation and transitional justice," in addition to aiding in the reconstruction of schools, clinics, and municipal buildings.[153] USAID's annual report, *A Year in Iraq*, further boasted, "More than 80 percent of Iraq's adult population had been engaged—either directly or indirectly—in democracy or governance at the local level because of U.S. programs."[154]

However, many local governance issues remained unresolved at the time of the June 2004 transfer of authority. In particular, provincial and city governments were only beginning to develop and follow such democratic procedures as holding meetings, taking notes, filing reports, creating and approving budgets, and preparing for elections. Moreover, some provinces were still struggling to create local governments at the time of the handover, and the governments in some were being contested. The government of Najaf, in particular, faced challenges to its legitimacy from Moqtada al-Sadr and his Mahdi Army. CPA officials and international aid workers in Iraqi noted that such challenges could cause local governments to collapse if not properly mentored over the next decade. Despite the departure of CPA, agencies such as USAID and RTI remained in the country to help work on the many challenges that continued to face local governance in Iraq.[155]

Lessons Learned

Efforts to address governance issues in Iraq offer valuable insights into nation building in general and the transition to democracy in particular. This chapter concludes by considering six issues that plagued coalition authorities in their 15-month effort to create an independent Iraqi political system: the importance of security for political development; the role of exiles in a new government; bringing all ethnic and religious groups into the political process; flexibility in sequencing political developments and constructing reasonable timelines; publicizing political events and informing the public; and recognizing that civil society and political culture take time to foster.

Security and Political Development

The security environment posed perhaps the single greatest obstacle for ORHA and CPA in their efforts to rebuild Iraq's political system. Beginning with rampant looting and violence throughout the country following the fall of Baghdad on April 9, coalition forces lost time and the trust of the population by failing to control the se-

---

[153] U.S. Agency for International Development, "Democracy in Iraq: Building Democracy from the Ground Up."

[154] U.S. Agency for International Development, *A Year in Iraq*, May 2004, p. 11. As of October 2007: http://www.usaid.gov/iraq/pdf/AYearInIraq.pdf

[155] Interview with USAID personnel, July 2004.

curity environment. The looting caused substantial damage to the government's infrastructure—including the destruction of most federal buildings and police stations, and many museums, schools, and hospitals—which, in turn, required coalition authorities to rebuild many of these institutions from the ground up, thus delaying political development. Furthermore, rampant unchecked looting made the population question coalition forces' capabilities and intentions in invading and occupying the country, including coalition plans to create a new Iraqi government.[156]

Throughout the occupation period, the lack of security in Iraq continued to hinder coalition authorities' efforts to establish a new Iraqi government. Following the appointment of the IGC and the establishment of a judiciary, Iraqis holding positions in either became targets of abduction and assassination. As noted earlier, two members of the IGC were assassinated while in office and several judges, mayors, and other officials were targeted between April 2003 and June 30, 2004. These assassinations were not only costly given the loss of Iraqi leadership, but they also created a more general atmosphere of insecurity. Most Iraqis blamed coalition authorities for failing to improve the security environment, further diminishing trust in the motives and intentions of the coalition's occupying powers.[157]

Finally, the UN named the poor security environment in Iraq as one reason for delaying elections until January 2005. The UN envoy argued that elections would not be feasible if security could not be ensured at voting stations.[158] Thus, continued violence helped to delay one of the key demands of most Iraqis—to select their leaders through popular votes. The failure of coalition forces to improve the security environment adversely affected the safety and productivity of the Iraqi Interim Government. It also damaged Iraqi perceptions of the occupying powers' motives and intentions toward the country.

The Role of Exiles in the New Government

Coalition authorities, and the United States in particular, emphasized the role of Iraqi exiles in creating a new political system in Iraq. Prior to the initiation of major combat operations on March 19, 2003, the United States spent considerable resources building a cadre of Iraqi exiles that could help with military operations in the country and then run Iraq's various ministries and other infrastructure. Relying on exiles, however, caused several problems. First, and most important, most of these exiles had been out of the country for at least a decade and were both out of touch with the situation in Iraq and unknown by the population. Many exiles were given positions of authority in the interim government, including in the IGC. The number

---

[156] Kifner and Burns; Williams, "Rampant Looting Sweeps Iraq."

[157] Rajiv Chandrasekaran, "'Our Heritage Is Finished,'" *Washington Post*, April 13, 2003.

[158] Wadhams.

of IGC leaders unknown to the public was named as one problem contributing to that body's lack of legitimacy with Iraqi citizens.[159]

Second, the trust that the U.S. government placed in some of the Iraqi exiles—such as Ahmed Chalabi—may have been misplaced. Chalabi, for example, was accused of offering bogus intelligence to the U.S. government leading up to the war.[160] In May 2004, Iraqi police raided Chalabi's offices in Baghdad because they suspected he was embezzling funds and leaking intelligence to Iran.[161]

Bringing All Groups into the Political Process

Iraq has a population that is both ethnically and religiously diverse. This includes two main ethnic groups, Arabs and Kurds, the majority of whom are either Shi'ite or Sunni Muslims, in addition to other ethnic and religious minorities. Such diversity presented challenges to coalition authorities in their efforts to forge a transitional government in Iraq.

The most important challenge resulted from the fact that ORHA and CPA were slow to bring Sunni Arabs into the political process. CPA's controversial policies, such as completely disbanding an Iraqi army whose officer corps was predominantly Sunni Arab and its strict de-Ba'athification program, hit Sunni Arabs the hardest, blocking many from participating in Iraq's new political and security infrastructure. These obstacles were exacerbated by the Sunni Arabs' own lack of political organization. Prior to major combat operations in Iraq, most Sunni Arabs were represented by the Ba'ath Party. Under occupation, however, the Ba'ath Party was banned, effectively dissolving the Sunnis' main vehicle for representation. It was not until six months after the fall of Saddam that Sunni Arabs succeeded in forming their own political party, the State Council for Sunnis, which brought together various ethnic and religious factions within the Sunni Arab population. This put Sunni Arabs at a perceived disadvantage relative to other ethnic and religious groups, most of which had political organizations prior to the fall of Saddam that coalition authorities allowed to participate in the new Iraqi political arena.

Starting in April 2004, partly in response to growing insurgency and violence throughout the country, efforts were made to soften the blow of the de-Ba'athification policy and to assure former Ba'athists that they, too, would have a

---

[159] At the end of October, for example, a poll showed that the majority of Iraqis still did not know that the IGC existed. Murphy, "Baghdad's Tale of Two Councils."

[160] See Jack Fairweather, "Chalabi Stands by Faulty Intelligence That Toppled Saddam's Regime," *Daily Telegraph* (London), February 19, 2004; "Notebook," *The New Republic*, March 8, 2004; Douglas Jehl, "Pentagon Pays Iraq Group, Supplier of Incorrect Spy Data," *New York Times*, March 11, 2004.

[161] Scott Wilson, "U.S. Aids Raid on Home of Chalabi," *Washington Post*, May 21, 2004; David E. Sanger, "Chalabi's Seat of Honor Lost to Open Political Warfare with U.S.," *New York Times*, May 21, 2004; Jim Hoagland, "Cutting Off Chalabi," *Washington Post*, May 21, 2004; David Johnston and Richard A. Oppel, Jr., "U.S. Steps Up Hunt in Leaks to Iraqi Exile," *New York Times*, May 24, 2004.

place in the new Iraq. While the policy had always been that even senior Ba'athists could appeal decisions excluding them from public service, the process was an onerous one. CPA officials promised to make it easier, particularly in order to attract former military personnel, teachers, engineers, and bureaucrats, whose skills and expertise were deemed necessary. In addition, CPA officials hoped this change in attitude might mitigate some of the Sunni hostility toward the coalition. From April to June 2004, CPA officials considered a range of initiatives to promote reconciliation and to include roles for former Ba'athists and the Sunni public as a whole.[162] By this time, however, many Sunnis did not trust the occupation authorities, and feared that they would be excluded from or underrepresented in the new Iraqi government.

Flexibility in Creating a New Government and Establishing Reasonable Timelines

Within months of the fall of Baghdad, ORHA and CPA identified a series of milestones they considered essential to creating democracy in Iraq. The first task was creating a new Iraqi constitution, followed by selecting an interim government via caucuses, and ending with full national elections. Interspersed between these milestones were other developments, such as the appointment of an interim cabinet, the creation of a judiciary, and the training and deployment of new Iraqi security forces.

Such progressive steps, however, proved to be unnatural for Iraq's political development. Specifically, coalition authorities' insistence that Iraqis draft a constitution before selecting an interim government stalled the political process. Before CPA and the Bush administration stepped in and reordered the sequence of events in the November 15 agreement, the IGC struggled for five months to secure an agreement on who should be allowed to draft the new constitution without reaching definitive conclusions. The new agreement placed greater emphasis on the selection of an interim government, calling on the UN to help resolve disputes between CPA and the IGC over the method of selection, and postponing the drafting of a constitution until later. In place of a permanent constitution, the November 15 agreement called for creation of an interim constitution, which Iraqis succeeded in drafting and ratifying in March 2004. Thus, the political process was stalled for five months by CPA's insistence that a constitution had to come first. If coalition authorities had demonstrated greater flexibility in the sequence of events leading toward Iraqi democracy, this time might not have been lost.

The November 15 agreement between CPA and the IGC not only rearranged priorities for what needed to be accomplished in Iraq's political development, it also

---

[162] Coalition Provisional Authority Press Briefing with Brigadier General Mark Kimmett, Deputy Director for Coalition Operations, and Dan Senor, Senior Advisor, CPA, April 22, 2004. As of February 2008:
http://www.globalsecurity.org/military/library/news/2004/04/mil-040422-dod01.htm
Also see Edward Wong, "Policy Barring Ex-Baathists from Key Iraqi Posts Is Eased," *New York Times*, April 23, 2004; Jeffrey Fleishman, "Ex-Baathists Offer U.S. Advice, Await Call to Arms," *Los Angeles Times*, April 27, 2004.

created a timeline for achieving these goals. While some welcomed the introduction of a timeline and, specifically, the selection of June 30, 2004 as the date for CPA to hand over authority to an Iraqi interim government, in practice the timeline was unrealistic given what still needed to be accomplished. In just over seven months, CPA and the IGC had to determine what body would inherit Iraq's political reins, how that body would be selected, and which basic laws would govern the country in lieu of a constitution.

Moreover, this timeline put pressure on the training and deployment of Iraqi security forces. As discussed in Chapter Nine, coalition forces accelerated the training of Iraqi forces, in some cases cutting training time in half.[163] As the June 30 date for political handover grew closer, violence in Iraq increased significantly, especially in April.[164] This put pressure on both coalition troops and Iraq's fledgling security forces. It also revealed that some Iraqi forces were unwilling to oppose their own people in battle, throwing into question their ability to contain Iraq's insurgency in the near future.[165]

Thus coalition authorities lost time in the first months of occupation by insisting on a sequence of events for a political process that was ill suited to Iraq's needs. The November 15 agreement, in contrast, reprioritized the sequence of events but established an unreasonably short period of time in which to accomplish monumental tasks necessary for Iraq's political viability.

Publicizing Political Developments and Informing the Population

Throughout their tenure in Iraq, ORHA and CPA struggled with informing the country of developments in the capital. This problem surfaced immediately following the fall of Baghdad when coalition forces attempted to issue a decree calling for new tactics toward looters, but there was no mechanism by which to publicize the announcement.[166] Following the selection of the IGC, in July 2003, CPA failed to circulate information either within Baghdad or throughout the country on this important development. As a result, many Iraqis had no idea what the IGC was or who its members were. At the end of October 2003, more than three months after its creation, a poll showed that the majority of Iraqis still did not know about the IGC.[167]

---

[163] Craig Gordon, "Iraq Security Plan Carries Own Risks," *Long Island Newsday*, November 4, 2003; Andrew Koch, "Training More Locals Is Key to U.S. Strategy for Iraq," *Jane's Defence Weekly*, November 5, 2003; Peter Grier and Faye Bowers, "The Risks of Rapid 'Iraqification,'" *Christian Science Monitor*, November 6, 2003; John Daniszewski, "Iraqi Security Forces Far from Ready," *Los Angeles Times*, November 16, 2003.

[164] Zoroya, "Iraq's Deadliest Month."

[165] Bradley Graham, "Iraqi Security Forces Fall Short, Generals Say," *Washington Post*, April 13, 2004; Tony Perry, "Authority of Iraqi General Questioned by Myers," *Los Angeles Times*, May 3, 2004.

[166] Rubin, "U.S. Struggles in Quicksand of Iraq."

[167] Murphy, "Baghdad's Tale of Two Councils"; "Iraq's Interim Leaders Review Transition Plans in Face of Shiite Poll Demand."

The lack of information on political developments within the federal government thus presented another obstacle to popular participation in the political development of Iraq.

Moreover, ORHA and CPA had to contend with competing sources of information. Specifically they had to vie with two Arab satellite networks, Al Jazeera and Al Arabiya, which had cameras and reporters on the ground throughout the country and could broadcast events in near real time. Both CPA and the IGC had run-ins with these news stations, including the IGC's decision to temporarily ban them in September 2003 on charges of inciting violence against coalition forces.[168] The U.S. government made several attempts to provide alternatives to these news sources; in particular, it established the Iraqi Media Network beginning in May 2003. That network included radio stations, a local TV station, a daily newspaper, and, beginning in January 2004, a satellite station called Al Hurra.[169] Yet Al Jazeera and Al Arabiya remained the primary sources of information for the Iraqi population.

ORHA and CPA were thus slow in: creating means for spreading information on developments in the capital; competing with Arab media sources—which were seldom complimentary toward the occupation; and putting down rumors when they arose. These informational problems did not help further Iraq's political progress. In particular, they did not bring the country's population on board by improving access to political developments in the capital.

The Importance of Civil Society and Political Culture

Finally, as argued by scholars of democracy theory,[170] democracy cannot flourish without civil society, which consists of voluntary, "organized collective participation in the public space between individuals and the state,"[171] including such groups as "associations, clubs, guilds, syndicates, federations, unions, [and] parties."[172] Nor can it flourish without a democratic political culture, defined as "values and behavioral codes of tolerating—if not accepting—others and a tacit or explicit commitment to

---

[168] Chandrasekaran, "Iraqi Council Denies Access to Two Arab Satellite Networks."

[169] Interview with ORHA official, October 2003. For more on Al Hurra, see Jim Rutenberg, "The Struggle for Iraq: Hearts and Minds," *New York Times*, December 17, 2004; William A. Rugh, testimony to the Senate Foreign Relations Committee, April 29, 2004.

[170] See Larry Diamond, Juan J. Linz, and Seymour Martin Lipset (eds.), *Politics in Developing Countries*, Boulder, CO: Lynne Rienner, 1990.

[171] Saad Eddin Ibrahim, "Liberalization and Democratization in the Arab World: An Overview," in Rex Brynen, Bahgat Korany, and Paul Noble (eds.), *Political Liberalization and Democratization in the Arab World, Volume I: Theoretical Perspectives*, Boulder and London: Lynne Rienner Publishers, 1995, p. 29.

[172] Augustus Richard Norton, "Introduction," in Augustus Richard Norton (ed.), *Civil Society in the Middle East, Volume I*, Leiden: E.J. Brill, 1995, p. 7.

the peaceful management of differences among individuals and collectivities sharing the same 'public space'—that is, the polity."[173]

Both civil society and political culture had languished in Iraq under Saddam's rule. Civil associations were highly regulated by the state or banned altogether. Moreover, key political leaders in Iraq not aligned with Saddam and his Ba'ath Party were subjected to arrest, torture, and execution, forcing many elites to flee the country, which drained Iraq of intellectual and political capital. Likewise, in addition to restrictions on civil associations, the Iraqi people have developed a political culture that corresponds to more than 30 years of oppressive, dictatorial rule in which the only real means of attempting to change the system was to violently overthrow it. In other words, the population is not accustomed to organizing and working nonviolently for political change through the system.

Thus, while Iraq successfully held its first set of national elections in January 2005, this milestone will not necessarily make Iraq a democracy. Rather, Iraq's transition to a truly healthy functioning democracy will require educating the population, both formally and informally, to accept democratic values, norms, and institutions and encouraging the growth of civil associations, which could take a decade or longer to foster. Iraq cannot establish these values and institutions alone. The task may require a sustained commitment from the United States, its partners, and the international community, if Iraq is to become a stable and lasting democracy.

---

[173] Ibrahim, p. 29.

CHAPTER ELEVEN

Economic Policy

This chapter describes economic conditions at the time of invasion, the economic problems facing the Coalition Provisional Authority (CPA), and the policies that CPA adopted to solve these problems. The chapter concludes with a discussion of tasks that were not fully implemented by the transition to Iraqi rule at the end of June 2004.[1]

Economic Conditions in Iraq at the End of Major Combat

Iraq's economy was in poor shape at the time of invasion, as a consequence of economic mismanagement on the part of the Iraqi regime, a quarter of a century of conflict, and more than a decade of economic sanctions. Output and incomes had fallen dramatically from Iraq's economic heyday of the oil boom years of 1979 and 1980. The Iran-Iraq War had saddled the country with debt. The 1991 Gulf War had destroyed key pieces of the country's infrastructure. Sanctions and economic mismanagement curbed the investment needed to repair the damage and replace depreciated equipment. Although the statistics are problematic,[2] real per capita GDP fell from a peak of $4,080 in 1980 to $1,021 in 2002, a quarter of the 1980 peak. In real terms, incomes had fallen even more dramatically. Infant mortality, closely related to eco-

---

[1] This chapter is based on the experience of RAND analyst Keith Crane and his colleagues in the Economic Section of CPA between May 2003 and June 2004, official CPA memoranda, various official reports, and an article on the experiences of the author and three of his colleagues, namely, Chris Foote, William Block, and Simon Gray, who co-authored "Economic Policy and Prospects in Iraq," *Journal of Economic Perspectives*, Vol. 18, No. 3, Summer 2004.

[2] Iraq has collected and published a very extensive series of national and regional statistics for a number of decades. Some series, such as inflation, population, and industrial output, are fairly good. However, in the 1990s the figures on gross domestic product (GDP) deteriorated. The underlying data on service sector output was poor. Most importantly, the use of artificial, centrally dictated exchange rates to calculate the value of foreign trade, including oil exports, seriously distorted the role of the oil sector in the economy, distorting figures on per capita GDP.

nomic well-being, rose from 54 deaths per 1,000 live births in 1979–1984 to 108 in 1994–1999.[3]

Some of the decline in economic output and increase in infant mortality has been blamed on economic sanctions imposed by the United Nations. The UN imposed a comprehensive trade embargo on Iraq and froze Iraqi assets held in foreign banks on August 6, 1990, in response to Iraq's invasion of Kuwait. Following the ceasefire in August 1991, the UN passed Resolution 687, which relaxed the embargo, permitting Iraq to export oil to generate revenues to pay for war reparations and humanitarian needs. However, the UN tied the end of sanctions to Iraq's compliance with agreements to eliminate weapons of mass destruction. Because Saddam refused to cooperate, sanctions remained in force. Consequently, oil export revenues fell from $23.3 billion in 1988 to just $2.2 billion in 1992 and 1993. In the face of declining food consumption and health, in April 1995 the UN passed Resolution 986, which set up the Oil for Food program. Under this resolution, all export revenues were to be channeled through accounts controlled by the UN. Revenues were to be used to provide food rations to all Iraqis and pharmaceuticals and other medical supplies for Iraq's state-run public health system. However, Saddam haggled over the proposal until 1996, when he finally acquiesced to UN terms and the program began to function.

Compounding the effects of sanctions, Saddam's economic policies contributed to the destruction of the Iraqi economy. The government, through the Ministry of Trade, controlled imports: only favored individuals were permitted to purchase imports.[4] They were also given permission to purchase dollars at the artificial official rate of 3.22 U.S. dollars per Iraqi dinar, while the black market exchange rate ran 1,900 dinars per dollar. Business owners who supported the regime were able to obtain the large number of licenses and registrations required to operate legally. The central bank was subordinated to the government. If Saddam needed more cash, he demanded that the requested amount be printed. Not surprisingly, inflation ran to 458 percent in 1994 and stayed in triple digits throughout the first half of the 1990s.[5]

The 2003 war further exacerbated Iraq's economic problems. When major combat ended on April 9, 2003, the Iraqi economy had come to a standstill. Oil exports had been halted, as pipelines were closed. As a consequence, oil production had to be severely curtailed. Iraq's rickety telecommunications system shut down when the major switches were destroyed in the war or damaged during looting. Although

---

[3] World Bank Data Sheet for Iraq, as of January 29, 2004. As of October 2007:
http://lnweb18.worldbank.org/mna/mena.nsf/Attachments/Datasheet/$File/iraqprototype.pdf

[4] In addition, substantial portions of the economy, notably in the import-export trade, came under the control of criminal gangs linked to key regime figures.

[5] "Central Statistical Organization of Iraq," *Statistical Yearbook of Iraq,* 2002.

the electric power system had not been attacked, looters destroyed important controls; organized gangs began to blow up power line pylons to steal the metal, further reducing the availability of electricity. Without electricity, pumps needed to operate the water systems failed, resulting in sharp reductions in the availability of water. Government ministries that managed most of the economy were also looted, destroying many of the records and accounts needed for the economy to function.

Economic Policies Pursued by CPA

It does not appear that much thought was given to economic policy before the war. ORHA focused primarily on humanitarian assistance and immediate reconstruction needs, as discussed in Chapter Five. Once CPA was set up, L. Paul Bremer created the Office of Economic Policy under Peter McPherson, a former Under Secretary of the Treasury and administrator of USAID. Initially, economic policies were formulated by the Office of Economic Policy and presented in the form of action memos that were to become the basis for orders signed by Bremer. Subsequently, some of the functions of the Office of Economic Policy were delegated to other units. Thomas Foley was recruited to head up the Office of Private Sector Development. David Oliver was given responsibility for developing budgets in a new unit, the Office of Management and Budget. Day-to-day operations proceeded independently within each of these offices. However, committees were set up to create common positions for such cross-cutting economic policy changes as price liberalization.

McPherson focused on reducing government controls and trying to create a climate for foreign direct investment. However, the Office of Economic Policy faced three major hurdles to making major economic policy changes. It was hampered by concerns about CPA's authority to enact irreversible reforms under Article 64 of the Geneva Convention of 1949; CPA wished to ensure that economic policies were politically acceptable so that they would not be reversed when an Iraqi government took over; and economic advisors had to contend with the security situation, which hampered CPA efforts to work with Iraqi ministries. In order to address the issue of political acceptability, the Iraqi ministers and the Iraqi Governing Council (IGC) had to agree on all economic policy proposals. Although some of the ministers were supportive of change, the IGC generally supported the status quo. Opposition from the IGC greatly slowed economic policymaking and derailed a number of initiatives.

Restarting the Economy

CPA's immediate task was to restart the economy. The Iraqi economy can be divided into three major components:

- the state-owned energy sector, which is dominated by the Ministry of Oil but also includes the Ministry of Electricity, formerly the Commission of Electric Power;

- the other state-sponsored economic activities, including education and public health, but also the nonenergy state-owned companies; and

- the private sector, most of which consists of the informal sector, small unincorporated businesses involved in trade, transport, and consumer services.

Different policies were adopted to stimulate activity in each of these segments.

A first priority was to get the oil sector and public utilities operating again. Here, a number of agencies were involved. The Army Corps of Engineers had been given responsibility and contracting authority to resuscitate the oil sector. It issued a sole-source contract to Halliburton on March 24, 2003, not only to put out oil fires but also to restart oil production and exports.[6] USAID took the lead on electric power and water systems, issuing a sole-source contract to Bechtel on April 17, 2003 to rehabilitate the electric power sector and rebuild government buildings, including schools.[7]

Resuming the Provision of Government Services

Another priority, almost as important as the first, was to resume the provision of government services and payments. First and foremost, CPA had to start paying Iraqi civil servants. As the interim authority in Iraq, CPA had responsibility for making payments to government employees, pensioners, and employees of state-owned enterprises (SOEs) after the war ended. In late 2002, U.S. Treasury officials worked out a payments strategy that would be funded with assets seized from the Ba'athist government during the Gulf War. On March 20, 2003, President Bush issued an order that these assets, of which $1.7 billion had been identified, were to be "vested" in a special account, the Development Fund for Iraq (DFI) in the Federal Reserve Bank of New York. These assets were to be used to pay for reconstruction costs, government salaries, and other expenditures that would directly benefit the Iraqi people.

Because of the breakdown in government operations, the Iraqi Ministry of Finance lacked cash to make salary payments immediately following major combat operations. Consequently, the coalition used dollars and what dinars were available in government coffers to pay salaries. Dollar bills (mostly in ones and fives) were flown from the United States to Camp Arifjan in Kuwait. Coalition economic advisors learned that Iraq's existing cash-based salary system could function in the post-

---

[6] Elizabeth Rosenberg, Adam Horowitz, and Anthony Alessandrini, "Iraq Reconstruction Tracker," *Middle East Report*. As of October 2007:
http://www.merip.org/mer/mer227/227_reconstruction.html.

[7] Rosenberg, Horowitz, and Alessandrini.

combat environment with only a few modifications. The main tactical issues would be how to move the cash around the country and confirm the accuracy of employee lists. CJTF-7 played a key role in setting up distribution centers and guarding payrolls so that civil servants, pensioners, teachers, and other government employees could be paid. The major modification to the payments system imposed by the coalition was to compress the previous salary schedule, widely perceived as unfair to those at the bottom, into a temporary schedule of four pay grades.[8]

Controlling Inflation

Reviving the Central Bank of Iraq (CBI). CPA and the U.S. Treasury were worried about controlling inflation in Iraq. They rightfully feared that rapid rates of inflation would be highly unpopular, generating political unrest.[9] Rapid inflation can be debilitating. It raises the cost of investing, as lenders have to charge increasingly higher interest rates to recoup principal as well as the cost of capital from loans. Buyers and sellers in high-inflation countries find it difficult to react judiciously to changes in relative prices. Governments have great difficulty in budgeting. As a consequence, high-inflation countries are prone to severe boom-bust cycles and have lower rates of economic growth than low-inflation countries.

A strong central bank is the best antidote to inflation. CPA immediately reinvigorated the Central Bank of Iraq. During the looting, the CBI had shut down. CPA called employees back to work, appointed an interim governor, and ordered the central bank to stop printing new currency immediately in order to forestall a surge in inflation. Over the course of the next several months, CPA helped introduce new accounting, statistics collection, and operating systems in the CBI. In addition, CBI branches were linked electronically with the headquarters in Baghdad, beginning the process of integrating activities of the branches with headquarters on a daily basis. These measures started the process of building a professional central bank in Iraq.

Currency exchange. The CPA quickly realized that the old Iraqi currency, the dinar, needed to be exchanged for a new currency. Immediately following the conflict, presses, plates, and paper for printing the 10,000-dinar note, worth about $5 in mid-summer 2003, were stolen from the central bank's Dar Al-Nahrain printing works.[10] As a consequence, counterfeit Iraqi 10,000-dinar notes were of the highest quality. To forestall massive counterfeiting, CPA needed to replace old currency with new as soon as possible.

---

[8] Foote, Block, and Gray.

[9] The Treasury team working on Iraq was very concerned about keeping inflation under control. A representative from the team on loan from the Federal Reserve's Board of Governors, Thomas Simpson, was sent to the October 2003 meeting between the Board of Governors of the Central Bank of Iraq and the IMF, which was held in Amman, Jordan.

[10] Foote, Block, and Gray.

Notes from the 1990s carried Saddam's visage, but earlier notes did not. Fortunately, plates for the earlier notes had not been stolen and were available. CPA contracted with a European firm to use these plates to print millions of new bills in a number of denominations. The new bills were airlifted to Jordan and Kuwait in 747s. Because U.S. military aircraft were completely engaged, CPA chartered aircraft and pilots from a Belarusian company to fly the new notes to bank branches around the country and return old notes that had been rendered worthless by dipping them in a red dye. The whole operation lasted three months, from mid-October 2003 to mid-January 2004. CJTF-7 played a key role in the success of the operation, providing security to banks and for local deliveries of cash.

Although the exchange was successful, a substantial amount of counterfeit cash appears to have been converted into the new currency, as the total amount converted exceeded the central bank's account of the old currency that had been issued. Either some of the new currency was exchanged for counterfeit as reported, or some old currency was not destroyed but instead returned to circulation and exchanged for new again.[11] Those who were able to make these illegal exchanges benefited handsomely, since they exchanged worthless paper for real money.

Foreign exchange auction. Aside from regulating the quantity of dinars it prints, the CBI had little in the way of policy instruments to control the supply of money. Iraqis have kept savings in dollars since before the Gulf War because they lack confidence in the domestic currency. Consequently, Iraq has been a dual-currency economy, using both dollars and Iraqi dinars for transactions. Thus, the total money supply in Iraq is determined by stocks of dollars as well as dinars and the rate of exchange between the two currencies. In dual-currency economies the rate of exchange is extraordinarily important for inflation. Households and businesses tend to hold most of their savings in dollars because dollars hold their value and can be used outside and inside Iraq. In dual-currency economies, even small declines in the exchange rate can trigger a flight to dollars, sometimes triggering a run on the local currency. Since imported goods are paid for in dollars, a decline in the value of the dinar immediately shows up in increased dinar prices for imported goods. Most consumption goods in Iraq, including food, clothing, and pharmaceuticals, are imported. Thus, an increase in the prices of imports has an immediate impact on consumer welfare.

Iraq did not have an official foreign exchange market under Saddam; the government controlled exchange rates by fiat. Dollars were sold unofficially by moneychangers; rates varied from city to city depending on local conditions. With the assistance of CPA, the CBI started a foreign exchange auction on October 4, 2003. By establishing a central foreign exchange market, the CBI could better monitor ex-

---

[11] "Counterfeit Trail Led to Chalabi," *The Sunday Times* (London), May 31, 2004.

change rate developments, provide the country with a clear, daily indicator of trends in the exchange rate (a key economic indicator), and attempt to influence the exchange rate to moderate inflationary forces. Volumes grew rapidly, in great part because the Ministry of Finance needed to convert a steady stream of the dollars it earned from oil exports into dinars to make salary and other payments in the domestic economy.

The auction became a success, with tens of millions of dollars exchanged on a daily basis. However, it did have some teething problems. Historically, Baghdad's money dealers congregated around the headquarters of the Rasheed Bank; an informal foreign exchange market among moneychangers had developed there. In order to eliminate settlement risk, only banks with accounts at the CBI were permitted to participate in the auctions. The moneychangers, who did not have banking licenses, had to bid for dollars through banks. Many of those in Baghdad met before the auction. Their aggregated bids went through the Rasheed Bank and frequently accounted for half of the total bids received, thus raising some concern about collusion on the part of those bidding through Rasheed. For its part, the CBI was not always transparent. Initially, the CBI chose a sell rate as much as 10 percent below the market rate to encourage participation. This rate often fell below both the buy and sell rates on the local market. Thus, banks had access to dollars sold at a discount. The discount was gradually eliminated, and by the first quarter of 2004 some dollars were sold as well as bought on the local market.

Tariffs and Taxes

Most of the Iraqi government's budget is financed from oil export revenues. The government does not tax oil: it takes all export revenues directly into government coffers. Revenues from traditional taxes were very small in that part of the country controlled by Saddam; the quasi-independent Kurdish area in the north had its own tax system, including tariffs and income taxes.

On June 8, 2003, CPA declared a tax holiday for the remainder of the calendar year. Tariffs and import licensing requirements were lifted. Because neither the tax administration nor the customs agency were functional at the time, the decision was partially one of necessity. However, the economic rationale was sound. The combination of UN sanctions and Saddam's tight control over foreign exchange and imports had prevented Iraqis from freely purchasing foreign goods. CPA's decision facilitated the beginning of the integration of Iraq into the world economy. Opening the country to imports was a key factor spurring economic activity, as wholesalers, transporters, and retailers expanded their operations to satisfy consumer demand.

Some observers criticized CPA's decision to eliminate tariffs, arguing that this would expose Iraqi manufacturers to unwarranted competition.[12] Because Iraq does not produce most of the imported products, these arguments are not very convincing. SOEs that manufactured products like clothing, which might compete with imports, were often sold only to the old Iraqi military because quality was so poor that consumers would not buy these goods.

The Budget

A large number of the few dozen CPA staff members involved in economic policy were engaged in writing and implementing the Iraqi budget. The first budget was for the second half of 2003. It was prepared using the traditional budgetary procedures of the Ministry of Finance. Expenditures and revenues were calculated in dollars. In tandem with the second-half 2003 budget, CPA and the Ministry of Finance began to write the 2004 budget. CPA was intent on restoring a more "normal" budgeting system for Iraq. Consequently, this budget was issued in the national currency, the dinar, not in dollars. Concurrently, CPA employed BearingPoint, a consulting firm, to introduce and train Ministry of Finance personnel in a modern Financial Management Information System (FMIS). The system was rolled out during the second half of 2003 and the first half of 2004. It improved control over budgetary expenditures, sped information flows, and provided a better basis for the development and implementation of future Iraqi budgets.

Like the second-half 2003 budget, the 2004 budget was put together quickly. Under the assumption that most of the costs of reconstruction would be financed from foreign assistance, most of the budget consisted of operating expenditures and social transfers. CPA counted on the U.S. supplemental appropriation and a donors' conference held in Madrid on October 23 and 24, 2003, to provide funding for reconstruction. Consequently, CPA staff were under the gun to complete the 2004 budget in time for the donors' conference so as to provide donors with information on areas that needed additional funding.

Law on Foreign Direct Investment

On September 21, 2003, Iraqi Finance Minister Kamal al-Kilani announced that CPA had issued an order on foreign investment, providing the same treatment for foreign-owned companies as for domestically owned companies. The order allowed foreign investors to own 100 percent of businesses outside of the natural resource sector. Investors faced no barriers or additional taxes on remitted profits. However, foreign investment in the oil industry remained prohibited, and strictures were also applied in banking and other financial services. In addition, foreigners were prohib-

---

[12] A.K. Gupta, "The Great Iraq Heist," *Z Magazine*, January 2004.

ited from purchasing land, and foreign retailers had to post a $100,000 bond with the government.

Although Iraq had allowed individuals and companies from Arab countries to invest in Iraq, going so far as to promulgate an "Arab investment law" in 2002, investors from non-Arab countries had been barred from legally investing in the country. This law was not universally popular. Many Iraqi businessmen feared foreign competition, and the populace remained concerned that the oil sector would be sold to foreign investors. The order was one of the most controversial reforms advanced by CPA and the IGC. To date, it has had a modest effect on Iraq's economy. Because of the security situation, foreign investment in Iraq has been very limited. Ironically, the most prominent investments (mobile telephones, banking) have been primarily undertaken by Arab investors.

Economic Policy Changes That Were Not Fully Implemented

Price Liberalization

Under Saddam, Iraq operated with a blend of market and controlled prices, some of which were highly distorted. In addition, food was distributed virtually gratis through the Public Distribution System, a food rationing system set up by Saddam immediately following the Gulf War. Outside of food, which is discussed in detail below, the most distorted prices were for energy. As of mid-year 2004, gasoline and diesel fuel were sold at less than a nickel a gallon.

The sale of refined oil products at such distorted prices had many pernicious effects on the Iraqi economy. Because fuel is so cheap, fuel consumption is much higher than it would be if prices were set by markets. To satisfy this additional consumption, the Oil Ministry had been pressed to invest in additional refining capacity that under market prices would be unnecessary. However, at these low prices, no amount of additional capacity can satisfy demand. As a consequence, gasoline and diesel fuel are allocated by queuing as customers attempt to obtain limited supplies before they run out.

Because domestic fuels are much cheaper than those sold at market prices in neighboring countries, gasoline and diesel is smuggled out of the country to markets where it can be sold at a profit. The Iraqi government has attempted to prevent smuggling by setting up cumbersome regulatory schemes and using customs agents to prevent the export of refined oil products. These schemes have not functioned well. Exporters have generally found it possible to avoid such restrictions through smuggling and bribing customs and other government officials.

The Iraqi government suffers enormous losses in forgone revenues from price controls, an estimated $4.9 billion in 2003, close to 25 percent of GDP. Moreover, the implicit fuel price subsidies go to the rich, not the poor. In Iraq, car ownership is concentrated in the top 20 percent of households by income, while poor Iraqis do not own cars. Consequently, upper-income households benefit more from fuel subsidies than lower-income households.

Although Bremer supported the concept of price liberalization, this policy change was never implemented even though CPA advisors presented a number of policy options for liberalizing prices. Opposition to immediate change came from a variety of sources. Regional coalition military commanders and civil affairs officers were concerned about the popular reaction to price increases. Riots in Basra in the summer of 2003—ironically over the length of lines for gasoline, a consequence of fuel price controls—had made commanders wary of tampering with the existing system. However, by early 2004, CJTF-7 was increasingly supportive of gradual price increases as the military recognized the security dangers of the long lines and massive imports of gasoline and diesel that were the result of not raising prices. Diplomats assigned to CPA also frequently took a negative view. Many had served in countries where price increases had been followed by riots. Although dozens of countries have successfully liberalized gasoline and diesel prices over the last three decades, diplomats focused on such countries as Nigeria and Zimbabwe where price increases had been followed by violence. Finally, the IGC had no stomach for difficult policy changes and generally opposed price increases of any kind.

Reforming the Food Rationing System

Under Saddam Hussein, Iraq had since 1991 had a system of publicly provided food rations, known as the Public Distribution System (PDS) for food. Saddam set up the system following the Gulf War to ensure that every Iraqi had a minimum amount of food. The system was needed because of the economic consequences of the UN trade sanctions. In August 1991, the UN relaxed the trade embargo; it passed Resolutions 706 and 712 permitting Iraq to sell oil through an escrow account to be used for food imports. Saddam rejected this arrangement. In April 1995 the UN and Iraq came to an agreement in the form of Resolution 986 to create the Oil for Food (OFF) Program, which with some modifications is the basis for financing the current system.

Under OFF, oil revenues deposited in escrow accounts abroad could be used to purchase food, medical supplies, and parts and equipment needed to keep the economy running.[13] In the Kurdish areas in the north, the UN's World Food Programme (WFP) hired local staff to distribute food and other supplies through leased warehouses. In the south, food and other supplies were shipped to the port of Umm Qasr; the Ministry of Trade handled distribution and transportation from the port to

---

[13] For more on OFF, see the UN's web site on the program. As of February 2008: http://www.un.org/Depts/oip/index.html

warehouses from which food was distributed to licensed food distributors (shops and bakeries).

Under the PDS, Iraqis received a monthly ration of rice, flour, sugar, legumes, soap, and tea. Families with infants received dried milk, baby food, and detergent. According to WFP vulnerability assessments, most Iraqi households continued to depend on the program in 2004. Based on surveys contracted by the WFP and program data, prior to OIF, 80 percent of Iraqis picked up their food rations and 60 percent depended on the ration for a major share of their food.[14] The Iraqi Ministry of Trade continued to operate under these assumptions. Initially, under the OFF agreement everything had to be imported so that the UN could monitor what was being purchased with the escrow monies. Subsequently, some home-grown foods were eligible for purchase. As of June 28, 2004, more food was being purchased domestically, but most products were still imported.

The program has successfully fed the Iraqi populace and contributed to reducing infant mortality. However, it has a number of negative side effects. First, it deprives local farmers of most of their market, retarding increases in agricultural output and penalizing some of the poorest people in Iraq, its farmers. Poor families who need cash for other expenditures are forced to sell part of their ration for money. Because Iraq is awash in rationed commodities, sellers receive only a pittance for these goods. The products are often of very low quality. In the north, which is a little better off, most families leave their ration at the local store, which then gives them a discount on better quality foods and soaps. Traders circulate among the local stores, purchasing this food at knockdown prices and then trucking it to Iran and Turkey for sale to poor people in those countries.[15] The program has also been subject to corruption up and down the entire supply chain.[16]

CPA failed to change this system. In September 2003, Bremer agreed that the in-kind system should be replaced with a cash payment. This policy change would permit Iraqis to purchase locally grown food of their own choice. It would also provide them with funds to obtain items other than food, such as shoes and clothing. Various options were debated for changing the system, and one option calling for a gradual replacement of food rations with a cash-based system, neighborhood by neighborhood, was presented to the Iraqi Minister of Trade in November 2003. However, the IGC did not wish to make major policy changes in 2004 because of its concerns that any policy change would serve to increase political discontent with

---

[14] Rajiv Chandrasekaran, "U.N. Official Warns of Iraqi Food Crisis," *Washington Post*, February 28, 2003.

[15] Interviews with Iraqi recipients, October 2003.

[16] U.S. General Accounting Office, "United Nations: Observations on the Oil for Food Program and Iraq's Security," testimony before the Committee on Agriculture, Statement of Joseph A. Christoff, Director, International Affairs and Trade, U.S. General Accounting Office, June 16, 2004, p. 1.

CPA, thereby fanning the insurgency. In addition, because of the complexity of moving to a new system, policy change in this area was not pursued further by CPA.

Rationalizing State-Owned Enterprises

Virtually all industrial assets in Iraq were and still are owned by the state: the oil industry, the electric power industry, and most manufacturing enterprises. As the representative of the occupying power, CPA was responsible for managing these SOEs. Under Saddam the SOEs were managed through ministries: the Ministry of Oil, the Commission of Electric Power, the Ministry of Trade, and the Ministry of Industry and Minerals. CPA maintained this management strategy, with some modifications. In the case of critical enterprises, such as those involved in oil production or electric power generation and distribution, CPA placed a priority on restoring former output levels, focusing the reconstruction effort on these industries. The Army Corps of Engineers, USAID contractors like Halliburton (oil) and Bechtel (electric power and other infrastructure), and the Iraqi ministries all worked to restore and improve operations. However, none of these entities tried to introduce organizational changes. In general, the approach was business as usual. However, improved systems of control and accounting were often installed along with new equipment.

Most of the remaining SOEs received relatively little attention. The Ministry of Industry and Minerals was responsible for 180 factories, all of which suffered under severe challenges. All contended with a lack of security. The Facilities Protection Services, the security force in charge of protecting equipment and employees, proved incapable of ensuring security at most factories, as discussed in Chapter Nine. Most enterprises, along with the rest of Iraq, did not have reliable power. Hence, they were unable to operate on a steady basis. Some had no power at all. Most of the top management was fired during de-Ba'athification. However, even if those managers had not been fired, it is not clear whether they would have managed the companies better than their successors. The Iraqi system of state control of enterprises was extraordinarily centralized: revenues went directly to the ministry, and the ministry, not the enterprise, was often the source of operating as well as investment expenditures. Management had never operated in the free market and had no idea how to respond. Top managers often did not even know the source of their raw materials or who their final customers were.

Under this system, neither the ministries nor local management provided employees with incentives to perform. Of more than 107,000 employees in the Ministry of Industry and Minerals, few cared about restarting operations. They were paid to stay home. When the Mishraq Sulfur Company staff was asked to go back to work, they destroyed the facility and set $40 million worth of sulfur on fire, completely destroying it.[17] They did not wish to return to work.

All the SOEs had suffered extensive damage from looting, the war, and 13 years of sanctions, during which time their capital stock deteriorated as production lines were cannibalized to keep the remaining lines open. Even before the war, these enterprises were not very productive. They survived on government subsidies and favorable treatment. There was little to suggest that such enterprises could be made productive. In some instances, enterprises destroyed value: the cost of imported components and materials needed to produce the final product were more valuable than the product that was manufactured.

Only a handful of people were assigned as advisors to the Ministry of Industry and Minerals. USAID money could have been used to pay for local advisors, but other tasks were given higher priority. Officers in the divisional commands ultimately became responsible for many of these enterprises. They petitioned CPA and the ministries for funds, fuels, and electric power. They attempted to restart production and protect the remaining assets of the facilities.[18]

Senior officials in CPA did examine a number of options for restructuring the SOEs. Because these enterprises offered little chance for success, CPA focused on other issues. A substantial amount of press attention was paid to suggestions that the SOEs be privatized. However, CPA never attempted to seriously implement such a policy. Most enterprises were in no shape to be sold: they had no accounts and no balance sheets. An investor would have had no way of knowing what he or she was buying. Giving the enterprise to employees or managers or using vouchers could have circumvented the immediate problem of defining the enterprise's assets, but there seemed to be little or no interest among Iraqis in privatization. Rather, opinion polls and public statements by Iraqis indicated that privatization was highly unpopular.

CPA intended to adopt a policy of downsizing SOEs, especially because of the way in which they were operated. SOEs were going to be forced to become self-sufficient in 2004. According to the original 2004 budget, the state would no longer provide SOEs with funds from the budget after December 31, 2003. Once cash ran out, unprofitable SOEs would have had to cease paying wages and close. Due to opposition from the IGC, however, this policy was never implemented. SOEs remain a drain on the budget and show little prospect of ever becoming competitive.

---

[17] Memo from CPA staff advising Ministry of Industry and Minerals.

[18] The British enjoyed more success in their sector with SOEs. The Department for International Development assigned a number of advisors to SOEs in that region. Many were restarted and some become profitable.

Lessons Learned

CPA had successes and failures in economic policy. CPA quickly restored the operations of the Ministry of Finance, especially the payment of government employees. Once government employees received their paychecks on a regular basis and ministries had access to operating budgets, ministers were able to reopen schools and clinics, pay pensioners and demobilized soldiers, and distribute food rations. CPA also successfully revamped operations of the Central Bank of Iraq, exchanging old currency for new and setting up a foreign exchange auction. CPA drew up two budgets (second-half 2003 and full-year 2004) and implemented the former. Iraqi households benefited from the influx of consumer goods, abetted by the policy of zero tariffs. In many instances, these goods had not been readily available to Iraqi households under Saddam. The revival of trade that zero tariffs helped facilitate contributed to a surge of economic activity in retailing, wholesaling, and transport.

CJTF-7 played an important role in many of these successes. Coalition forces played primary roles in guarding, transporting, and distributing wages and pensions to government employees, pensioners, and demobilized soldiers. CJTF-7 also guarded banks from the day the currency exchange began. Without CJTF-7's support, CPA would not have been able to restart the provision of government services so quickly.

CPA failed to implement a number of key policies. The failure to liberalize energy prices was the most significant. Price controls and subsidies on refined oil products remain one of the most debilitating economic policies in Iraq. They are extraordinarily expensive, equal to almost a quarter of GDP in 2003; they encourage smuggling and corruption; and their benefit primarily accrues to higher-income individuals. Shortages of gasoline due to price controls have resulted in riots and deaths of coalition soldiers. The persistence of these subsidies will retard economic growth and severely distort Iraq's economy in the years ahead. In this instance, resistance to raising prices within CPA, CJTF-7, and Washington had deleterious implications for the economy.

The failure to provide a better alternative to the Public Distribution System for food was also unfortunate. The existing system functioned; there was a consensus that any new system had to be in place and functioning before the existing system was dismantled. Most proposals for new systems entailed setting up processes to distribute cash. These new systems would involve setting up new distribution procedures and making major changes in supply chains. CPA and CJTF-7 staff needed to see that the new systems worked before they would be willing to dismantle the old one.

The failure to improve the operations of state-owned enterprises was less understandable. Some attempt was made to categorize the enterprises into salvageable and unsalvageable, but virtually nothing was done to provide enterprises with the accounting, billing, and purchasing systems they needed to become independent operating units. Managers did not have access to training or advisors. CPA also failed to set up employee buyout and restructuring programs that might have encouraged employees to exit the wage rolls so that enterprises could be either officially closed or restructured to become profitable.

The failure of CPA to make progress on liberalizing prices, revamping the food ration system, or restructuring state-owned enterprises left a legacy of misguided economic policies to the new Iraqi government. Although a number of Middle Eastern countries suffer from the ill effects of similar programs, Iraq has some of the most dysfunctional economic policies and institutions in the world. Although the deteriorating security situation made it increasingly difficult to change these policies and institutions, the new Iraqi government has been left to face the pernicious legacies of Saddam Hussein. Based on the poor track record of other Middle Eastern governments in tackling problems posed by fuel subsidies, food supports, and SOEs, it is doubtful that these issues will be addressed by any Iraqi government for the foreseeable future.

The misallocation of resources caused by implicit subsidization of gasoline, diesel fuel, and the food ration system will impose a heavy economic price on Iraq. As noted above, roughly $5 billion dollars of potential government revenues are squandered by keeping gasoline and diesel prices at 100 Iraqi dinars per liter, less than a nickel a gallon. This is equivalent to more than a quarter of the U.S. supplemental funding and almost a quarter of Iraq's entire budget. In light of the many needs for reconstruction and government services, these subsidies constitute an enormous waste of resources.

Although continued provision of food rations has successfully forestalled widespread hunger and malnutrition, this program is on the verge of outliving its usefulness. The program centralizes food procurement in the hands of the Ministry of Trade. Most purchases consist of imported items, many of poor quality. Iraqi farmers, retailers, and wholesalers find that most of their potential market is supplied through imported products provided through the food ration program, depriving them of their primary markets and forestalling the development of agriculture and retailing.

CHAPTER TWELVE

Essential Services and Infrastructure

This chapter describes the condition of Iraq's infrastructure at the time of the invasion, the problems facing CPA and coalition forces in reconstructing this infrastructure and restoring essential services, the means that CPA and the U.S. government employed to achieve these goals, and their accomplishments in terms of restoring basic services through June 28, 2004.[1]

Status at the End of Major Combat

Before the onset of combat operations in 2003, Iraq's infrastructure was in such disrepair that electricity was available for only a few hours a day, and many Iraqis no longer had access to piped, potable water. Before the 1991 Gulf War, times were better. In 1990, Iraq had a functional installed generating capacity of 9,295 megawatts (MW) with peak demand of about 5,100 MW.[2] At that time, 87 percent of the population had access to electric power and 95 percent of the urban population and 75 percent of the rural population had access to potable water.

By the end of the Gulf War, wartime damage had reduced generating capacity to 2,325 MW. In the immediate aftermath of that conflict, the Iraqi government was unable to invest in repairing infrastructure because of the UN trade embargo. However, by 1996 the Saddam government had agreed to the Oil for Food program, which placed oil export revenues in an offshore account under the control of the UN. Iraq, with UN approval, could use funds not only for food imports, but also to purchase equipment for the repair and construction of critical infrastructure. On the eve of the 2003 war, $8.9 billion of contracts were outstanding, of which about $4.9 billion were for equipment and other nonfood items.[3]

---

[1] This chapter is based on the experiences of RAND analyst Keith Crane and his colleagues at CPA between May 2003 and June 2004, official CPA memoranda, various official reports, and press accounts.

[2] United Nations and World Bank, *Joint Iraq Needs Assessment*, October 2003, pp. 19 and 28.

[3] "Secretary General to Administer Iraq's Oil-For-Food Program," International Information Programs, U.S. Department of State, March 28, 2003.

Despite these investments, Iraq's infrastructure was severely degraded by 2003 compared to its condition in 1990. U.S. engineers were both amazed and horrified at the stopgap measures devised by the Iraqis to keep the electric power system up and running, from ingenious substitutes for important components to running power lines around circuit breakers, endangering the entire system. Needs assessments conducted by U.S. contractors and the United Nations and World Bank found multibillion-dollar backlogs of maintenance, refurbishment, and new investment needed to restore power supplies, increase supplies of potable water, and treat sewage. Investment in telecommunications and the oil sector had also been very limited. As a consequence, on a per capita basis the provision of electricity and water had fallen sharply. Although the population has grown by half since the end of the Gulf War, by 2002 functioning generating capacity had only been restored to 4,500 MW.

As noted in Chapter Eleven, OIF and the subsequent looting resulted in additional damage to Iraq's infrastructure and reductions in output of electricity and oil, and in the provision of water. After the looting ended in mid-April, the Iraqi economy had come to a standstill. Oil exports had been halted because pipelines were closed. Oil production had to be severely cut back. Iraq's rickety telecommunications system had shut down because the major switches had been destroyed in the war or damaged during looting. Although the electric power system had not been attacked by the coalition, looters destroyed important controls. To compound the problems posed by looting, organized gangs knocked down power line pylons to steal the metal in the cables, further reducing the availability of electricity. In early May 2003, electric power generation was one-third of output levels in the previous year; for the entire month, output was 56 percent of May 2002 levels.[4] Without electricity, pumps needed to operate the water systems failed, resulting in sharp reductions in the availability of potable water. Sewage treatment plants were also unable to operate properly.

Prewar Assumptions

Although substantial planning was undertaken to forestall a humanitarian catastrophe and to prevent destruction of the oil fields, there was little planning prior to the war for the restoration of essential services.

An interagency team started working on humanitarian assistance and reconstruction in the fall of 2002, as discussed in Chapter Three. Plans were developed for water and sanitation, electricity, telecommunications, shelter, and transportation, in addition to health, education, governance and the rule of law, agriculture and rural development, and economic and financial policy. Among other agencies, USAID was

---

[4] Data provided by Iraqi Ministry of Electricity.

involved in this process. Subsequently, that planning process and some of the associated staff were transferred to ORHA when it was set up early in 2003.[5]

ORHA put a great deal of effort into planning for a potential humanitarian disaster, as discussed in Chapters Five and Six. The U.S. government estimated that 800,000 Iraqis were internally displaced and 740,000 were refugees, primarily in Iran, at the beginning of 2003, and that as many as 2,000,000 other Iraqis might flee abroad or internally.[6] ORHA planned to provide water, food, and shelter to refugee populations; to counteract the effects of chemical and biological weapons attacks on the civilian population; to provide emergency health care; and to restore water and sewage services.

ORHA did attempt to identify infrastructure before the war that would be needed to keep the country functioning. In contrast to the 1991 Gulf War, CENTCOM chose not to target critical infrastructure (except for telephone switches) so that electric power and other services could be quickly restored to the Iraqi population following the war. Before OIF, ORHA knew that electrical generation plants, transmission lines, and water treatment and sewage plants would not be struck.[7]

Before the war, the Department of Defense engaged in detailed planning to avoid damage to the oil industry. Units were assigned to quickly seize and guard oil fields and installations to prevent sabotage. DoD was especially concerned about Iraqi agents setting fire to oil wells as they had done in the Kuwaiti fields during the Gulf War. In the event, only nine fires were lit in the important Southern Rumaila field,[8] in part because of the speed at which U.S. forces moved and in part because Iraqi oil engineers were loath to destroy their own facilities.[9]

Aside from planning to protect the oil industry and avoid targeting utility installations, little effort was spent planning to restart and rehabilitate Iraqi infrastructure. One reason for the lack of planning for post-conflict reconstruction was that DoD underestimated the magnitude of the reconstruction program needed to meet its prewar goals. Prior to the conflict, the Deputies Committee set the goal of restoring essential services to their prewar levels, and it subsequently added the goal of rehabilitating 6,000 schools. However, planners assumed restoration of pre-conflict service levels would be feasible by repairing damage to infrastructure suffered during the 2003 war or still stemming from the Gulf War. Since CENTCOM was not

---

[5] U.S. General Accounting Office, *Rebuilding Iraq: Fiscal Year 2003 Contract Award Procedures and Management Challenges*, GAO-04-605, Washington, D.C., June 2004, p. 6.

[6] Woodward, p. 276.

[7] Interview with ORHA official, July 2004.

[8] Christina Reed, "Burning Assets: Oil Fires in Iraq," *Geotimes*, May 2003.

[9] Interviews with CPA officials and CENTCOM staff, October 2003.

planning on targeting infrastructure and planners did not envisage damage from looting, this task did not seem that daunting. DoD did not intend to rehabilitate Iraq's entire electric power or water systems.

When CPA was created in May 2003, it took on a number of responsibilities for governing the country, as dictated by the Geneva Convention.[10] CPA interpreted its responsibilities to include the provision of public services that had been provided by the former regime: electricity, water, garbage pickup, and sewage, where municipal facilities existed. CPA set itself the task of increasing the average hours of electric power available by geographic region and the share of population with potable water and adequate sewage compared to conditions under Saddam.

Political imperatives forced CPA to focus on turning on the lights and taps and restoring oil production. Demonstrations and opinion polls all pointed to high expectations from Iraqis for a speedy restoration of services, followed by considerable improvement. When these expectations were not met, levels of dissatisfaction were high. During the summer of 2003, shortages of gasoline and diesel fuel were blamed for triggering riots and disturbances, including a major riot in Basra that led to a number of deaths.[11] In short, CPA's goal shifted from repairing damage inflicted during the two wars to providing Iraqi citizens with a basic level of services, especially electricity and water.

CPA's ability to achieve its goals was hampered by poor information. Information on the extent and condition of Iraq's electric power and water systems was fragmentary and often wrong. U.S. policymakers operated under the assumption that the electric power and water systems were in much better shape than was the case and that Iraqi personnel were technically proficient.[12] They also assumed that the Iraqis would be willing and able to bring the systems back on line quickly. As CPA attempted to restore power and water, engineers found that the infrastructure for them was operating on shoestrings and sealing wax; it would need massive investments just to maintain previous levels of service.

The use of erroneous information for planning appears to have stemmed from ignorance of other information sources, which was compounded by attempts to limit the number of individuals involved in the planning process.[13] Better information was available from government sources, individuals who had visited Iraq, and scholars of

---

[10] RAND staff seconded to CPA were involved in monitoring and reporting on the provision of essential services to the Iraqi population. This section is based on RAND staff experiences and interviews with CPA staff in Baghdad as well as on published documents.

[11] Charles Recknagel, "Iraq: Al-Basrah Riot Underlines Frustration with Energy Shortages," *Radio Free Europe*, August 11, 2003.

[12] Interviews with ORHA officials, July 2004.

[13] Interviews with ORHA officials, July 2004.

Iraq and the Middle East all resident in the United States.[14] Interviewees claimed that knowledgeable individuals had been excluded from the planning process and that many of the Iraqi expatriates who provided much of the information about the condition of Iraqi infrastructure had not been in Iraq for a considerable period of time.[15] To compound matters, there was no contingency planning in the event that the initial assumptions did not hold.

One of the coalition's major failures was to not plan adequately to safeguard facilities from looters. Far more damage was done to Iraq's infrastructure by looting than by military action.[16] Using lists compiled by the U.S. military and the CIA, ORHA put together a roster of facilities to be safeguarded following major combat operations and sent it to CENTCOM. CENTCOM did not use this list to deploy guard units to designated facilities after they had been seized. U.S. military officers said that they lacked sufficient troops to guard all the selected critical facilities.

A number of U.S. officers noted that looting should not have been a surprise and that a number of measures could have been adopted to suppress it. Units should have been prepared to combat looters and given the appropriate rules of engagement to do so. One officer stated that, according to the Geneva Convention, Iraqi forces could have been mustered into a constabulary and assigned to guard facilities under the command of U.S. forces. Briefings had been prepared by military staff to address these issues, but senior commanders turned down opportunities to hear them in the run-up to the war.[17]

Contracting for the Resumption of Essential Services

Not surprisingly, the coalition put a high priority on resuming the provision of water and electric power for political and humanitarian reasons, as well as on oil production and exports for economic reasons. Without water, Iraq was vulnerable to outbreaks of cholera and other infectious diseases. Without electricity, water treatment plants could not operate. And without fuel, electric power generators could not run. Consequently, the resumption of electric power generation and oil production was imperative.

To restore the provision of government services and repair Iraqi infrastructure, the U.S. government turned to contractors and went through three stages of contracting for the reconstruction of infrastructure in Iraq.

---

[14] Interviews with National Defense University officials, July 2004.

[15] Interviews with National Defense University officials, July 2004.

[16] Interviews with Army Corps of Engineers personnel, October 2003.

[17] Interviews with CJTF-7 officers, October 2003.

Initial Contracts

Even though the interagency planning process started in August 2002, most of the agencies involved were not requested to start procurement actions for reconstruction until early 2003. Consequently, agencies had only a few weeks before the war to set up contracts for reconstruction.[18] Facing such tight deadlines, agencies took two approaches: letting new contracts or issuing task orders to existing contracts.

The U.S. government let several large contracts prior to or during the war, to ensure the rapid resumption of essential services (especially water and electric power), restore transportation systems, resume oil production, and minimize damage to oil fields. On March 24, 2003, the Defense Department, through the Army Corps of Engineers, gave a sole-source contract to Kellogg, Brown and Root (KBR), a subsidiary of Halliburton, to put out oil fires and to restart oil production and exports.[19] USAID let two contracts prior to the end of major combat operations for the repair of infrastructure. One went to Bechtel for electric power, reconstruction of government buildings (including schools), dredging and repairing the Umm Qasr seaport, and other construction projects.[20] A third contract to Bechtel was let on April 17, 2003, after major combat ended—which included financing for an assessment of Iraqi infrastructure needs. The other contract went to SkyLink Air and Logistic Support for repairing airports. A third USAID contract went to Stevedoring Service of America to operate the port of Umm Qasr; it covered the operation of the port rather than its reconstruction, which was handled by Bechtel.

In addition, both USAID and DoD wrote task orders under existing contracts for early work on reconstruction. For example, a task order was written under a Logistics Civil Augmentation Program (LOGCAP) contract to commission work by KBR on contingency planning for the reconstruction of Iraq's oil industry.

Initially, none of the contracts were let under standard contracting procedures that call for open competition. As noted above, the Army Corps of Engineers issued a sole-source contract, and USAID employed limited competition. The Corps of Engineers has been heavily criticized for giving Halliburton a sole-source contract for the reconstruction of Iraq's oil industry.[21] A U.S. General Accounting Office (GAO) report stated that the Corps properly justified the award under U.S. legal provisions

---

[18] U.S. General Accounting Office, *Rebuilding Iraq: Fiscal Year 2003 Contract Award Procedures and Management Challenges*, p. 7.

[19] Rosenberg, Horowitz, and Alessandrini.

[20] Rosenberg, Horowitz, and Alessandrini.

[21] See, for example, "Democratic Policy Committee Hearing on Iraq Contracting Abuses," February 13, 2004. As of October 2007:
http://democrats.senate.gov/dpc/hearings/hearing12/transcript.pdf

for circumventing normal competitive procedures.[22] USAID's contracts received less criticism in the press. USAID may have been better able to deflect such criticism because its contracts were awarded on the basis of limited competition rather than sole-source procurements. Because of the press and public focus on Halliburton as a result of Vice President Cheney's past position as the company's CEO, it is not clear whether criticism of the Army Corps of Engineers would have been any less had it opted for limited competition rather than sole-source procurement.

Virtually all of the contracts and task orders issued before and during CPA's existence were on an indefinite quantity/indefinite cost basis. Contractors were to be reimbursed for permitted costs plus a fee. To encourage contractors to operate efficiently, the U.S. government provides incentives for them to provide goods and services at lower cost or more efficaciously. These incentives are tied to detailed descriptions of tasks and the terms under which they are to be completed. The process of defining tasks and performance metrics is called "definitization."

Post-Conflict Contracts in FY2003

Through the end of FY2003, 100 contracts or task orders were awarded or issued for the reconstruction of Iraq or the provision of services for CPA. Of these, eight were awarded by the Army Corps of Engineers and three by USAID for reconstruction of infrastructure. These 11 contracts and task orders were the most important by value, running $2.755 billion in total, three-quarters of the total obligated in FY2003.[23]

Following the end of major combat operations, more contracts were let using full and open competition. Of the 100 contracts and task orders, 14 new contracts were awarded using other than full and open competition: five sole-source contracts and nine limited-competition contracts. Only in one instance, a contract issued by the State Department, did GAO find that agency officials had not adequately justified awarding contracts on a basis other than full and open competitive procedures.[24] GAO was more critical about the process of issuing task orders. Of eleven task orders reviewed by GAO, seven were not within the scope of the initial contract and two were questionable. The task order issued by the Army Field Support Command under its contract with KBR was deemed to have been outside the scope of the original contract and should not have been granted.[25] GAO also criticized the way that the

---

[22] U.S. General Accounting Office, *Rebuilding Iraq: Fiscal Year 2003 Contract Award Procedures and Management Challenges*, p. 11.

[23] U.S. General Accounting Office, *Rebuilding Iraq: Fiscal Year 2003 Contract Award Procedures and Management Challenges*, pp. 36 and 39.

[24] U.S. General Accounting Office, *Rebuilding Iraq: Fiscal Year 2003 Contract Award Procedures and Management Challenges*, p. 12.

[25] U.S. General Accounting Office, *Rebuilding Iraq: Fiscal Year 2003 Contract Award Procedures and Management Challenges*, pp. 4–5.

Corps of Engineers handled three major contracts. In March 2003, the Corps let three contracts for construction-related activities in CENTCOM's area of responsibility. The contracts were let on the basis of limited competition and had a maximum value of $100 million each. GAO found that task orders issued under each of these contracts for rehabilitating the Iraqi electric power system violated legal provisions: the task orders exceeded the size of the contracts, they were issued in a noncompetitive manner, and they were issued without written justification.[26]

The Project Management Office

In November 2003, CPA set up a Project Management Office (PMO) to handle contracting. This occurred shortly before passage of the Emergency Supplemental Appropriations Act for Defense and for the Reconstruction of Iraq and Afghanistan. The PMO was headed by Admiral David Nash (ret.). The supplemental legislation stipulated that Congress be informed no later than seven calendar days before any contract of $5 million or more was awarded and that all contracts be awarded on the basis of full and open competition procedures.[27] Nash was intent on ensuring that CPA satisfy these legislative stipulations and that the newly appropriated $18.4 billion for Iraq be spent in an accountable fashion. CPA had been criticized for not being able to account for the $3.7 billion obligated for contracts in FY2003.

The PMO adopted a strategy of issuing one program management support contract to support its own office and to oversee reconstruction efforts in specific sectors. In addition, six program management contracts were let to coordinate the reconstruction efforts in the six sectors: electricity, oil, public works and water, security and justice, transportation and communications, and buildings and health. Finally, 15 to 20 design-build contracts were to be awarded to execute specific tasks.[28]

These contracts were to be let through a variety of channels. By March 2004, 17 contracts had been awarded on behalf of CPA by various offices in DoD. The 17 contracts included all 7 program management contracts and 10 design-build contracts. In addition, the Corps of Engineers and USAID continued to receive funds for reconstruction, which were obligated through their own contracting mechanisms.

---

[26] U.S. General Accounting Office, *Rebuilding Iraq: Fiscal Year 2003 Contract Award Procedures and Management Challenges*, pp. 20–21.

[27] U.S. General Accounting Office, *Rebuilding Iraq: Fiscal Year 2003 Contract Award Procedures and Management Challenges*, p. 9.

[28] U.S. General Accounting Office, *Rebuilding Iraq: Fiscal Year 2003 Contract Award Procedures and Management Challenges*, pp. 27–28.

The Players: Who Was Involved in Reconstruction

Iraqi Ministries and State-Owned Enterprises

A number of actors have been involved in the reconstruction of Iraq. On the Iraqi government side, the Ministry of Oil, the Ministry of Electricity (formerly the Commission of Electric Power), the Ministry of Municipalities and Public Works (water and sewage), the Ministry of Transport, the Ministry of Communications, and the Ministry of Water (irrigation) played the primary roles in managing Iraq's infrastructure. In conjunction with the Finance Ministry, ministries made decisions on operating expenditures. Under Saddam, decisions on major investments were made in a highly centralized fashion. The Ministry of Planning took the lead on major infrastructure projects, with the Ministry of Finance and the line ministries participating. Stronger line ministries like Oil and Electricity played major roles in decisions on key investments. Weaker ministries found their priorities receiving less attention.

All of the major utilities and the entire oil sector are state-owned in Iraq. They form major components of the line industries: enterprises in the oil industry are part of the Ministry of Oil, water treatment plants are part of the Ministry of Municipalities and Public Works, etc. The state-owned sector in Iraq is extraordinarily centralized. Enterprise managers have very little decisionmaking authority: managers of operating units have no control over revenues, have little authority to make payments for operating expenditures, and have virtually no control over investment. Investment decisions are still taken at the ministry level. Operating units play a limited role in investment decisions, primarily as providers of information concerning bottlenecks and unsatisfied demand.

CPA

The CPA senior advisors played a major role in the day-to-day operations of the ministries and in planning. Until ministers were appointed by the Iraqi Governing Council (IGC) and the CPA Administrator in September 2003, the CPA senior advisors to these ministries had executive authority. Even after ministers were appointed, the senior advisor had veto authority over the minister's decisions, until the ministry was officially "graduated." The graduation process only began in early 2004. Thus, the senior advisors played important roles not only in day-to-day operations, but in establishing priorities.

To establish cross-ministry expenditure priorities, CPA set up a Program Review Board (PRB). This board consisted of senior CPA staff, including individuals from CPA's Office of Management and Budget, which was responsible for the Iraqi budget, the Economics Group, and other senior staffers. The PRB made decisions on all major CPA expenditures involving Iraqi money—Iraqi assets that had been frozen, as well as revenues from oil sales through the Oil for Food program.

Regional Military Commanders

Much of the day-to-day governance of Iraq took place through the regional military commands. Given the need for day-to-day contact with civil servants in the ministries, senior advisors spent most of their time in Baghdad. Because of the security situation and shortages of staff, CPA had little presence at the regional or governorate level. Moreover, regional and local commanders played a very important role in government at the local level because military units were widely deployed throughout the country, and because CPA lacked the regional and local presence necessary to do this work. Regional commanders and their subordinate officers were in day-to-day contact with their local communities. These commanders and officers, primarily through their civil affairs officers, engaged with the local community. They conveyed local concerns and priorities to CPA, though they often believed that CPA's presence inside the Green Zone and its focus on Baghdad made it unresponsive to the needs of other parts of the country. In addition, through the use of Commanders' Emergency Response Funds (CERF) and other resources, local commanders were able to repair and rehabilitate local infrastructure, especially buildings.

U.S. Government Contracting Institutions

Although the CPA Administrator was the governing authority in Iraq, CPA did little of the actual contracting for Iraqi reconstruction. Initially, USAID and the Army Corps of Engineers had the money and the contracting authority for reconstruction projects, subject to CPA oversight. Subsequently, the Project Management Office was given this authority, although it continued to work with and through the Corps of Engineers, USAID, and other agencies that had the authority to issue contracts. These agencies wrote the task orders and handled the contracting. Although all task orders and contracts had to pass CPA's PRB and later the PMO, the contracting agencies played an important implicit role in setting priorities.

International Institutions

Despite their absence from Iraq following the August 2003 bombing of the UN mission in Baghdad, the World Bank and the United Nations Development Program (UNDP) played roles in reconstruction. They conducted initial needs assessments that were used as a basis for funding requests at the October 2003 Madrid donors' conference. They were also charged with creating and managing two trust funds that would be used for foreign assistance.

Contractors

Contractors are doing the actual work of reconstruction and rehabilitation in Iraq. Contractors design projects; organize teams, usually involving a number of subcontractors, to build or repair a facility; manufacture or purchase equipment; install the equipment; and begin operations. Contractors were also responsible for providing for their own security. Because of the scale of the reconstruction effort, the U.S. government chose to give very large contracts to a few lead contractors who then broke out individual projects. Table 12.1 shows a partial list of major contractors for infrastructure projects as of September 30, 2003.

Financing

A tremendous amount of money is being dedicated to the reconstruction of Iraq. As of April 30, 2004, the GAO estimated that $58 billion in grants, loans, assets, and revenues from various sources had been pledged or made available for the provision of government services and reconstruction. Of this amount, the United States appropriated about $24 billion, of which $4.5 billion was appropriated in FY2003, primarily through the Emergency Wartime Supplemental Appropriations Act enacted in April 2003. Of the remainder, $18.4 billion was appropriated through the Emergency Supplemental Appropriations Act passed on November 6, 2003.[29] The second-largest source of funds is Iraq's earnings from oil exports, totaling $18 billion

Table 12.1
Major Reconstruction Contracts for Iraq as of September 30, 2003

Contractor	Sector	Amount Obligated ($U.S. millions)
Corps of Engineers		
Brown and Root Services	Oil	1,390.1
Washington International Corporation	Electricity	111.0
Fluor Intercontinental, Inc.	Electricity	102.5
Perini Corporation	Electricity	66.6
IAP Worldwide Services	Supplied diesel generators	11.9
Michael Baker Jr., Inc.	Program management support	4.4
Stanley Consultants, Inc.	Program management support	4.4
USAID		
Bechtel National Inc.	Reconstruction, including electricity	1,029.8
Stevedoring Service of America	Port management	14.3
SkyLink Air and Logistic Support USA Inc.	Airports	17.5

SOURCE: U.S. General Accounting Office, *Rebuilding Iraq: Fiscal Year 2003 Contract Award Procedures and Management Challenges,* pp. 36 and 39.

---

[29] U.S. General Accounting Office, *Rebuilding Iraq: Resource, Security, Governance, Essential Services, and Oversight Issues,* GAO-04902R, Washington, D.C., June 2004, p. 11.

since oil exports were restarted after OIF ended. These funds have been deposited in the Development Fund for Iraq (DFI), and they constitute the principal source of Iraqi government revenues. In addition to ongoing earnings from oil exports, Iraq has accrued assets to draw upon. Some of these are assets formerly controlled by Saddam or the Ba'ath Party that have been seized abroad and deposited in the DFI. Other assets include funds transferred from the Oil for Food account set up and run by the United Nations. A number of rehabilitation and repair projects had been started before the invasion. Oil for Food funds were obligated to pay for these projects. Non-U.S. donors promised an additional $3.6 billion in grant aid and about $10 billion in the form of subsidized loans ($9.6 billion to $13.3 billion).[30] Table 12.2 provides a breakdown of estimated funds as of April 2004.

U.S. assistance to Iraq has been extraordinary. The November 2003 supplemental appropriation was 2.5 times the entire budget request for USAID in 2002. Total U.S. appropriations through April 2004 exceeded Iraq's estimated GDP for 2003.

The magnitude of foreign assistance to Iraq is also unusual in that the Iraqi government has a ready source of funds in the form of oil revenues. In other postconflict situations where foreign assistance has been a large share of GDP, such as Bosnia, Kosovo, and East Timor, the local governments lacked ready sources of tax revenue. Foreign assistance helped jump-start the economy by supporting government operations during a period when traditional tax revenues were impossible to collect. In the case of Iraq, the government has large streams of export earnings that it can tap for expenditures.

Actual figures on planned spending on infrastructure are difficult to come by because of the many sources of funding and differing definitions of infrastructure.

Table 12.2
Total Funds Available, Obligated, and Disbursed for Iraq by Source as of April 2004 (in billions of U.S. dollars)

Source	Made Available	Obligated	Disbursed
United States	24.00	8.2	3.0
Development Fund for Iraq	18.00	13.0	8.3
Vested and seized assets	2.65	2.5	2.4
International pledges	13.60	N/A	N/A
Total	58.25	23.7	13.7

SOURCE: U.S. General Accounting Office, *Rebuilding Iraq: Resource, Security, Governance, Essential Services, and Oversight Issues*, p. 10.

---

[30] U.S. General Accounting Office, *Rebuilding Iraq: Resource, Security, Governance, Essential Services, and Oversight Issues*, p. 20.

According to CPA, of the U.S. supplemental appropriation, two-thirds ($12.4 billion) was earmarked for construction projects. In addition to these monies, some CERP funds and funds for the Rapid Regional Response Program (RRRP), a DoD-funded program for commanders to use for local and regional projects, were used for infrastructure projects. Total spending on the CERP and the RRRP were $550 million and $430 million, respectively, through April 2004.[31]

Most of the Iraqi monies were being spent on government operations and transfers to the Iraqi people. The Iraqi state budget for 2004 earmarked $2.5 billion for capital projects, financed from oil revenues and Iraqi assets, out of a total budget of $22.4 billion.[32] As noted above, according to the GAO, about three-fourths of the U.S. contribution, or $2.755 billion total obligated assistance for Iraq in FY2003, was for infrastructure projects.[33] Total funding for spending on infrastructure from the end of OIF to the end of 2004 was $18.8 billion, excluding non-U.S. donor funding for infrastructure projects. Of the $13.6 billion in non-U.S. pledges, roughly $10 billion (the subsidized credits) or more went to infrastructure. This funding has been slow to arrive. Thus, out of $58.25 billion dollars in pledged funding, more than half, $30 billion, may be designated for infrastructure. Of this, $5.447 billion was obligated for the reconstruction of infrastructure and the provision of essential services as of April 2004, more than 70 percent of which was provided by the United States.[34] For the next few years, more than 90 percent of projected spending on infrastructure is to be financed from foreign assistance.

U.S. assistance for the reconstruction of infrastructure is extraordinary not only for its size, but also because of the form it has taken: grants rather than loans. The use of grant aid to finance infrastructure projects is highly unusual. Grant aid is generally considered best spent on projects for which it is difficult to arrange loans, such as education, improving government operations, and developing health care systems. Long-term investments in electric power, water systems, and the oil sector are almost invariably financed through equity or loans. These projects are very expensive, and they produce goods and services that are sold to consumers, generating revenue streams that can be used to make debt payments. The use of loan financing generally imposes a discipline on project design and implementation that grant aid often does not. The process of obtaining a loan to finance a project forces the borrower to

---

[31] U.S. General Accounting Office, *Rebuilding Iraq: Resource, Security, Governance, Essential Services, and Oversight Issues*, p. 15.

[32] Coalition Provisional Authority, "Working Papers: Iraq Status," unclassified briefing, June 29, 2004, slides 14 and 18.

[33] U.S. General Accounting Office, *Rebuilding Iraq: Fiscal Year 2003 Contract Award Procedures and Management Challenges*, pp. 36 and 39.

[34] U.S. General Accounting Office, *Rebuilding Iraq: Resource, Security, Governance, Essential Services, and Oversight Issues*, p. 14.

evaluate the project in terms of rate of return. In the course of this appraisal, borrowers are forced to tailor the size and design of the project to projected demand and, hence, revenue streams, which can lead to more efficient use of funds and more attention to revenue generation and collection, a deficiency that plagues many projects in Iraq.

Coordinating and Implementing Reconstruction Projects

The diverse sources of funding and the many actors engaged in reconstruction complicated the process of coordinating and implementing reconstruction projects in Iraq. Participants had to reach decisions on the selection of projects, the allocation of funds, and the choice of contractors. Contractors had to design projects, order and obtain equipment, materials, machinery, and men, and complete the projects. Governments and agencies needed to perform financial and project audits to ensure that the promised goods and services were provided and that charges conformed to the law.

Project Selection

CPA had to start project selection virtually from scratch. Although the Ministry of Planning had official responsibility for planning investments under Saddam, little planning took place in the 1990s because of the dearth of funds. Consequently, a new list of projects needed to be identified. What little investment was made in infrastructure during the 1990s was determined primarily by the line ministries as they struggled to maintain services. These ministries had identified a number of rehabilitation and investment projects for which imported components had been approved by the UN under the Oil for Food program. Because contracts under the Oil for Food program had already been let, these projects generally proceeded, although in some instances CPA canceled them if they seemed unwarranted.

Sectoral needs assessments completed in the summer of 2003 provided the initial basis to select projects for reconstruction. USAID and DoD tasked two contractors, Bechtel and KBR, to assess Iraqi investment needs. KBR assessed the oil industry, while Bechtel assessed the electric power, transportation, communications, and other infrastructure. In October 2003, the World Bank and UNDP published a broader joint assessment of needs that covered many of the same sectors as the Bechtel report but excluded the oil sector so as not to duplicate the KBR report.[35] In addition to evaluations of physical infrastructure needs, the report also included assessments of human needs that Bechtel was not asked or equipped to address. Because the World Bank and UNDP effectively withdrew from Iraq after the August

---

[35] United Nations and World Bank, *Joint Iraq Needs Assessment*.

2003 bombing of UN headquarters and because Bechtel and KBR won major reconstruction contracts, the Bechtel and KBR needs assessments became the operational documents for Iraqi reconstruction. Forward Engineering Support Teams (FESTs) from the Corps of Engineers also conducted assessments. However, the UNDP/World Bank report played a major role in requests for donor funding at the Madrid conference. The Bechtel report also drew upon and referred to some of the analysis and findings in the UNDP/World Bank studies.

In some ministries, detailed project plans were developed by the senior advisors and Iraqi civil servants, the Corps of Engineers, USAID, and contractors. For example, in October 2003, the Ministry of Electricity and the CPA Electricity Advisory Team published a detailed plan for increasing power output.[36] The plan was based on the Bechtel and UNDP/World Bank electric power sector needs assessments, but it also incorporated ministry analysis of crucial needs. It included detailed plans for scheduling fall and spring maintenance of power plants and transmission and distribution systems, plans that had not been developed for more than a decade. This was an operational plan, and it included a level of detail that neither Bechtel nor the UNDP/World Bank teams had addressed. The Bechtel and UNDP/World Bank teams noted that they sometimes could not make site visits during their surveys because of the security situation. Consequently, the Commission of Electric Power's October plan incorporated information to which Bechtel and UNDP/World Bank teams did not have access.

Reconstruction and Project Management

Because of separate sources of funding, reconstruction sometimes proceeded on separate tracks. In the case of the oil and electric power industries, two special task forces were set up outside the Iraqi ministries: Project Restore Iraqi Oil (RIO) started in the fall of 2002, and Project Restore Iraqi Electricity (RIE) commenced in late summer 2003. Both projects were under the auspices of the Army Corps of Engineers.

Reconstructing the oil industry. Project RIO was set up under the auspices of the Corps of Engineers. It was initially designed to repair damage to Iraq's oil industry stemming from the war, including putting out oil well fires. Subsequently, it was given responsibility for reconstructing the entire oil industry. Project RIO maintained responsibility for supervising the KBR oil contract. Project RIO staff played a major role in project selection, using the KBR report but also incorporating additional information from the Ministry of Oil. Project managers and contractors cooperated with the Iraqi Ministry of Oil and the senior advisor on choices of projects. However, Project RIO took the lead in managing the reconstruction and rehabilitation of the oil sector. It, not the Oil Ministry and senior advisor, monitored contrac-

---

[36] U.S. General Accounting Office, *Rebuilding Iraq: Resource, Security, Governance, Essential Services, and Oversight Issues*, p. 86.

tors in terms of progress, charges, and quality of work, because it controlled funding through the Corps of Engineers, which selected the contractors. The Corps of Engineers' Project RIO was generally credited with successfully restoring oil output and exports.

Reconstructing the electric power industry. Project RIE was more complicated. By late summer 2003, the CPA Administrator and CJTF-7 had become quite concerned about progress in increasing the output of electricity. Popular dissatisfaction with CPA was ascribed to power outages. Senior policymakers hoped that if electric power provision improved, popular dissatisfaction, and hence support for the nascent insurgency, would decline.

General John Abizaid sponsored a conference on this topic in late summer 2003 at CENTCOM headquarters in Tampa. Based on the favorable experience with Project RIO, the Corps of Engineers was charged with creating Task Force Restore Iraqi Electricity (RIE). Brigadier General Steven Hawkins from the Corps of Engineers was brought back to Iraq run Project RIE.[37] Project RIE was expected to accelerate the reconstruction of the electric power sector because the Corps had the ability to let emergency contracts.

As noted above, the Corps had awarded three construction contracts in March 2003, to Washington International Corporation, Fluor Intercontinental, Inc., and Perini Corporation on the basis of limited competition. Once Project RIE was approved, the Corps issued task orders under these contracts for reconstruction and rehabilitation work in the electric power sector.

Some acrimony arose among the various participants over the coordination of reconstruction efforts for the electric power sector. In contrast to Project RIO, Project RIE was not the only organization working on electricity. Bechtel had a large contract through USAID to rehabilitate the power sector, and Iraq's Commission of Electric Power had more than a billion dollars in equipment and supplies on order and in the process of delivery for projects begun under Saddam's regime. The commission had its own engineers and project planners. In March 2004, the Program Management Office awarded three additional contracts for expanding electric power generating capacity—outside of Project RIE.[38]

Iraqi engineers from the Commission of Electric Power complained that their expertise had not been recognized or utilized to the extent it should have been. Knowledgeable Americans, however, claimed that the commission did not have the expertise or capability to successfully manage new projects. Meanwhile, GAO criticized the Corps of Engineers for improperly issuing task orders to the three main

---

[37] Hawkins had worked on electricity reconstruction activities for a few months after Task Force IV disbanded. See Chapter Four.

[38] U.S. General Accounting Office, *Rebuilding Iraq: Resource, Security, Governance, Essential Services, and Oversight Issues*, p. 87.

contractors under Project RIE, for not establishing detailed work assignments, and for not opening the task orders to competition.[39]

The Department of Defense initially allocated $350 million for RIE, with a promise of an additional appropriation of $730 million. RIE made obligations on the assumption that its total budget would be close to $1 billion, but the additional funds never materialized. Bremer was forced to use DFI funds to cover RIE's expenditures once the initial $350 million was gone, which was very frustrating for CPA officials.[40]

Some participants believed that USAID's contract with Bechtel was too expensive. Because USAID has no in-house engineers, it contracted with the Corps of Engineers to provide quality assurance and quality control for its contract with Bechtel. Consequently, the chain of institutions involved was quite long: CPA, USAID, Corps of Engineers, Bechtel, major subcontractors such as General Electric and Siemens, and so on. This complicated structure raised concerns that unnecessary costs were being added to the contract. The PMO apparently found neither Project RIE nor USAID totally satisfactory, because it chose to separately fund new contracts for the rehabilitation of the electric power sector. Despite these problems, participants stated that turf wars were rare, because all parties realized that there was more than enough work to go around. Each organization and contracting group reportedly focused on areas where it had particular expertise.[41]

Reconstructing public buildings, bridges, and roads. CPA gave reconstruction of public buildings, bridges, and roads lower priority than the big three: telecommunications, electric power, and water. However, while CPA struggled to restore power, Iraqis complained to local commanders that the Americans had made no improvements or, in some instances, that life had been better under Saddam. Commanders attempted to address these complaints by using funds from the CERP and RRRP to repair local buildings and other community projects. These projects were much more visible and also employed local people. Commanders also complained that school reconstruction under the Bechtel contract was proceeding too slowly, that work was sometimes substandard, and that subcontractors were overpaid.

CERP and RRRP deserve praise for providing local commanders with the flexibility and the resources to address immediate local problems. However, these programs and the school reconstruction projects under Bechtel have also been criticized. CPA had set itself the goal of creating a secure, democratic Iraq with a flourishing market economy. To achieve this vision, CPA needed to set up mechanisms for reducing corruption and encouraging free and fair competition. Commanders fre-

---

[39] U.S. General Accounting Office, *Rebuilding Iraq: Fiscal Year 2003 Contract Award Procedures and Management Challenges*, pp. 20–21.

[40] Interviews with USAID official, July 2004, and CPA official, September 2004.

[41] Interview with USAID official, July 2004.

quently did not set up competitive bidding mechanisms and, in a number of instances, awarded contracts to politically important locals even after other Iraqis underbid them. Civil affairs officers stated that this approach to contracting left local Iraqis with the impression that it was "business as usual." Political connections often determined awards, not the lowest bid. In addition, Iraqis claimed that some of the reconstruction work was shoddy. In light of the amount of money being paid, communities thought they could have done a better job themselves.[42]

Allocation of Funds

CPA did not have a good system for making decisions on the allocation of funds for infrastructure and other projects. The request for the largest source of funding, the November 2003 supplemental, was drawn up by a few people in CPA's Office of Management and Budget. Senior advisors and USAID staff complained that they were excluded from the process of putting together the final funding request. They argued that as a consequence, the original supplemental request did not allocate resources in a manner best designed to improve security, foster democracy, and create a basis for sustained economic growth.

Critics point to the $2 billion request for funding to create an Iraqi army, while the request for the Iraqi Civil Defense Corps was $76 million. The police were allocated $150 million for equipment and refurbishing stations, while $400 million was requested for a maximum security prison, $100 million for a witness protection program, and $100 million for investigative support for crimes against humanity.

As noted above, supplemental funds were skewed toward investment in infrastructure, especially electric power. Out of a $20.3 billion total request for Iraq, $5.675 billion was designated for electric power. This allocation was made without addressing the question of solvent demand for power in Iraq—the amount that people would consume if they knew they had to pay for it. Because users are not compelled to pay for power, nor do they face realistic prices, demand for power in Iraq is unconstrained by price. Increases in capacity to be financed under the supplemental would move power availability in Iraq above per capita levels in other Middle Eastern states, including states with substantially higher per capita incomes and much more developed industrial structures.

Meanwhile, movement toward creating a regulatory structure and a proper metering and billing system languished. Not surprisingly, because of the lack of constraints on demand, power availability (the number of hours per day power was available) was no higher at the end of May 2004 than it had been prior to the war, despite expenditures of $1.38 billion by Project RIE alone.[43]

---

[42] Interview with civil affairs officer, August 2004.

[43] U.S. General Accounting Office, *Rebuilding Iraq: Resource, Security, Governance, Essential Services, and Oversight Issues*, pp. 85–86.

Results as of June 28, 2004

Resuming the Provision of Essential Services

Despite large sums committed to the reconstruction of Iraq, the coalition had mixed success in restoring Iraqi services. After rising sharply from May 2003 to October 2003, electric power generation stopped rising rapidly: production fell in April 2004.[44] By June 28, 2004, output still ran at about the same level as before the war. Hours of availability in May 2004 were 90 percent of what they were in May 2002.[45]

Oil production had failed to reach its pre-conflict peak by mid-year 2004. At the end of June 2004, sabotage, an ongoing problem, had reduced output to just 1.1 million barrels per day (mbd), a level last seen just after the war, compared to the prewar peak of 2.5 mbd.[46]

On the other hand, the provision of water did improve. Municipal water systems were being expanded throughout the country as of June 28, 2004. Sewage systems were also operating more efficiently and were being expanded. After three mobile telephone licenses were granted in 2003, mobile telephone service was being rapidly introduced. Although the number of subscribers to fixed line services still lagged pre-conflict numbers, which consisted of a little more than 800,000 lines, Iraq soon signed up 457,000 cell phone subscribers.

The slower than planned pace at which electricity and oil production were restored stems from a variety of reasons. First, the coalition underestimated the magnitude of the task. Not only did the coalition have to repair war damage, but years of poor maintenance and underinvestment had caught up with Iraq. In many instances, electric power-generating equipment was not worth repairing; it had to be replaced. This slowed the process of reconstruction. Second, looting immediately after the war and ongoing sabotage knocked out refurbished installations, resulting in frequent setbacks in restoring service levels. Third, the continued use of subsidies and price controls has exacerbated problems. Iraqis have used their rising incomes to purchase refrigerators and air conditioners that have increased the demand for electric power, power for which they do not pay. Price controls on diesel fuel have encouraged the Ministry of Electricity to use diesel for power generation and discouraged the development of natural gas. Not surprisingly, diesel fuel is in short supply. The Ministry of Electricity has not been able to operate newly installed generating capacity because of shortages of fuel. Fourth, the lack of security has slowed contractor work and added to costs. Project RIE missed many of its deadlines for providing additional ca-

---

[44] Coalition Provisional Authority, "Working Papers: Iraq Status," slide 20.

[45] It is worth noting, however, that the levels in May 2004 exceeded the levels reached in 1999–2000 by 20 percent. Interview with CPA official, September 2004.

[46] Coalition Provisional Authority, "Working Papers: Iraq Status," slide 22.

pacity. The inability of contractors to work because of violence and threats contributed to these delays.[47]

Expenditures

The PMO has been criticized for not spending the supplemental more rapidly. Despite Nash's experience in government contracting procedures, it took some months before sizeable numbers of contracts could be awarded and funds spent. When CPA ceased to exist on June 28, 2004, $5.291 billion of the $18.439 billion supplemental had been obligated; only $366 million had been spent. Of the $12.406 billion of the supplemental to be spent on construction, $3.562 billion had been obligated by that date.[48]

Reconstruction fell behind the initially optimistic timelines set for it, stemming in part from the difficulty in contracting for and implementing projects. U.S. congressional concerns about properly monitoring contract expenditures cannot be satisfied without a reasonable period of time for issuing requests for proposals, evaluating bids, and choosing winners. Only after the winning bidders have been selected does the process of developing specific projects, contracting for equipment, and installing equipment begin. The time needed to manage this process correctly stretched out the reconstruction timeline. Finally, the high degree of violence has raised contractor costs, slowed reconstruction, and generally retarded the effort.

It is not clear that a more rapid rate of expenditures would have been in the best interest of Iraq or U.S. taxpayers. As noted above, total U.S. appropriations for assistance to Iraq since the end of the war have exceeded the best estimates of Iraq's GDP for 2003. Trying to spend that amount of money in one or two years would be equivalent to pushing $11 trillion, roughly the size of the United States' GDP at that time, through the U.S. economy in 24 months. Sums of this magnitude have highly distorting effects on demand and prices for construction inputs: labor, land, materials, transport, and so on.

Costs

Because of the desire to provide assistance quickly and the highly uncertain nature of the venture, all major contracts for reconstructing Iraqi infrastructure are based on cost reimbursement plus a fee. These types of contracts do not lend themselves to encouraging cost savings. DoD has addressed the inherent lack of incentives to save by providing financial incentives for exceeding performance criteria set down in task orders.

---

[47] U.S. General Accounting Office, *Rebuilding Iraq: Resource, Security, Governance, Essential Services, and Oversight Issues*, pp. 90–92.

[48] Coalition Provisional Authority, "Working Papers: Iraq Status," slide 14.

We do not have enough information at this point to make a judgment on the costs of infrastructure projects in Iraq. We do know that Project RIE has experienced several significant cost overruns, in part because of the increased costs of security.[49] The GAO report criticizes the Army Corps of Engineers for failing to come to early agreements with KBR on definitions of task orders. These definitions are the basis for rewarding contractors for achieving cost savings and providing services in an efficient manner. The Corps failed to define criteria for either the March 2003 contract or one of its contracts for rehabilitating the Iraqi electric power industry.[50] In part because of the lack of incentives and oversight, GAO "saw very little concern for cost considerations."[51]

Quality Control

Neither official government reports nor interview data suggest that quality has been a problem in reconstruction projects involving electric power, the oil sector, or water systems. Some interviewees and press reports have been critical of the quality of work under the Bechtel contract pertaining to school rehabilitation. In some instances, local subcontractors used poor-quality materials and provided shoddy workmanship.[52] Although this contract does not cover major reconstruction and repairs, anecdotal reports suggest that quality was inadequate in certain cases.

---

[49] U.S. General Accounting Office, *Rebuilding Iraq: Resource, Security, Governance, Essential Services, and Oversight Issues*, pp. 90–92.

[50] U.S. General Accounting Office, *Rebuilding Iraq: Fiscal Year 2003 Contract Award Procedures and Management Challenges*, p. 21.

[51] U.S. General Accounting Office, *Rebuilding Iraq: Fiscal Year 2003 Contract Award Procedures and Management Challenges*, p. 10.

[52] Dan Murphy, "Quick School Fixes Won Few Iraqi Hearts," *Christian Science Monitor*, June 28, 2004.

CHAPTER THIRTEEN

Assessing Postwar Efforts

The United States did not plan well for the complexities and violence of post-Saddam Iraq. Although many U.S. government agencies invested a lot of time and effort in identifying possible reconstruction requirements for postwar Iraq, the basic plan for war largely pushed these requirements aside. Indeed, in overlooking the need to enforce security in the immediate aftermath of Saddam's fall, the warplan may well have contributed to the problems U.S. forces face.

How did this happen? More important, what can be done to avoid the flaws of the OIF warplan in the future? After all, Iraq and Afghanistan may not be the only state-sized battlegrounds of the war on terrorism. Insofar as terrorists find safe haven in states like Afghanistan, or exploit the weaknesses of failing states generally, regime change, state-building, reconstruction, governance—the missions U.S. forces have taken on in Iraq—are likely to recur in future military operations. Although these missions call on the forces of all services as well as those of other states, they fall mainly to ground forces, for the obvious reason that only ground forces can seize, hold, and control territory and people.

The challenge to the Army is critical but especially vexing. As they do in Iraq today, Army forces play a major role in post-conflict operations. Yet they are only part of a much larger planning process[1] that includes large sections of the government and is driven, rightly, by the nation's and the Defense Department's civilian leadership. The Army can shape doctrine and forces to better take on counterinsurgency and stability operations, but the larger issue raised in this report is how it can help to better shape a planning process much larger than itself.

Shaping the Plan

If myriad voices within and outside the U.S. government sought to call attention to the possible challenges of post-Saddam Iraq, why did the planning process fail to ac-

---

[1] Operational planning is largely the purview of the combatant commands, while strategic planning is controlled by the Chairman of the Joint Chiefs of Staff.

count for these possibilities? The story begins at a fairly high level, where key assumptions about post-Saddam Iraq remained largely unchallenged. But even at this high level, military voices could have sounded a louder warning than they did. In fact, the plan was shaped by a rough convergence of civilian assumptions about post-Saddam Iraq and military views about responsibility for handling that part of the operation that together downplayed post-Saddam challenges and the role military forces should play in meeting them.

Unchallenged Assumptions and Expectations

U.S. government planning was driven by a particular view, held by senior policymakers in key positions in the government, of what would emerge as a result of combat operations and what would be required thereafter. While alternative views and outcomes were examined by the various organizations described in previous chapters, none of them formed a consistent basis for planning throughout the U.S. government. The high-level civilian view that prevailed was based on the following assumptions:

- **The military campaign would have a decisive end and would produce a stable security environment.** As discussed in Chapter Two, the civilian leadership believed that military operations would end once Saddam Hussein was removed from power, giving rise to a largely stable situation during the reconstruction phase. Local forces, particularly the police and the regular army, would be capable of providing law and order, so U.S. and coalition military forces could be withdrawn rapidly from Iraq. Administration officials had hoped to shrink the U.S. military presence to two divisions—between 30,000 and 40,000 troops—by the fall of 2003.[2] Deputy Secretary of Defense Paul Wolfowitz succinctly expressed this assumption in his testimony to Congress on February 27, 2003, when he stated "it's hard to conceive that it would take more forces to provide stability in post-Saddam Iraq than it would take to conduct the war itself and to secure the surrender of Saddam's security forces and his army."[3]

- **U.S. and coalition forces would be greeted as liberators.** After Saddam was removed from power, the Iraqi population was expected to support the U.S. presence. Three days before the war, Vice President Cheney clearly articulated this

---

[2] Interview with V Corps official, January 2005. These plans called for a third division, from a yet-to-be-determined coalition country, to join the two U.S. divisions in Iraq. Michael R. Gordon with Eric Schmitt, "U.S. Plans to Reduce Forces in Iraq, with Help of Allies," *New York Times*, May 3, 2003; Esther Schrader and Paul Richter, "U.S. Delays Pullout in Iraq," *Los Angeles Times*, July 15, 2003.

[3] Paul Wolfowitz, testimony to the House Budget Committee, February 27, 2003. It is worth noting that this assumption stood in stark contrast to the force estimates developed by the military. General Franks believed that the number of troops would have to be increased to 250,000 before being reduced, and General Shinseki, Task Force IV, and the Army staff estimated that the reconstruction phase would involve multiple hundreds of thousands of troops. See Chapters Two and Four.

view by stating "my belief is we will, in fact, be greeted as liberators."[4] Iraqi exiles emphasized that the Iraqis would greet U.S. forces with "sweets and flowers,"[5] and CENTCOM commander General Franks worked from the assumption that the Iraqis would support U.S. forces, and perhaps even join them in combat, once they believed that the United States was serious about removing Saddam from power.[6] It may be that U.S. officials insisted on the vision of the United States as a liberator as a way to avoid the responsibilities of occupying powers under the Geneva Conventions.[7] Yet regardless of whether this view was ideological or instrumental, it was one of the reasons why U.S. forces on the ground after the fall of Baghdad did not take steps to restore public order.[8]

- **Government ministries would continue to function.** Because the Ba'ath regime maintained a tightly centralized grip on power throughout the country, U.S. officials assumed that government ministries were largely effective state structures. As discussed in Chapter Five, they assumed that the top leadership of each ministry could be replaced, leaving the remaining technocrats and civil servants to continue running the state.

- **Humanitarian relief requirements would be extensive.** Humanitarian relief was the one area where U.S. officials assumed Phase III and Phase IV might overlap. They prepared for the possibility that more than a million people would flee their homes to avoid the effects of combat, and they worried that those numbers might be even higher if Saddam Hussein used weapons of mass destruction during the military campaign. They also assumed that hunger would be widespread, due to disruptions in the food distribution system, and that Iraq might experience significant sanitation problems as well.

- **Infrastructure throughout the country would remain largely intact.** The military campaign was designed to damage as little of Iraq's infrastructure as possi-

---

[4] Vice President Richard Cheney, remarks made on *Meet the Press*, televised March 16, 2003.

[5] Kanan Makiya, one of the Iraqi participants in the Future of Iraq project, acknowledged after the war that this had been his message to President Bush, and he stated "I admit I was wrong." Joel Brinkley and Eric Schmitt, "Iraqi Leaders Say U.S. Was Warned of Disorder After Hussein, but Little Was Done," *New York Times*, November 30, 2003. See also Woodward, p. 259.

[6] Woodward, p. 81.

[7] Sandra Mitchell, vice president of the International Rescue Committee, recounted that she and other NGO representatives discussed the responsibilities of occupying powers under the Fourth Geneva Convention with representatives of the U.S. Agency for International Development. She noted, "we were corrected when we raised this point. The American troops would be 'liberators' rather than 'occupiers,' so that the obligations did not apply." Quote from Fallows, p. 63.

[8] As the 3rd Infantry Division noted in its after action report, "Because of the refusal to acknowledge occupier status, commanders did not initially take measures available to occupying powers, such as imposing curfews, directing civilians to work and controlling the local governments and populace. The failure to act after we displaced the regime created a power vacuum, which others immediately tried to fill." Quote from John J. Lumpkin and Dafna Linzer, "Army Says Policy Choice Led to Chaos in Iraq," *Philadelphia Inquirer*, November 28, 2003.

ble, focusing instead on regime centers of power. U.S. officials were concerned that Saddam might torch the oil fields during the war, so they took special precautions to prevent that. Generally speaking, however, they expected the oil sector, the power grid, and other key aspects of Iraq's infrastructure to remain mostly unaffected by the war, with only minimal reconstruction required afterward.

Actual postwar events proved most of these assumptions to be faulty. Phase III combat operations did not end neatly, and the United States was not greeted as a liberator.[9] An insurgency started developing almost immediately, suggesting that General Shinseki's estimate that postwar operations would require "something on the order of several hundred thousand soldiers"[10] may have been closer to the mark than the administration's optimistic assumption had been. Humanitarian relief requirements were minimal, which meant that the one contingency for which detailed plans had been developed never arose. Although the military campaign left most of Iraq's infrastructure intact, extensive looting in the aftermath of the conflict severely damaged infrastructure throughout the country. Moreover, U.S. analysts had underestimated the level of debilitation to Iraq's infrastructure after more than a decade of sanctions.[11] Government ministries turned out to be hollow, without the capabilities and resources necessary to run the country once the Ba'athists were removed from power. Wolfowitz later acknowledged that defense officials had erred by making assumptions that "turned out to underestimate the problem" in postwar Iraq.[12] Iraq's reconstruction became the major undertaking described in Chapters Eight through Twelve.

Although the prevailing assumptions proved to be wrong, they were not unreasonable—or at least they were no less reasonable than a variety of other, less optimistic sets of assumptions. The problem was that the prevailing assumptions were never seriously challenged. Despite a predilection for questioning virtually all operational

---

[9] Ten days after the war started, an unnamed senior administration official was already quoted in the press as questioning this assumption: "We underestimated their capacity to put up resistance. We underestimated the role of nationalism. And we overestimated the appeal of liberation." Bob Drogin and Greg Miller, "Plan's Defect: No Defectors," *Los Angeles Times*, March 28, 2003.

[10] General Eric Shinseki, testimony to the Senate Armed Services Committee, February 25, 2003.

[11] For more on the failure to correctly assess the status of Iraq's infrastructure, see Rajiv Chandrasekaran, "Crossed Wires Kept Power Off in Iraq," *Washington Post*, September 25, 2003.

[12] Wolfowitz identified three conditions that were worse than defense officials had anticipated: the failure of Iraqi army units to fight alongside the United States and assist in the reconstruction; the requirement to rebuild the police forces; and the difficulty of imagining that Ba'ath Party remnants would continue to fight. See U.S. Department of Defense News Transcript, "Deputy Secretary Wolfowitz Briefing on His Recent Trip to Iraq," July 23, 2003. As of October 2007:
http://www.defenselink.mil/transcripts/transcript.aspx?transcriptid=2872

military assumptions from several directions,[13] and despite the existence of alternative analyses within the government, those charged with planning for Iraq assumed that one particular scenario would play out and did not plan for other possible contingencies that had been identified both inside and outside the U.S. government.[14]

What explains the failure to challenge assumptions and develop alternate scenarios? One factor was the context in which the decision to go to war was being debated. Proponents of the intervention, within the administration and without, tended to argue that post-conflict stabilization and reconstruction would be manageable in terms of cost, time, and effort, while opponents argued the contrary. Once the basic decision to intervene had been made, reservations about the scale of effort needed for stabilization and reconstruction may have been interpreted as a form of bureaucratic obstruction—as obstacles to be overcome rather than valid concerns to be addressed.

Meanwhile, in the aftermath of the September 11, 2001 terrorist attacks, Congress and the country as a whole were disinclined to challenge the administration on national security judgments. As a result, the decision to intervene received less congressional and media scrutiny than might have been the case under other circumstances. The administration's early and rapid success in installing a successor to the Taliban regime in Afghanistan only weeks after the initial entry of U.S. troops also engendered optimism that a similarly quick transition might be made in Iraq.

Ineffective Interagency Coordination
The dominance of a single set of assumptions about postwar Iraq suggests the absence of a robust interagency coordination process. Several U.S. government organizations, particularly the Office of the Secretary of Defense (OSD), the State Department, the U.S. Agency for International Development (USAID), the National Security Council (NSC), and the military commands conducted separate studies of postwar possibilities. Looking back, some of these studies appear to have been reasonably prescient. The problem, therefore, was not that the U.S. government failed to plan for the postwar period. Instead, it was the failure to effectively coordinate and integrate these various planning efforts.

Those functions normally fall to the National Security Council staff, which has overall responsibility for coordinating U.S. foreign and defense policies. In fact, as outlined in previous chapters, the NSC staff oversaw several interagency working groups that brought together representatives from DoD, State, CIA, and other orga-

---

[13] See Woodward and Franks for more detailed discussions.

[14] The Defence Committee of the British House of Commons reached the same conclusion after an extensive inquiry into postwar planning and operations. After stating that a "wide range of predictions for the post-conflict situation in Iraq were made in advance of the conflict," the committee notes that "there is some evidence that the extensive planning, which we know took place in both the U.S. and the U.K., did not fully reflect the extent of that range." See United Kingdom House of Commons Defence Committee, *Iraq: An Initial Assessment of Post-Conflict Operations, Volume I*, Sixth Report of Session 2004–05, March 16, 2005, p. 3.

nizations. Most of these working groups focused on the conduct of the war, but the Iraq Relief and Reconstruction (IR+R) Working Group did focus on postwar plans. This group produced fairly detailed humanitarian relief plans, while its reconstruction plans remained vague—reflecting the assumptions described earlier in this chapter. These assumptions appear to have dominated the NSC staff's approach to planning as well.

If the NSC staff failed to consider alternative scenarios that might pose differing requirements, neither did it provide strategic guidance on various aspects of U.S. policy during the postwar period. Repeated requests for policy guidance from CENTCOM, Task Force IV, ORHA, and others went unanswered, leaving each agency to make its own assumptions about key aspects of the postwar period.[15] Key questions, such as whether the U.S. postwar authority would be military or civilian in nature, went unanswered throughout the planning process. When the NSC finally did issue strategic guidance in late March 2003 (as discussed in Chapter Three), the war was already under way. As a result, the various planning processes that occurred throughout the U.S. government were neither coordinated nor guided by a set of consistent goals and objectives.

Above all, the NSC seems not to have mediated the persistent disagreement between the Defense Department and the State Department that existed throughout the planning process. Secretary of State Powell influenced a few key diplomatic decisions—notably the decision to take the case for war with Iraq to the United Nations in September 2002[16]—but the Defense Department controlled most planning decisions. Richard Haass, then the Director of Policy Planning at the State Department, later stated that he realized the decision to confront Iraq had already been made in July 2002, despite continuing opposition from State.[17] As discussed in Chapter Three, State's main postwar planning effort, the Future of Iraq project, was largely ignored by the Defense Department throughout 2002. And as noted in Chapter Five, in February 2003 DoD prevented Tom Warrick, the study's leader, from working for ORHA.[18]

The Department of Defense was named the lead agency for postwar Iraq in January 2003, on grounds that the civilian and military authorities in postwar Iraq would coordinate more effectively if they both reported to the Secretary of Defense,

---

[15] See, for example, the discussion of CENTCOM's frustration with the lack of policy guidance in Fineman, Wright, and McManus.

[16] Woodward, especially pp. 148–153.

[17] Haass stated that during a conversation with National Security Advisor Condoleezza Rice held in the first week of July 2002, "I raised this issue about were we really sure that we wanted to put Iraq front and center at this point, given the war on terrorism and other issues. And she said, essentially, that that decision's been made, don't waste your breath . . . it was that meeting with Condi that made me realize it was farther along than I had realized." Nicholas Lemann, "How It Came to War," *The New Yorker*, March 31, 2003.

[18] For more on tension between DoD and State, see Rieff, "Blueprint for a Mess."

rather than having the civilian authorities reporting to State and the military authorities reporting to DoD. While this may have made sense in theory, it did not work in practice. DoD's understandable emphasis on military operations led it to form ORHA, the civilian planning agency, only eight weeks before combat operations began, and more than a year after CENTCOM began military planning for the war. But DoD lacked the expertise and personnel necessary to address the civilian aspects of reconstruction, and it did not possess enough bureaucratic leverage to compel other U.S. agencies to provide experienced personnel. DoD's lack of capacity for civilian reconstruction planning and execution continued to pose problems throughout the occupation period.

Security as the Key Postwar Task

The biggest failure of both military planning and the interagency process was the failure to assign responsibility and resources for providing security in the immediate aftermath of the war. Clearly, this failure stemmed directly from the assumptions identified at the beginning of this chapter: the Iraqi population would generally support the U.S. military presence, and the Iraqi police would maintain law and order throughout the country. Postwar planning did not account for the insurgency that emerged as soon as major combat ended and did not anticipate that all of Iraq's security structures, including the police, would essentially disintegrate and prove incapable of providing security.

The failure is all the more glaring for the presence of countering advice available to planners. The question of post-conflict security was addressed explicitly at the February 2003 rock drill, as discussed in Chapter Five, but was never satisfactorily resolved. Ron Adams, the deputy director of ORHA, later recalled that "There were some of us saying, right from the get-go, 'We think there's a troops-to-task mismatch here—I'm not sure there are enough troops here to maintain security.'"[19] Meanwhile, General Anthony Zinni (ret.), the CENTCOM commander before General Franks, noted after the war that the warplans he developed for potential operations against Saddam did include enough forces for post-combat stability operations. Zinni argues that DoD leaders were "very proud that they didn't have the kind of numbers [of troops] my plan had called for. The reason we had those two extra divisions was the security situation. Revenge killings, crime, chaos—this was all foreseeable."[20]

In the event, Secretary Rumsfeld and General Franks chose to stop the flow of forces into Iraq sooner than their planners had envisioned, further reducing the forces available to provide security. As discussed in Chapter Two, senior commanders believed that forces would continue to flow into theater until after end-state condi-

---

[19] Michael Elliott, "So, What Went Wrong?" *Time*, October 6, 2003.

[20] Fallows, p. 65.

tions were met for both Phases III and IV. The decision to stop the flow of forces (notably the 1st Cavalry Division) into theater prior to that point was, in effect, a change to OPLAN 1003V. It further exacerbated a shortfall in the number of troops required to simultaneously complete Phase III and begin Phase IV operations.

Significantly, few military voices besides that of Army Chief of Staff Shinseki protested these force levels. Clearly one reason was that the military operated within the prevailing assumptions outlined earlier in this chapter, which did not identify security as a problem. Yet another part of the answer lies in the reluctance of the military to take responsibility for security and stabilization missions in the aftermath of major combat—though that reluctance was, in part, due to the failure to adequately resource Phase IV operations.

The CENTCOM commander, General Tommy Franks, unintentionally shows this reluctance in his memoirs. He mentions the importance of Phase IV throughout his book, but never identifies the specific mission that U.S. forces should have had during that time. To the contrary, he expresses the strong statement that his civilian superiors should focus on the postwar while he focused on the war itself.[21] He goes on to argue that civic action sets the preconditions for security rather than the other way around.[22] And he justifies his decision to retire right after combat ended because the mission was changing and a new commander should be there throughout Phase IV.[23] These statements reveal a mindset that sees major combat operations during Phase III as fundamentally distinct from Phase IV stability and reconstruction requirements.

We know now, of course, that the failure to plan for and adequately resource stability operations had serious repercussions that affected the United States throughout the occupation period and continue to affect U.S. military forces in Iraq. Because U.S. forces were not directed to establish law and order—and may not have had sufficient capabilities to do so—they stood aside while looters ravaged Iraq's infrastructure and destroyed the facilities that the military campaign took great pains to ensure remained intact, creating greater reconstruction requirements than existed when major combat ended. Because U.S. forces have had to focus far more on providing security for U.S. personnel (both military and civilian) than on providing se-

---

[21] Franks states, "While we at CENTCOM were executing the warplan, Washington should focus on policy-level issues . . . I knew the President and Don Rumsfeld would back me up, so I felt free to pass the message along to the bureaucracy beneath them: *You pay attention to the day* after and *I'll pay attention to the day* of." Franks, p. 441. Emphasis in the original.

[22] Franks writes, "As I had said throughout our planning sessions, civic action and security were linked— *inextricably* linked. There was a commonly held belief that civil action would not be possible in Iraq without security. I would continue to argue that there could be no security without civic action." Franks, p. 526. Emphasis in the original.

[23] Franks, p. 530.

curity for Iraqis, ordinary Iraqis started growing frustrated with the lack of law and order in their country soon after Saddam was removed from power.

This trend has only gotten worse since the insurgency began, as U.S. forces have had to assume that ordinary citizens may be potential belligerents and that civilians are often caught in the crossfire. A consistent majority of the Iraqi population identified security and safety as the most urgent issue facing Iraq throughout the occupation period.[24] The failure to stabilize and secure Iraq has therefore had the inadvertent effect of strengthening the insurgency, as Iraqis witness many of the negative effects of the U.S. military presence without seeing positive progress on the issues that matter to them most. The insurgency has also been aided by the failure of U.S. military forces to prioritize the mission of sealing the country's borders during the occupation period, which enabled critical foreign support to flow into Iraq.

Lessons for the Army

This experience provides several lessons for how the military in general, and the nation's ground forces in particular, can seek to avoid or at least ameliorate in the future the problems that it now faces in Iraq. Arguably, three of those lessons are particularly helpful in shaping a new approach to military planning aimed at improving post-conflict operations.

- First, it should be clear from U.S. interventions not just in Iraq, but in Afghanistan, Kosovo, and Bosnia, that wars do not end when major conflict ends. Wars emerge from an unsatisfactory set of political circumstances, and they end with the creation of new political circumstances that are more favorable to the victor—in this case, circumstances more favorable to U.S. interests. Creating those new circumstances may not involve continuing conflict, and even if conflict is involved, it may not be as intense as the counterinsurgency operations confronting U.S. forces in Iraq. But given the likely security vacuum in the immediate aftermath of major conflict, planners cannot avoid considering a variety of possible missions and scenarios.

- Second, these post-conflict missions will almost unavoidably fall to forces present on the ground at the time. To some extent the security missions that follow

---

[24] This trend continued after the June 28, 2004 transfer of authority. Results do vary somewhat by city. Between January and August 2004, the percentage of the population identifying safety and security as the most urgent issue averaged 63 percent in Baquba; 60 percent in Mosul; 53 percent in Baghdad; 47 percent in Najaf; and 30 percent in Basra. When asked "How safe do you feel in your neighborhood?" the number of respondents who answered "not very safe" or "not safe at all" averaged 63 percent in Basra; 58 percent in Baquba; 57 percent in Baghdad and Najaf; 46 percent in Mosul; and 33 percent in Karbala. See "Opinion Analysis," U.S. Department of State Office of Research, M-106-04, September 16, 2004, Appendix 6A.

major conflict are legitimate tasks for ground forces that, by virtue of their possession of the instruments of violence, can impose security in such situations. But the absence of security makes it unlikely that the civilian organizations that would normally handle reconstruction tasks will be available quickly to take on those roles. In the immediate aftermath of major conflict, and perhaps for a good deal longer, "civilian" as well as "military" missions will fall to forces on the ground.

- Finally, it should be clear that the way the actual conflict unfolds greatly influences the situation that emerges and evolves after the major conflict ends. In order to provide security in the aftermath of Saddam's fall, the invading force needed more troops. These observations testify to the dangerous artificiality of the distinction between Phase IV and the phases that preceded it. They are not distinct phases; planning for them sequentially can produce unhappy outcomes.

These lessons have significance for the U.S. Army's Title 10 role of organizing, training, and equipping forces for use by combatant commanders in major conflicts. The Army must put real meaning into the phrase "full spectrum force." It must be able to fight and dominate an adversary in a major conflict. But as Iraq demonstrates, Army forces must also be prepared to provide security to a civilian populace, reconstitute and retrain local security forces, reconstruct infrastructure as necessary, escort children safely to school, perhaps even help clear raw sewage from the streets. They will usually do so in a cultural environment foreign to them, yet those missions will require them to have at least enough cultural awareness to avoid undermining the mission.

But the more crucial significance of these basic lessons comes at the level of military and strategic planning. Clearly these lessons produce a very different view of the military planning process than the one recorded in this report. Military planning must start with a view of the desired outcome of the war—not the outcome of major conflict, but the creation of the desired political circumstances that signal the true end of the war. They must do so both because their forces, and especially forces on the ground, will be intimately involved in creating those circumstances, and because the way in which military action unfolds will heavily shape subsequent developments.

One way to capture this lesson is to say that military planners must start with "Phase IV." A still more productive approach would be to dispense with phases, which inevitably produce sequenced plans that risk missing crucial connections from phase to phase. Planners must start with strategic guidance from the civilian leadership on where they want to be, strategically, when the war ends. They can then work backward to points of major conflict, shaping plans for those in ways that contribute to the larger and longer-term strategic goal.

Starting to plan in this way will ensure that "Phase IV" will not be ignored or underplayed in the planning process. As planning for OIF makes clear, it is essential that planners entertain a full array of possible scenarios for getting to that strategic end point. Even the most reasonable assumptions must be challenged, and hedging actions must be an integral part of the plan. Recognizing that military forces—largely U.S. Army forces—will play a role in these activities should give the combatant commander good reason to force this conversation into the planning process.

They may not have to "force" anything, of course. Every planning process is different, and many may be open to the kind of questions that were not seriously considered in planning for Phase IV of OIF. The point is that military commanders and planners have a right and a responsibility to raise questions about so-called post-conflict activity, the "stability operations" assumed to follow a major combat operation. At the very least, the way these questions are answered will shape plans for major combat operations. It may also be the case, as it has been in Iraq, that military forces will be engaged in fashioning a favorable political outcome to a particular conflict long after major fighting has ended.

Strategic Studies Institute's Mission Matrix for Iraq

As described in Chapter Three, the Strategic Studies Institute of the U.S. Army War College issued a report in February 2003 called *Reconstructing Iraq*. It contains the single most comprehensive list of postwar reconstruction tasks produced either inside or outside the U.S. government before the war. This mission matrix, as it is called in the report, is reproduced here. It divides the reconstruction period into four phases: providing security (up to six months); stabilization (six months to one year); building institutions (one to two years); and handing power to local authorities (more than two years). It then identifies 135 tasks that must be accomplished during reconstruction, and names the organizations or agencies that should be responsible for each task during each of the four phases. It also assigns a priority to each task, defined as follows:

> **Critical:** "If the commander of coalition military forces does not put immediate emphasis and resources on these activities he risks mission failure."

> **Essential:** "These tasks also require quick attention and resources from the commander of coalition military forces, though they are generally not as time sensitive as the critical tasks. However, failure in accomplishing them will have significant impact on the overall mission."

> **Important:** "These tasks must still be performed to create and maintain a viable state, but they are more important in later phases of transition and/or primarily the responsibility of nonmilitary agencies."

Based largely on this matrix, the report concluded that military leaders, particularly within the U.S. Army, should conduct more detailed analyses of their responsibilities and missions in postwar Iraq before major combat commences.

Table A.1
Mission Matrix for Iraq

Category and Task	Priority	Security Phase (0–6 Months)	Stabilize (6–12 Months)	Build Institutions (1–2 Years)	Handover Phase (2+ Years)
Major Security Activities					
1A. Secure/Destroy WMD	Critical	Coalition	Coalition, State, UNMOVIC	State, UNMOVIC	Transparent Iraqi government
1B. Stop Intra- and Inter-factional Fighting	Critical	Coalition	Coalition	Coalition, Iraqi army	Iraqi army
1C. Train New Iraqi Army	Essential	Coalition	Coalition	Coalition (broadened)	U.S. Army
1D. Round Up Regime	Critical	Coalition	Coalition, CIA, DOJ	Coalition, CIA, DOJ	—
1E. Eliminate Pockets of Resistence	Critical	Coalition	Coalition	Coalition	—
1F. Process Detainees and POWs	Essential	Coalition, DOJ	Coalition, Interagency Task Force	Coalition, Interagency Task Force	Iraqi institutions
1G. Secure Borders	Critical	Coalition	Coalition	Coalition, Iraqi army	Iraqi army
1H. Seize and Secure Oil Facilities	Critical	Coalition	Coalition	Coalition, Iraqi army	Iraqi army
1I. Plan and Conduct Conse-quence Management	Critical	Coalition	Coalition	Coalition, Iraqi army	Iraqi army
1J. Plan and Conduct Theater Information Operations	Critical	Coalition	Coalition	Coalition, Iraqi army	Iraqi army
1K. Maintain Freedom of Movement	Critical	Coalition	Coalition	Coalition, Iraqi army	Iraqi army
1L. Regulate Freedom of Movement	Essential	Coalition	Coalition	Coalition, Iraqi army	Iraqi army
Public Administration					
2A. Establish and Assist Regional and Local Governments	Critical	Coalition	Coalition, USAID, NED, State, IOs	USAID, NED, State, UNDP, IOs	UNDP, Iraqi institutions
2B. Establish and Assist National Legislative System	Important	Coalition	USAID, NED, State, IOs, Coalition	USAID, NED, State, UNDP, IOs	UNDP, Iraqi institutions
2C. Establish and Assist National Executive Office	Important	Coalition, State	USAID, NED, State, IOs, Coalition	USAID, NED, State, UNDP, IOs	UNDP, Iraqi institutions
2D. Establish and Assist Ministries	Important	Coalition, State	USAID, NED, State, IOs, Coalition	USAID, NED, State, IOs, Coalition	UNDP, Iraqi Institutions
2E. Preserve and Improve Public Records System	Essential	Coalition, DOJ	Coalition, DOJ	Iraqi institutions	—

Table A.1 (continued)

Category and Task		Priority	Security Phase (0–6 Months)	Stabilize (6–12 Months)	Build Institutions (1–2 Years)	Handover Phase (2+ Years)
Legal						
3A.	Operate Criminal Court System	Essential	Coalition, Iraqi courts	USAID, AOUSC, DOJ, Arab League, Iraqis	USAID, AOUSC, DOJ, Arab League, Iraqis, State	Arab League, Iraqis
3B.	Operate Civil Court System	Essential	Coalition, Iraqi courts	USAID, AOUSC, DOJ, Arab League, Iraqis	USAID, AOUSC, DOJ, Arab League, Iraqis, State	Arab League, Iraqis
3C.	Establish and Operate System to Enact and Publicize Laws	Essential	Coalition	USAID, DOJ, AOUSC, Arab League	USAID, DOJ, AOUSC, Arab League, State	Arab League, Iraqis
3D.	Operate Judicial Administrative System	Essential	Coalition, Iraqi courts	USAID, AOUSC, DOJ, Arab League, Iraqis	USAID, AOUSC, DOJ, Arab League, Iraqis, State	Arab League, UNDP, ICM, Iraqis
3E.	Support and Conduct War Crimes Tribunals	Essential	Coalition	Coalition, DOJ, State	Coalition and ICC or Tribunal	Iraqi institutions
3F.	Provide Legal Education	Important	—	USAID, Education, Arab League, NGOs	USAID, Education, Arab League, NGOs	Arab League, Iraqis
3G.	Protect Human Rights	Important	Coalition	State, USAID, Coalition, NGOs	State, USAID, ICM, NGOs	ICM, NGOs
Public Finance						
4A.	Stabilize Currency	Essential	Coalition, Treasury	Treasury, USAID	Treasury, USAID, World Bank, IMF	World Bank, IMF, Iraqis
4B.	Maintain and Operate Government Finance System	Important	Coalition, Treasury	Treasury, USAID	Treasury, USAID, World Bank, IMF	World Bank, IMF, Iraqis
4C.	Establish Private Financial Institutions	Important	—	Treasury, USAID	Treasury, USAID, World Bank, IMF	World Bank, IMF, Iraqis
4D.	Conduct Foreign Currency Exchange	Important	Treasury	Treasury, USAID	Treasury, USAID, World Bank, IMF	World Bank, IMF, Iraqis
4E.	Pay Government Civil and Military Employees	Essential	Coalition	USAID, State, Coalition	USAID, State, Coalition, World Bank	World Bank, Iraqis
4F.	Collect Customs and Duties	Important	Coalition	Treasury, USAID, Coalition	Treasury, USAID, Coalition, Iraqis	Iraqis

Table A.1 (continued)

Category and Task		Priority	Security Phase (0–6 Months)	Stabilize (6–12 Months)	Build Institutions (1–2 Years)	Handover Phase (2+ Years)
Civil Information						
5A.	Restore and Maintain Newspapers and Print Media	Essential	Coalition	USAID, State, Coalition	USAID, State, Coalition	Iraqis
5B.	Restore and Maintain Government Radio System	Critical	Coalition	USAID, BBG, Coalition, State, FCC	USAID, BBG, Coalition, State, FCC	Iraqis
5C.	Restore and Maintain Government Television System	Critical	Coalition	USAID, BBG, Coalition, State, FCC	USAID, BBG, Coalition, State, FCC	Iraqis
5D.	Establish Private TV System	Important	—	USAID, FCC	USAID, FCC	Iraqis
5E.	Establish Private Radio System	Important	—	USAID, FCC	USAID, FCC	Iraqis
5F.	Develop Censorship and Libel Laws	Important	—	USAID, DOJ, Coalition, State, Arab League	USAID, DOJ, Coalition, State, Arab League	Arab League, Iraqis
5G.	Restore and Maintain Cable Systems	Important	—	USAID, Coalition, State	USAID, Coalition, State	Iraqis
5H.	Coordinate U.S. Gov./ Iraq Government Info	Critical	Coalition, State	State, Coalition	State, Coalition	State
Historical, Cultural, and Recreational Services						
6A.	Maintain Art and Cultural Institutions	Important	Iraqis	State, USAID, Iraqis	State, USAID, Iraqis, UNESCO	UNESCO, Iraqis
6B.	Protect Historical Artifacts	Important	Coalition, Iraqis	State, DOJ, USAID, Iraqis	State, DOJ, USAID, Iraqis, UNESCO	UNESCO, Iraqis
6C.	Maintain Sports and Recreational Systems	Important	Iraqis	USAID, State, Iraqis	USAID, State, Iraqis, UNESCO	UNESCO, Iraqis
6D.	Protect Religious Sites and Access	Critical	Coalition, Iraqis	Coalition, Iraqis	Iraqi Forces	—
Public Safety						
7A.	Establish and Maintain Police Systems and Operations	Critical	Coalition	DOJ, State, Coalition, APA	DOJ, State, APA, UN Police	APA, Iraqis
7B.	Train Police	Important	—	DOJ, State, APA	DOJ, State, APA, UN Police	APA, Iraqis
7C.	Maintain Penal System	Essential	Coalition	USAID, DOJ, Coalition, Arab League	USAID, DOJ, Arab League	Arab League, Iraqis
7D.	Provide and Support Fire Fighting Systems	Important	Coalition	USAID, FEMA, Coalition	USAID, FEMA	Iraqis
7E.	Conduct Explosive Ordnance Disposal and Demining	Essential	Coalition	Coalition, USAID, State, NGOs	USAID, State, UN, NGOs	UN, Iraqis, NGOs
7F.	Protect Foreign Residents	Important	Coalition	Coalition, State	Coalition, State, Iraqi Forces	Iraqis

Table A.1 (continued)

Category and Task		Priority	Security Phase (0–6 Months)	Stabilize (6–12 Months)	Build Institutions (1–2 Years)	Handover Phase (2+ Years)
Disarmament, Demobilization, and Reintegration						
8A.	Demobilize and Reorganize Army, Security Forces, and Militias	Essential	Coalition, State	Coalition, State	Coalition, State	—
8B.	Transfer and Reorient to Integrate into Civil Sector	Important	—	USAID, Coalition	USAID, Coalition	—
8C.	Reintegrate Demobilized Persons Into Civil Sector	Important	—	USAID, State	USAID, State	—
8D.	Restructure and Reorganize New Civil Security Forces	Important	Coalition	DOJ, State, Coalition, APA	DOJ, State, Coalition, APA	APA
8E.	Dismantle Ba'ath Party	Essential	—	NED, USAID, State, CIA	NED, USAID, State, CIA	—
8F.	Disarm and Secure Weapons	Critical	Coalition	Coalition	—	—
Electoral Process for More Participatory Government						
9A.	Plan Local Elections	Important	Coalition	NED, USAID, State	NED, USAID, State, NGOs	NGOs
9B.	Plan National Elections	Important	—	NED, USAID, State	NED, USAID, State, NGOs	NGOs
9C.	Prepare Local Elections	Important	Coalition	NED, USAID, State	NED, USAID, State, NGOs	NGOs
9D.	Prepare National Elections	Important	—	NED, USAID, State	NED, USAID, State, NGOs	NGOs
9E.	Assist Conduct of Local Elections	Important	Coalition	NED, USAID, State, Coalition	NED, USAID, State, Coalition, NGOs	NGOs
9F.	Assist Conduct of National Elections	Important	—	NED, USAID, State, Coalition	NED, USAID, State, Coalition, NGOs	NGOs
9G.	Provide Post Local Election Support	Important	—	NED, USAID, State	NED, USAID, State, NGOs	NGOs
9H.	Provide Post National Election Support	Important	—	NED, USAID, State	NED, USAID, State, NGOs	NGOs
9I.	Plan for Constitutional Convention	Important	—	NED, USAID, State	NED, USAID, State, NGOs	NGOs
9J.	Assist Conduct of Constitutional Convention	Important	—	NED, USAID, State, Coalition	NED, USAID, State, Coalition NGOs	NGOs
9K.	Assist Conduct of Constitutional Referendum	Important	—	—	NED, USAID, State, Coalition NGOs	—

Table A.1 (continued)

Category and Task	Priority	Security Phase (0–6 Months)	Stabilize (6–12 Months)	Build Institutions (1–2 Years)	Handover Phase (2+ Years)
Disaster Preparedness and Response					
10A. Provide Emergency Warning Systems	Important	Coalition	USAID, Coalition, FEMA	USAID, Coalition, FEMA	Iraqis
10B. Provide Emergency Evacuation and Treatment	Important	Coalition	USAID, Coalition, FEMA	USAID, Coalition, FEMA	Iraqis
10C. Provide Post Disaster Recovery	Important	Coalition	USAID, Coalition, FEMA	USAID, Coalition, FEMA	Iraqis
10D. Conduct Pre-Disaster Planning	Important	Coalition	USAID, Coalition, FEMA	USAID, Coalition, FEMA	Iraqis
Public Works					
11A. Repair Roads and Streets	Critical	Coalition	USAID, Coalition, State	USAID, Coalition, State, UNDP	UNDP, Iraqis
11B. Repair Bridges	Critical	Coalition	USAID, Coalition, State	USAID, Coalition, State, UNDP	UNDP, Iraqis
11C. Repair Port Facilities	Critical	Coalition	USAID, Coalition, State	USAID, Coalition, State, UNDP	UNDP, Iraqis
11D. Repair Airports	Critical	Coalition	USAID, FAA, Coalition, State	USAID, Coalition, State, Interagency Task Force, UNDP	Interagency Task Force, UNDP, Iraqis
11E. Repair Railroads	Critical	Coalition	USAID, Coalition, State	USAID, Coalition, State, UNDP	UNDP, Iraqis
11F. Repair Dams	Important	Coalition	USAID, Coalition, State	USAID, Coalition, State, UNDP	UNDP, Iraqis
11G. Repair Canal System	Important	Coalition	USAID, Coalition, State	USAID, Coalition, State, UNDP	UNDP, Iraqis
Public Utilities					
12A. Restore and Maintain Power Systems	Critical	Coalition	USAID, DOE	USAID, DOE, UNDP	UNDP, Iraqis
12B. Restore and Maintain Water Systems	Critical	Coalition	USAID, HHS	USAID, HHS, UNDP, WHO	UNDP, WHO, Iraqis
12C. Restore and Maintain Gas Systems	Important	Coalition	USAID, DOE	USAID, DOE, UNDP	UNDP, Iraqis
12D. Restore and Maintain Sewage Systems	Critical	Coalition	USAID, HHS	USAID, HHS, UNDP, WHO	UNDP, WHO, Iraqis
12E. Restore and Maintain Garbage Collection	Essential	Coalition	USAID	USAID, UNDP	UNDP, Iraqis

Table A.1 (continued)

Category and Task	Priority	Security Phase (0–6 Months)	Stabilize (6–12 Months)	Build Institutions (1–2 Years)	Handover Phase (2+ Years)
Telecommunications and Public Communications					
13A. Restore and Maintain Telecommunications System	Essential	Coalition	USAID, FCC	USAID, FCC, IPU	IPU, Iraqis
13B. Restore and Maintain Broadcasting Systems	Critical	Coalition	USAID, FCC	USAID, FCC, IPU	IPU, Iraqis
13C. Restore and Maintain Postal System	Important	Coalition	USAID, USPS	USAID, USPS, IPU	IPU, Iraqis
Education					
14A. Operate Public School System	Essential	Coalition	USAID, Education	USAID, Education, UN, World Bank	UN, World Bank, Iraqis
14B. Operate Private School System	Important	—	USAID, Education	USAID, Education, UN, World Bank	UN, World Bank, Iraqis
14C. Provide Adult Education Services	Important	—	USAID, Education	USAID, Education, UN	UN, Iraqis
14D. Provide Job Training Programs	Important	USAID	USAID, UNDP, NGO	USAID, UNDP, NGO	UNDP, NGOs
14E. Provide University Education	Important	—	State, Education, USAID	State, Education, USAID, UNESCO	UNESCO, Iraqis
Public Health					
15A. Provide Emergency Medical Service	Essential	Coalition, Iraqi technocrats (IT)	USAID, FEMA, IT	USAID, FEMA, IT, UN, ICRC, NGOs	UN, ICRC, NGO, Iraqis
15B. Operate Hospitals	Essential	Coalition, IT	USAID, HHS, WHO, NGOs, IT	USAID, HHS, WHO, NGOs, IT	WHO, NGOs, Iraqis
15C. Provide Doctors and Health Professionals	Essential	Coalition, IT	USAID, HHS, WHO, NGOs, IT	USAID, HHS, WHO, NGOs, IT	WHO, NGOs, Iraqis
15D. Provide and Distribute Pharmaceutical Supplies	Essential	Coalition	USAID, HHS, WHO, NGOs, Coalition	USAID, HHS, WHO, NGOs, Coalition	WHO, NGOs, Iraqis
15E. Provide and Distribute Non-Pharmaceutical Medical Supplies	Essential	Coalition	USAID, HHS, WHO, NGOs, Coalition	USAID, HHS, WHO, NGOs, Coalition	WHO, NGOs, Iraqis
15F. Dispose of Medical Waste	Essential	Coalition	USAID, HHS, WHO	USAID, HHS, WHO	WHO, Iraqis
15G. Provide Vector Control Systems	Essential	Coalition	USAID, HHS/CDC, WHO	USAID, HHS/CDC, WHO	WHO, Iraqis
15H. Provide Garbage Disposal System	Essential	Coalition	USAID	USAID, UNDP	UNDP, Iraqis
15I. Ensure Proper Sanitation	Essential	Coalition	USAID, HHS, WHO, NGOs	USAID, HHS, WHO, NGOs, UNDP	WHO, UNDP, Iraqis
15J. Perform Preventive Medicine	Important	Coalition	USAID, HHS, WHO, NGOs	USAID, HHS, WHO, NGOs, UNDP	UNDP, Iraqis
15K. Provide Mortuary Services	Important	Coalition	USAID, HHS	USAID, HHS, UNDP	UNDP, Iraqis

Table A.1 (continued)

Category and Task		Priority	Security Phase (0–6 Months)	Stabilize (6–12 Months)	Build Institutions (1–2 Years)	Handover Phase (2+ Years)
Public Welfare and Humanitarian Relief						
16A.	Provide Assistance to Poor	Important	Coalition	USAID, USDA, WFP, NGOs	USAID, USDA, WFP, NGOs, UNDP	UNDP, WFP, NGOs, Iraqis
16B.	Provide Emergency Relief	Important	Coalition	USAID, DoD	USAID, DoD, ICRC	ICRC, Iraqis
16C.	Operate Orphanages	Important	Coalition	USAID, HHS	USAID, HHS, UNICEF, NGOs	UNICEF, NGOs, Iraqis
16D.	Provide Care for Aged	Important	Coalition	USAID, HHS	USAID, HHS, WHO, NGOs	WHO, NGOs, Iraqis
16E.	Provide Psychological Assistance	Important	Coalition	USAID, HHS	USAID, HHS, WHO, NGOs	WHO, NGOs, Iraqis
16F.	Care for and Relocate Refugees	Critical	Coalition	State, Coalition, USAID, UN, NGOs	State, Coalition, USAID, UN, NGOs, OCHA	OCHA, NGOs, Iraqis
16G.	Care for and Relocate Displaced Persons	Critical	Coalition, State	USAID, Coalition, State, UN, NGOs	USAID, Coalition, State, UN, NGOs,OCHA	OCHA, NGOs, Iraqis
16H.	Care for and Relocate Displaced Persons	Critical	Coalition, State	USAID, Coalition, State, UN, NGOs	USAID, Coalition, State, UN, NGOs,OCHA	OCHA, NGOs, Iraqis
16I.	Administer Oil for Food Program	Critical	Coalition, UN	UN	UN	—
Economics and Commerce						
17A.	Revitalize Commercial Sector	Important	—	USAID, Commerce, Treasury, IT	USAID, Commerce, Treasury, IT, World Bank	World Bank, Iraqis
17B.	Revitalize Industrial Sector	Important	—	USAID, Commerce, Treasury, IT	USAID, Commerce, Treasury, IT, World Bank	World Bank, Iraqis
17C.	Repair and Maintain Oil Facilities	Essential	Coalition, IT	USAID, Education, IT	USAID, UNDP, IT	UNDP, Iraqi
17D.	Manage Oil Revenues	Essential	Coalition, UN	USAID, State, Treasury, DOE, UN	USAID, State, Treasury, DOE, UN. World Bank, Iraqis	UN, World Bank, Iraqis
17E.	Implement Wage and Price Controls	Important	—	USAID, Treasury	USAID, Treasury, World Bank	World Bank, Iraqis
17F.	Maintain Foreign Trade System	Important	—	USAID, State, Treasury	USAID, State, Treasury, World Bank, IMF	World Bank, IMF, Iraqis
17G.	Set Customs and Duties	Important	—	State, Treasury, Commerce	Iraqis	—
17H.	Implement Oil Fire Contingencies	Critical	USAID, Coalition	USAID	—	—

Table A.1 (continued)

Category and Task		Priority	Security Phase (0–6 Months)	Stabilize (6–12 Months)	Build Institutions (1–2 Years)	Handover Phase (2+ Years)
Labor						
18A.	Establish and Provide Employment Services and Benefits	Important	—	USAID, Labor	USAID, Labor, ILO	ILO, Iraqis
18B.	Establish and Maintain System to Resolve Management-Labor Disputes	Important	—	USAID, Labor, State	USAID, Labor, State, ILO	ILO, Iraqis
18C.	Establish and Monitor Worker Safety Programs	Important	—	USAID, Labor, OSHA	USAID, Labor, OSHA, ILO	ILO, Iraqis
Property Control						
19A.	Establish and Enforce Ownership System for Real Property	Important	—	USAID, Commerce, Justice	USAID, Commerce, Justice	Iraqis
19B.	Establish and Enforce Ownership System for Personal Property	Important	—	USAID, Commerce, Justice	USAID, Commerce, Justice	Iraqis
Food, Agriculture, and Fisheries						
20A.	Maintain Production System	Important	Coalition, IT	USAID, USDA, IT	USAID, USDA, IT, UNFAO	UNFAO, Iraqis
20B.	Maintain Processing System	Important	Coalition, IT	USAID, USDA, IT	USAID, USDA, IT, UNFAO	UNFAO, Iraqis
20C.	Maintain Distribution System	Important	Coalition, IT	USAID, USDA, IT	USAID, USDA, IT, UNFAO	UNFAO, Iraqis
20D.	Maintain Retail Sales System	Important	—	USAID, USDA, Commerce	USAID, USDA, Commerce	Iraqis
20E.	Establish and Execute Inspection System	Important	Coalition	USAID, USDA	USAID, USDA	Iraqis
20F.	Maintain Irrigation System	Important	Coalition, IT	USAID, USDA, IT	USAID, USDA, IT, UNFAO	UNFAO, Iraqis
20G.	Support Harvest	Essential	Coalition	USAID, Coalition, USDA	—	—
Transportation						
21A.	Operate Ports	Critical	Coalition, IT	USAID, Coalition, DOT, IT, State	USAID, Coalition, DOT, IT, State, UNDP	UNDP, Iraqis
21B.	Operate Rail System	Critical	Coalition, IT	USAID, Coalition, DOT, IT, State	USAID, Coalition, DOT, IT, State, UNDP	UNDP, Iraqis
21C.	Maintain Intercity Road Network	Critical	Coalition	USAID, Coalition, DOT, State	USAID, Coalition, DOT, State, UNDP	UNDP, Iraqis
21D.	Maintain Municipal Roads	Essential	Coalition	USAID, Coalition	USAID, Coalition, UNDP	UNDP, Iraqis

Table A.1 (continued)

Category and Task		Priority	Security Phase (0–6 Months)	Stabilize (6–12 Months)	Build Institutions (1–2 Years)	Handover Phase (2+ Years)
Transportation (continued)						
21E.	Operate Air System (Including Airspace Management)	Critical	Coalition	USAID, FAA, Coalition, State	USAID, FAA, Coalition, State, UNDP	UNDP, IATA, Iraqis
21F.	Operate Pipelines	Critical	Coalition, IT	USAID, DOE, IT, State	USAID, DOE, IT, State, UNDP	UNDP, Iraqis

SOURCE: Conrad C. Crane and W. Andrew Terrill, *Reconstructing Iraq: Insights, Challenges, and Missions for Military Forces in a Post-Conflict Scenario,* Carlisle, PA: U.S. Army War College, Strategic Studies Institute, February 2003.

Bibliography

"40,000 Police Recruits to Train in Jordan," *USA Today*, October 14, 2003.

Allen, Mike, "Expert on Terrorism to Direct Rebuilding," *Washington Post*, May 2, 2003.

Anderson, Gary, "On Patrol With Iraq's 'Homeboys,'" *Washington Post*, January 28, 2004.

Andrews, Edmund L., "Allied Officials Now Allow Iraqi Civilians to Keep Assault Rifles," *New York Times*, June 1, 2003.

———, "U.S. Military Chief Vows More Troops to Quell Iraqi Looting," *New York Times*, May 15, 2003.

———, "Why Wheels of Recovery Are Spinning in Iraq," *New York Times*, June 15, 2003.

———, and Patrick E. Tyler, "As Iraqis' Disaffection Grows, U.S. Offers Them a Greater Political Role," *New York Times*, June 7, 2003.

"Arab League Nations Agree to Grant Seat to Iraq's Council," *New York Times*, September 9, 2003.

Baker, Gerard, and Stephen Fidler, "The Best Laid Plans?" *Financial Times* (London), August 4, 2003.

Baker, Peter, "Top Officers Fear Wide Civil Unrest," *Washington Post*, March 19, 2003.

Barnard, Anne, "U.S. Trains 'New Breed' of Iraqi Border Guards," *Boston Globe*, January 13, 2004.

Barrionuevo, Alexei, "In Critical Matter of Security, U.S. Gives Iraq a Bigger Role," *Wall Street Journal*, July 22, 2003.

Barry, John, and Evan Thomas, "The Unbuilding of Iraq," *Newsweek*, October 6, 2003.

Barton, Frederick D., and Bathsheba N. Crocker, *A Wiser Peace: An Action Strategy for a Post-Conflict Iraq*, Washington, D.C.: Center for Strategic and International Studies, January 2003.

Basu, Monica, "Iraqi Political Parties Vie to Fill Postwar Void," *Atlanta Journal-Constitution*, June 8, 2003.

Behn, Sharon, "Chalabi Says No to UN Oversight," *Washington Times*, April 29, 2004.

———, "Governing Council Huddles over Power Transfer," *Washington Times*, April 19, 2004.

———, "Iraqi Leaders Meet to Plan Reconstruction," *Washington Times*, April 30, 2003.

———, and Paul Martin, "Chalabi's Forces Enter Baghdad," *Washington Times*, April 17, 2003.

Bennett, Brian, "The World's Toughest Beat," *Time*, March 1, 2004.

Berenson, Alex, "For Iraqi Police, a Bigger Task but More Risk," *New York Times*, November 9, 2003.

————, "Iraqis' New Army Gets Slow Start," *New York Times*, September 21, 2003.

————, "Security: Use of Private Militias in Iraq Is Not Likely Soon, U.S. Says," *New York Times*, November 6, 2003.

Blanford, Nicholas, "Iraqi Force Elicits Hope—And Fear," *Christian Science Monitor*, December 9, 2003.

————, "Iraqi Police Walk Most Perilous Beat," *Christian Science Monitor*, November 25, 2003.

Booth, William, "Iraqis in Basra Chafe at Authority," *Washington Post*, May 30, 2003.

————, "U.S.-Run Local Election Touted as 'First Step' for Iraqis," *Washington Post*, May 25, 2003.

————, and Rajiv Chandrasekaran, "Occupation Forces Halt Elections Throughout Iraq," *Washington Post*, June 29, 2003.

Borger, Julian, "The Spies Who Pushed for War," *The Guardian* (London), July 17, 2003.

Bowman, Tom, "U.S. Examining Plan to Reconstitute Iraqi Army," *Baltimore Sun*, November 10, 2003.

Branigin, William, and Rick Atkinson, "Anything, and Everything, Goes," *Washington Post*, April 12, 2003.

"Bremer: U.S. in Tough Fight in Iraq," *BBC News Online*, November 17, 2003. As of October 2007: http://news.bbc.co.uk/2/hi/middle_east/3275485.stm

"Briefing on Humanitarian Reconstruction Issues," Elliott Abrams, Andrew Natsios, et al. Transcript released by the White House, Office of the Press Secretary, February 24, 2003. As of October 2007: http://www.state.gov/g/prm/rls/2003/18402.htm

Brinkley, Joel, "Iraqi Council Agrees on National Elections," *New York Times*, December 1, 2003.

————, "Some Members Propose Keeping Iraqi Council After Transition," *New York Times*, November 25, 2003.

————, "U.S. Rejects Iraq Plan to Hold Census by Summer," *New York Times*, December 4, 2003.

————, and Eric Schmitt, "Iraqi Leaders Say U.S. Was Warned of Disorder After Hussein, but Little Was Done," *New York Times*, November 30, 2003.

————, and ————, "The Struggle for Iraq: Postwar Planning," *New York Times*, November 30, 2003.

Burns, John F., "In Hussein's Shadow, New Iraqi Army Strives to Be Both New and Iraqi," *New York Times*, January 7, 2004.

"Bush Doesn't Expect Shiite Theocracy," *Los Angeles Times*, May 21, 2004.

Cambanis, Thanassis, "Forces Get First Test of Building New Order," *Boston Globe*, April 4, 2003.

————, "Stability of Iraqi Police Eroding," *Boston Globe*, April 18, 2004.

Campbell, Matthew, "Quiet Britons Outpace U.S. in Taming Iraq," *The Sunday Times* (London), December 28, 2003.

Cass, Connie, "General: 10 Percent of Iraqi Forces Turned on U.S. During Attacks," *USA Today*, April 22, 2004.

Cha, Ariana Eunjung, "Crash Course in Law Enforcement Lifts Hopes for Stability in Iraq," *Washington Post*, December 9, 2003.

————, "Flaws Showing in New Iraqi Forces," *Washington Post*, December 30, 2003.

————, "In Iraq, the Job Opportunity of a Lifetime," *Washington Post*, May 23, 2004.

Chandrasekaran, Rajiv, "Appointed Iraqi Council Assumes Limited Role," *Washington Post*, July 14, 2003.

———, "Baghdad Figure Won't Vacate Post," *Washington Post*, April 23, 2003.

———, "Bremer Supports Iraqi-Led Force," *Washington Post*, November 5, 2003.

———, "Crossed Wires Kept Power Off in Iraq," *Washington Post*, September 25, 2003.

———, "Envoy Bowed to Pressure in Choosing Leaders," *Washington Post*, June 3, 2004.

———, "Envoy Urges U.N.-Chosen Iraqi Government," *Washington Post*, April 15, 2004.

———, "Exile Finds Ties to U.S. a Boon and a Barrier," *Washington Post*, April 27, 2003.

———, "How Cleric Trumped U.S. Plan for Iraq," *Washington Post*, November 26, 2003.

———, "In Iraqi City, A New Battle Plan," *Washington Post*, July 29, 2003.

———, "Interim Leaders Named in Iraq," *Washington Post*, June 2, 2004.

———, "Iraqi Council Agrees to Ask U.N. for Help," *Washington Post*, March 18, 2004.

———, "Iraqi Council Denies Access to Two Arab Satellite Networks," *Washington Post*, September 24, 2003.

———, "Iraqi Council Postpones Selection of a Leader," *Washington Post*, July 15, 2003.

———, "Iraqi Minister Assembling Security Force," *Washington Post*, September 18, 2003.

———, "Iraqi Police Now Targets of Choice," *Washington Post*, November 2, 2003.

———, "Iraqis Assail U.S. Plans for Council," *Washington Post*, June 3, 2003.

———, "Iraqis Call U.S. Goal on Constitution Impossible," *Washington Post*, September 30, 2003.

———, "Iraqis Say U.S. to Cede Power by Summer," *Washington Post*, November 15, 2003.

———, "Iraqis Set Timetable to Take Power," *Washington Post*, April 29, 2003.

———, "Iraqis Vow to Hold National Assembly," *Washington Post*, June 4, 2003.

———, "Iraq's Barbed Realities," *Washington Post*, October 17, 2004.

———, "'Our Heritage Is Finished,'" *Washington Post*, April 13, 2003.

———, "Shiites Massacred in Iraqi Blasts," *Washington Post*, March 3, 2004.

———, "Shiites to Seek Changes," *Washington Post*, March 9, 2004.

———, "U.N. Envoy Backs Iraqi Vote," *Washington Post*, February 13, 2004.

———, "U.N. Official Warns of Iraqi Food Crisis," *Washington Post*, February 28, 2003.

———, "U.S. Presses New Iraqi Council to Begin Tackling Major Issues," *Washington Post*, July 16, 2003.

———, "U.S. Sidelines Exiles Who Were to Govern Iraq," *Washington Post*, June 8, 2003.

———, "U.S. to Appoint Council in Iraq," *Washington Post*, June 2, 2003.

———, "U.S. to Form New Iraqi Army," *Washington Post*, June 24, 2003.

———, and Monte Reel, "Iraqis Set Timetable to Take Power," *Washington Post*, April 29, 2003.

———, and Anthony Shadid, "Shiites Massacred in Iraq Blasts," *Washington Post*, March 3, 2004.

Chesterman, Simon, *You, The People*, Oxford: Oxford University Press, 2004.

Chives, C.J., "Marines Keep Order in Saddam Hussein's Home Town," *New York Times*, April 16, 2003.

Coalition Provisional Authority, *Achieving the Vision: Taking Forward the CPA Strategic Plan for Iraq*, July 18, 2003.

———, "Ministry of Interior—Security Forces Information Packet." As of February 2008: http://cpa-iraq.org/security/MOI_Info_Packet.html

———, Orders. As of October 2007: http://www.cpa-iraq.org/regulations/#Orders

———, Regulations. As of October 2007: http://www.cpa-iraq.org/regulations/#Regulations

———, *Vision for Iraq*, July 11, 2003.

Cody, Edward, "Influential Cleric Backs New Iraqi Government," *Washington Post*, June 4, 2004.

Collins, Joseph J., Deputy Assistant Secretary of Defense for Stability Operations, "DoD News Briefing," February 25, 2003. As of June 2004: http://www.centcom.mil/centcomnews/transcripts/20030225.htm

Constable, Pamela, "New Policy Unveiled for Ex-Baathists," *Washington Post*, January 12, 2004.

———, and Khalid Saffar, "Blasts at Iraqi Police Facilities Kill 68," *Washington Post*, April 22, 2004.

Cooper, Christopher, "Troop Training in Afghanistan Outpaces Similar Efforts in Iraq," *Wall Street Journal*, September 8, 2003.

Cordesman, Anthony H., "One Year On: Nation Building in Iraq," CSIS Working Paper, April 6, 2004.

Corera, Gordon, "Iraq Provides Lessons in Nation Building," *Jane's Intelligence Review*, January 2004.

Cornwell, Rupert, "U.S.-Backed Iraqis Launch Bid for Power," *Independent on Sunday* (London), April 13, 2003.

"Counterfeit Trail Led to Chalabi," *The Sunday Times* (London), May 31, 2004.

Crane, Conrad C., and W. Andrew Terrill, *Reconstructing Iraq: Insights, Challenges, and Missions for Military Forces in a Post-Conflict Scenario*, Carlisle, PA: U.S. Army War College, Strategic Studies Institute, February 2003.

Crocker, Bathsheba N., *Post-War Iraq: Are We Ready?* Washington, D.C.: Center for Strategic and International Studies, March 25, 2003.

Daniszewski, John, "Baghdad Is Asking, Where Are the Police?" *Los Angeles Times*, May 27, 2003.

———, "Hundreds Line Up to Join New Iraqi Army" *Los Angeles Times*, July 22, 2003.

———, "Iraqi Security Forces Far from Ready," *Los Angeles Times*, November 16, 2003.

———, "New Iraqi Army Makes Its Debut," *Los Angeles Times*, September 16, 2003.

———, "Policing Isn't Black and White in Baghdad," *Los Angeles Times*, May 20, 2003.

———, and Jailan Zayan, "Iraqi Council's Foreign Minister Takes a Seat at the Arab League's Table," *Los Angeles Times*, September 10, 2003.

Dao, James, and Eric Schmitt, "President Picks a Special Envoy to Rebuild Iraq," *New York Times*, May 7, 2003.

Daragahi, Borzou, "Disbanded Army Draws Praise, Condemnation," *Washington Times*, January 6, 2004.

———, "Use of Private Security Firms in Iraq Draws Concern," *Washington Times*, October 6, 2003.

Dempsey, Judy, "Several European States Ready to Help Train 5,000 Iraqi Police," *Financial Times* (London), November 18, 2003.

Dewan, Shaila K., "G.I.'s Turn Over Policing of Iraqi Town to Local Force," *New York Times*, July 12, 2003.

DeYoung, Karen, "U.S. Sped Bremer to Iraq Post," *Washington Post*, May 24, 2003.

———, and Glenn Kessler, "U.S. to Seek Iraqi Interim Authority," *Washington Post*, April 24, 2003.

———, and Dan Morgan, "U.S. Plan for Iraq's Future Is Challenged," *Washington Post*, April 6, 2003.

———, and Daniel Williams, "Training of Iraqi Exiles Authorized," *Washington Post*, October 19, 2002.

Diamond, Larry, Juan J. Linz, and Seymour Martin Lipset (eds.), *Politics in Developing Countries*, Boulder, CO: Lynne Rienner, 1990.

Dilanian, Ken, "Iraqi Police Force Taking Shape, but It Still Lacks Some Basics," *Philadelphia Inquirer*, September 24, 2003.

Dinmore, Guy, and James Harding, "Bremer in Talks on Timing of Iraq Pullout," *Financial Times* (London), October 29, 2003.

Dixon, Robyn, "Iraq Council picks 9 Leaders Instead of 1," *Los Angeles Times*, July 30, 2003.

Djerejian, Edward, and Frank G. Wisner, Co-Chairs, *Guiding Principles for U.S. Post-Conflict Policy in Iraq*, Washington, D.C.: Council on Foreign Relations and the James A. Baker III Institute for Public Policy, Rice University, December 2002.

Dobbins, James et al., *America's Role in Nation-Building: From Germany to Iraq*, Santa Monica, CA: RAND Corporation, MR-1753-RC, 2003.

Dorning, Mike, "Policing Baghdad Not for the Meek," *Chicago Tribune*, November 2, 2003.

Douglas, William, and John Walcott, "U.S. Focuses on Faster Handover to the Iraqis," *Philadelphia Inquirer*, November 13, 2003.

Dreazen, Yochi J., "Disarming Iraq's Many Militias Poses Major Challenge for U.S.," *Wall Street Journal*, April 28, 2003.

———, "Insurgents Turn Guns on Iraqis Backing Democracy," *Wall Street Journal*, December 10, 2003.

Drechsler, Donald R., "Reconstructing the Interagency Process After Iraq," *Journal of Strategic Studies*, Vol. 28, No. 1 (February 2005), pp. 3–30.

Elliott, Michael, "So, What Went Wrong?" *Time*, October 6, 2003.

Fairweather, Jack, "Chalabi Stands by Faulty Intelligence That Toppled Saddam's Regime," *Daily Telegraph* (London), February 19, 2004.

———, "Jordan to Train Iraqi Police Force," *Daily Telegraph* (London), September 30, 2003.

Fallows, James, "Blind into Baghdad," *The Atlantic Monthly*, January/February 2004.

Fan, Maureen, "Trying to Restore Order in Iraq, with Little Help from Law," *Philadelphia Inquirer*, June 8, 2003.

Farley, Maggie, and Sonni Efron, "U.N. Envoy May Provide the Key to a Transfer of Power in Iraq," *Los Angeles Times*, April 14, 2004.

Fathi, Nazila, "One Iraq City Tries on Democracy and Finds It Fits Well," *New York Times*, May 18, 2003.

"A 'Federal System' with Leaders Chosen by the People," *New York Times*, April 16, 2003.

Filipov, David, "Harsh Legacy Slows Training of Iraq Police," *Boston Globe*, October 19, 2003.

Filkins, Dexter, "In Baghdad, Free of Hussein, A Day of Mayhem," *New York Times*, April 12, 2003.

———, "Iraq Government Considers Using Emergency Rule," *New York Times*, June 21, 2004.

———, "Iraqi Ayatollah Insists on Vote by End of Year," *New York Times*, February 27, 2004.

———, "Iraqi Council Picks a Cabinet to Run Key State Affairs," *New York Times*, September 2, 2003.

———, "Iraqi Leadership Gains Agreement on Constitution," *New York Times*, March 1, 2004.

———, "Iraqis Name Team to Devise Way to Draft Constitution," *New York Times*, August 12, 2003.

———, "Iraqis Receive U.S. Approval of Constitution," *New York Times*, March 2, 2004.

———, "Panel Starts to Draw Up Constitution for Short Run," *New York Times*, February 3, 2004.

———, "Top Shiites Drop Their Resistance to Iraqi Charter," *New York Times*, March 8, 2004.

———, "A U.S. General Speeds the Shift in An Iraqi City," *New York Times*, November 18, 2003.

———, "U.S. to Send Iraqis to Site in Hungary for Police Course," *New York Times*, August 25, 2003.

———, and Ian Fisher, "U.S. Is Now in Battle for Peace After Winning the War in Iraq," *New York Times*, May 3, 2003.

———, and John Kiefner, "U.S. Troops Move to Restore Order in Edgy Baghdad," *New York Times*, April 13, 2003.

———, and Richard A. Oppel, Jr., "Truck Bombing; Huge Suicide Blast Demolishes U.N. Headquarters in Baghdad," *New York Times*, August 20, 2003.

Fineman, Mark, "Arms Plan for Iraqi Forces Is Questioned," *Los Angeles Times*, August 8, 2003.

———, "U.S. Forces Eager to Relinquish Baghdad Beat," *Los Angeles Times*, May 3, 2003.

———, Warren Veith, and Robin Wright, "Dissolving Iraqi Army Seen by Many as a Costly Move," *Los Angeles Times*, August 24, 2003.

———, Robin Wright, and Doyle McManus, "Preparing for War, Stumbling to Peace," *Los Angeles Times*, July 18, 2003.

Finer, Jonathan, "Marines Get Hands-on Civics Lesson," *Washington Post*, April 14, 2003.

———, "Marines Get New Mission: Restoring Law and Order," *Washington Post*, April 12, 2003.

Finn, Peter, "New Force Moves to Gain Sway," *Washington Post*, April 12, 2003.

———, "Pentagon's Iraqi Proteges Go Home to Mixed Welcome," *Washington Post*, April 17, 2004.

"First Battalion of 700 Troops Completes Its U.S. Training," *USA Today*, October 6, 2003.

Fisher, Ian, "Iraqi Army Takes Shape as Recruits End Training," *New York Times*, October 5, 2003.

———, and John Kifner, "G.I.'s and Iraqis Patrol Together to Bring Order," *New York Times*, April 15, 2003.

Fleishman, Jeffrey, "Ex-Baathists Offer U.S. Advice, Await Call to Arms," *Los Angeles Times*, April 27, 2004.

———, "Several Blasts Kill 30 in Basra," *Los Angeles Times*, April 21, 2004.

Fontenot, Colonel Gregory (USA, ret.) and Lieutenant Colonel Edmund J. Degen, *On Point: The United States Army in Operation Iraqi Freedom*, Fort Leavenworth, Kansas, 2004.

Foote, Christopher, William Block, Keith Crane, and Simon Gray, "Economic Policy and Prospects in Iraq," *Journal of Economic Perspectives*, Vol. 18, No. 3, Summer 2004.

Ford, Peter, "Democracy Begins to Sprout in Iraq," *Christian Science Monitor*, April 23, 2003.

———, "Iraqi Holy Men Leap into Postwar Politics," *Christian Science Monitor*, May 14, 2003.

———, "U.S. Troops in Perilous Police Work," *Christian Science Monitor*, May 19, 2003.

Frankel, Glenn, "British Troops Bring Their Brand of Civic-Minded Peacekeeping to Iraq," *Washington Post*, April 4, 2003.

Franks, Tommy, *American Soldier*, New York: Regan Books, 2004.

Fuller, Thomas, "Hungary Is Cool to U.S. Idea to Train Iraqis," *International Herald Tribune*, September 3, 2003.

Garner, Jay, interview transcript, dated July 17, 2003, for *Frontline,* "Truth, War & Consequences," first aired October 9, 2003. As of October 2007:
http://www.pbs.org/wgbh/pages/frontline/shows/truth/interviews/garner.html

Gerlin, Andrea, "Iraqi Opposition Groups Agree on Power-Sharing Plan," *Philadelphia Inquirer*, December 18, 2002.

Gettleman, Jeffrey, "G.I.'s Padlock Baghdad Paper Accused of Lies," *New York Times*, March 29, 2004.

———, "U.S. Training a New Iraqi Military Force to Battle Guerrillas at Their Own Game," *New York Times*, June 4, 2004.

Glain, Stephen J., "Chaos Thrives in Baghdad Despite Prewar Planning," *Boston Globe*, June 10, 2003.

Glasser, Susan B., and Rajiv Chandrasekaran, "Reconstruction Planners Worry, Wait and Reevaluate," *Washington Post*, April 2, 2003.

Globalsecurity.org, "Facility Protection Service." As of February 2008:
http://www.globalsecurity.org/intell/world/iraq/fps.htm

———, "Iraqi Border Police." As of February 2008:
http://www.globalsecurity.org/intell/world/iraq/ibp.htm

———, "Iraqi Civil Defense Corps." As of February 2008:
http://www.globalsecurity.org/military/world/iraq/icdc.htm

———, "Iraqi Military Reconstruction." As of February 2008:
http://www.globalsecurity.org/military/world/iraq/iraq-corps3.htm

Goldberg, Jeffrey, "A Little Learning," *The New Yorker*, May 9, 2005.

Gordon, Craig, "Iraq Security Plan Carries Own Risks," *Long Island Newsday*, November 4, 2003.

Gordon, Michael R., "101st Airborne Scores Success in Northern Iraq," *New York Times*, September 4, 2003.

———, "Allies to Begin Seizing Weapons from Most Iraqis," *New York Times*, May 21, 2003.

————, "Baghdad's Power Vacuum Is Drawing Only Dissent," *New York Times*, April 21, 2003.

————, "Between War and Peace," *New York Times,* May 2, 2003.

————, "Iraqi Opposition Gets U.S. Pledge to Oust Hussein for a Democracy," *New York Times*, August 11, 2002.

————, "Iraqi Opposition Groups Meet Bush Aides," *New York Times*, August 10, 2002.

————, "One Hundred and Twenty Degrees in the Shade," *New York Times*, August 8, 2003.

————, "U.S. Planning to Regroup Armed Forces in Baghdad, Adding to Military Police," *New York Times*, April 30, 2003.

————, with Eric Schmitt, "U.S. Plans to Reduce Forces in Iraq, with Help of Allies," *New York Times*, May 3, 2003.

Graham, Bradley, "Iraqi Security Forces Fall Short, Generals Say," *Washington Post*, April 13, 2004.

————, and Rajiv Chandrasekaran, "Iraqi Security Crews Getting Less Training," *Washington Post*, November 7, 2003.

Grier, Peter, and Faye Bowers, "The Risks of Rapid 'Iraqification,'" *Christian Science Monitor*, November 6, 2003.

Gupta, A.K., "The Great Iraq Heist," *Z Magazine*, January 2004.

Halchin, L. Elaine, *The Coalition Provisional Authority: Origin, Characteristics, and Institutional Authorities,* CRS Report for Congress, RL32370, Congressional Research Service, Washington, D.C., April 29, 2004.

Hammer, Joshua, and Colin Soloway, "Who's in Charge Here?" *Newsweek*, May 26, 2003.

"Handing Over the Keys in Iraq," *BBC Online*, November 13, 2003. As of October 2007: http://news.bbc.co.uk/2/hi/middle_east/3266617.stm

"Handover Completed Early to Thwart Attacks, Officials Say," *New York Times on the Web*, June 28, 2004.

Harman, Danna, "Baathists Need Not Apply: Kirkuk Elects a New Mayor," *Christian Science Monitor*, May 28, 2003.

Harris, Kent, "Army Training Iraqis for Guard Duty," *European Stars and Stripes*, June 25, 2003.

Harris, Paul, Martin Bright, and Ed Helmore, "U.S. Rivals Turn on Each Other as Weapons Search Draws a Blank," *The Observer* (London), May 11, 2003.

Healy, Patrick, "Policing Iraq's Police," *Boston Globe*, May 24, 2003.

————, "Security Proves a Daunting Task for U.S. in Iraq," *Boston Globe*, May 18, 2003.

Hendawi, Hamza, "Expanded Council Likely When U.S. Cedes Power," *Miami Herald*, March 5, 2004.

————, "Iraqi Police Desert Stations," *Washington Times*, April 9, 2004.

Hersh, Seymour M., "Selective Intelligence," *The New Yorker*, May 12, 2003.

Hider, James, "One Local Police Officer Has Job of Finding Embassy Bombers," *The Times* (London), August 9, 2003.

Higgins, Andrew, "As It Wields Power Abroad, U.S. Outsources Law and Order Work," *Wall Street Journal*, February 2, 2004.

Hoagland, Jim, "Cutting Off Chalabi," *Washington Post*, May 21, 2004.

Hodge, Nathan, "Pentagon Agency May Train Iraqi War-Crimes Prosecutors," *Jane's Defence Weekly*, September 15, 2003.

Hoge, Warren, "U.N. Envoy Seeks New Iraqi Council by Close of May," *New York Times*, April 28, 2004.

Hong, Peter Y., "U.N. Hands Task of Organizing Vote to Iraqi Panel," *Los Angeles Times*, June 5, 2004.

Horan, Deborah, "U.S. Scales Back Hopes for Iraqi Security Forces," *Chicago Tribune*, April 22, 2004.

Howard, Michael, "Conference Delegates Vie for Political Role in New Iraq," *The Guardian* (London), December 16, 2002.

Hundley, Tom, and E.A. Torriero, "U.S. Boosts Troop Strength in Baghdad to Help Fight Crime," *Chicago Tribune*, May 18, 2003.

Hurst, Steven R., "Shi'ite Picked to Be Iraq's First President," *Washington Times*, July 31, 2003.

———, "U.S. Sends 400 Iraqi Recruits to Basic Training for New Army," *Philadelphia Inquirer*, August 5, 2003.

Ibrahim, Saad Eddin, "Liberalization and Democratization in the Arab World: An Overview," in Rex Brynen, Bahgat Korany, and Paul Noble (eds.), *Political Liberalization and Democratization in the Arab World, Volume I: Theoretical Perspectives*, Boulder and London: Lynne Rienner Publishers, 1995.

"The Interim Government of Iraq from 2003–06," *Middle East Reference.* As of October 2007: http://middleeastreference.org.uk/iraqministers.html

International Crisis Group, "Iraq: Building a New Security Structure," ICG Middle East Report No. 20, December 23, 2003.

———, "Iraq's Constitutional Challenge," ICG Middle East Report No. 19, November 13, 2003.

"Iraq Constitution Delayed, Leaders Call for Calm," *New York Times on the Web*, March 3, 2004.

"Iraq Defector Information from Chalabi of Little Use," *The Independent* (London), September 30, 2003.

"Iraq Militias Said Threat to Civil Order," *New York Times on the Web*, January 8, 2004.

"Iraq Police Training a Flop," *Associated Press*, June 10, 2004.

"Iraqi Battalion Loses a Third of Its Recruits," *Los Angeles Times*, December 11, 2003.

"The Iraqi Council," *Washington Post*, July 14, 2003.

"Iraqi Governing Council Members," *BBC News Online*, July 14, 2003. As of October 2007: http://news.bbc.co.uk/2/hi/middle_east/3062897.stm

"The Iraqi Reconstruction and Development Council," *Middle East Reference.* As of October 2007: http://middleeastreference.org.uk/irdc.html

"Iraqis Assume Control of Oil Industry," *Washington Times*, June 9, 2004.

"Iraqis Protest over Running of Iraq," *The Times* (London), June 16, 2003.

"Iraq's Interim Leaders Review Transition Plans in Face of Shiite Poll Demand," *Associated Foreign Press*, November 29, 2003.

"Islamic Group Endorses Iraqi Government," *Los Angeles Times*, June 17, 2004.

Jaffe, Greg, "Military Plans Pullback in Major Iraq Cities," *Wall Street Journal*, September 18, 2003.

———, "On a Remote Base, U.S. Drill Sergeants Train Iraqi Exiles," *Wall Street Journal*, February 24, 2003.

———, "U.S. Seeks Lower Profile in Iraq," *Wall Street Journal*, August 19, 2003.

Jehl, Douglas, "CIA Report Suggests Iraqis Are Losing Faith in U.S. Efforts," *New York Times*, November 13, 2003.

———, "Pentagon Pays Iraq Group, Supplier of Incorrect Spy Data," *New York Times*, March 11, 2004.

———, "U.S. Considers Private Iraqi Force to Guard Sites," *New York Times*, July 18, 2003.

———, and Dexter Filkins, "Rumsfeld Eager for More Iraqis to Keep Peace," *New York Times*, September 5, 2003.

———, with Eric Schmitt, "U.S. Reported to Push for Iraqi Government, with Pentagon Prevailing," *New York Times*, April 30, 2003.

Johnson, Keith, and Alexei Barrionuevo, "U.S. Declares Local Force Will Police," *Wall Street Journal*, June 26, 2003.

Johnson, Kevin, "Attacks on Iraqi Police Increase," *USA Today*, March 31, 2004.

Johnston, David, and Richard A. Oppel, Jr., "U.S. Steps Up Hunt in Leaks to Iraqi Exile," *New York Times*, May 24, 2004.

Joint Task Force Planning Guidance and Procedures, Joint Publication 5-00.2, January 13, 1999.

Jontz, Sandra, "Soldiers Face Dangerous Situation at Weapons Turn-In Stations in Iraq," *European Stars and Stripes*, June 3, 2003.

———, "U.S. Gives Iraqi Police 3,000 Radios," *European Stars and Stripes*, June 5, 2003.

———, "Weapons Sweep Through Baghdad Marketplace Not Very Productive," *European Stars and Stripes*, June 13, 2003.

Josar, David, "MPs Start Training Police in Baghdad," *European Stars and Stripes*, May 5, 2003.

Kahwaji, Riad, "Forging a New Iraqi Army—In Jordan," *Defense News*, February 9, 2004.

———, and Barbara Opall-Rome, "Rebuilding Iraq's Military," *Defense News*, June 2, 2003.

Kessler, Glenn, "Rumsfeld: No Need for More U.S. Troops," *Washington Post*, November 3, 2003.

Keysor, Jason, "U.S. Trained Civil Defense Corps Has Started to Show Some Cracks," *St. Louis Post-Dispatch*, January 4, 2004.

Khalaf, Roula, "Support Surges for Rebel Iraqi Cleric," *Financial Times* (London), May 20, 2004.

Kifner, John, and John F. Burns, "As Tanks Move in, Young Iraqis Trek Out and Take Anything Not Fastened Down," *New York Times*, April 9, 2003.

King Jr., Neil, "Bush Has an Audacious Plan to Rebuild Iraq Within a Year," *Wall Street Journal*, March 17, 2003.

Knight, Molly, "Exile Groups Frustrated by U.S. Role in Meeting," *Baltimore Sun*, April 17, 2003.

Koch, Andrew, "Iraq Occupation: Questions Remain," *Jane's Defence Weekly*, July 16, 2003.

———, "Training More Locals Is Key to U.S. Strategy for Iraq," *Jane's Defence Weekly*, November 5, 2003.

———, "U.S. Central Intelligence Agency Forces: Covert Warriors," *Jane's Defence Weekly*, March 12, 2003.

Komarow, Steven, "Iraqis, U.S. Split on New Leaders," *USA Today*, June 1, 2004.

Krane, Jim, "Iraq Takes Over Last of Ministries," *Philadelphia Inquirer*, June 25, 2004.

———, "Iraq's Security Forces Not Ready," *Denver Post*, June 10, 2004.

LaFranchi, Howard, "New Fast Track for Iraqi Sovereignty," *Christian Science Monitor*, November 17, 2003.

Labbe, Theola, "Iraq's New Military Taking Shape," *Washington Post*, September 16, 2003.

Lacey, Marc, "British Give Port Control to the Iraqis," *New York Times*, May 23, 2003.

———, "Plans for a British-Appointed Ruling Council in Basra Go Awry," *New York Times*, June 2, 2003.

———, "Their Jobs in Jeopardy, Iraqi Troops Demand Pay," *New York Times*, May 25, 2003.

Lasseter, Tom, "Top Cleric Spurns U.S. Plans," *Miami Herald*, January 26, 2004.

LeBor, Adam, "Exiles Prepare for a Happy Return at Camp Freedom," *The Times* (London), February 24, 2003.

Lemann, Nicholas, "How It Came to War," *The New Yorker*, March 31, 2003.

Lieberman, Brett, "Marines Uneasy About Shift to Humanitarian Mission," *Newhouse.com*, April 15, 2003.

Loeb, Vernon, "Occupation of Iraq Has No Time Limit," *Washington Post*, May 10, 2003.

———, "U.S. Force Pulling Back in the North," *Washington Post*, October 23, 2003.

———, "U.S. to Appoint Council to Represent Iraqis," *Washington Post*, July 9, 2003.

Luhnow, David, "Shortages, Lack of Time to Plan Bedevil New U.S. Agency in Iraq," *Wall Street Journal*, June 5, 2003.

Lumpkin, John J., and Dafna Linzer, "Army Says Policy Choice Led to Chaos in Iraq," *Philadelphia Inquirer*, November 28, 2003.

Lynch, Colum, "Brahimi to Be U.N. Adviser on Iraq," *Washington Post*, January 12, 2003.

———, "U.N. Plan for Iraq Transition Released," *Washington Post*, February 24, 2004.

———, and Robin Wright, "U.N. Chief to Urge Delaying Elections, Senior Officials Say," *Washington Post*, February 19, 2004.

MacFarquhar, Neil, "Gains by Kin in Iraq Inflame Kurds' Anger at Syria," *New York Times*, March 24, 2004.

Magnier, Mark, and Sonni Efron, "Arrested Development on Iraqi Police Force," *Los Angeles Times*, March 31, 2004.

Marshall, Mike, "Training of Iraqi Police Seen as Key to U.S. Success," *Newhouse.com*, December 2, 2003.

Marshall, Tyler, "Iraqi Police Struggle to Hold Back a Crime Wave," *Los Angeles Times*, October 28, 2003.

———, and Edmund Sanders, "Iraqi Advisors Are Left Cooling Their Heels," *Los Angeles Times*, May 27, 2003.

Martin, Paul, "Democracy Starts to Take Shape as Councilors Meet in Baghdad," *Washington Times*, July 8, 2003.

———, "General Promises a 'Big Tent' for Iraqis," *Washington Times*, April 13, 2003.

McDonnell, Patrick J., "Iraqi Police Chief Paid for Friendship with U.S.," *Los Angeles Times*, November 26, 2003.

———, and Alissa J. Rubin, "Iraq Paramilitary Force Is Weighed," *Los Angeles Times*, November 6, 2003.

McElroy, Darien, "New Iraqi Government Accuses Iran and Syria of Backing Insurgents," *The Sunday Telegraph* (London), July 4, 2004.

McEntee, Marni, "173rd Airborne Brigade Takes to the Streets of Kirkuk," *European Stars and Stripes*, May 7, 2003.

———, "Joining Forces: Iraqis, U.S. on Patrol," *European Stars and Stripes*, May 20, 2003.

McMillian, Joseph, "Building an Iraqi Defense Force," *Strategic Forum*, No. 198, June 2003, pp. 1–7.

Michaels, Jim, "U.S. Plans Elite Iraqi Force for Security," *USA Today*, April 23, 2004.

Miller, Judith, "Displaced By the Gulf War: Five Million Refugees," *New York Times*, June 16, 1991.

———, "Ending Conference, Iraqi Dissidents Insist on Self-Government," *New York Times*, March 3, 2003.

———, "Iraqi Leadership Team to Prepare for a Transition to Democracy," *New York Times*, February 28, 2003.

Moaveni, Azadeh, "Thousands of Ex-Soldiers in Iraq Demand to Be Paid," *Los Angeles Times*, June 3, 2003.

Morello, Carol, "'Nucleus' of Iraqi Leaders Emerge," *Washington Post*, May 6, 2003.

Mroue, Bassem, "Iraqis Raid Offices of Television Network," *Associated Foreign Press*, November 24, 2003.

Murphy, Dan, "Baghdad's Tale of Two Councils," *Christian Science Monitor*, October 29, 2003.

———, "The G.I.'s Weapon of Choice in Iraq: Dollars," *Christian Science Monitor*, January 29, 2004.

———, "Quick School Fixes Won Few Iraqi Hearts," *Christian Science Monitor*, June 28, 2004.

———, "Why Iraq Governing Council Failed," *Christian Science Monitor*, April 29, 2004.

Murphy, Richard, Chair, *Winning the Peace: Managing a Successful Transition in Iraq*, Washington, D.C.: The Atlantic Council of the United States and the American University Center for Global Peace, January 2003.

Nedawi, Hamza, "Iraqi Police Desert Stations," *Washington Times*, April 9, 2004.

"'Nobody Said This Was Going to Be Easy,'" *National Journal*, October 25, 2003.

"Northrop Grumman to Train New Iraqi Army," *Jane's Defence Weekly*, July 7, 2003.

Norton, Augustus Richard, "Introduction," in Augustus Richard Norton (ed.), *Civil Society in the Middle East, Volume I*, Leiden: E.J. Brill, 1995.

Oliker, Olga et al., *Aid During Conflict: Interaction Between Military and Civilian Assistance Providers in Afghanistan, September 2001–June 2002*, Santa Monica, CA: RAND Corporation, MG-212-OSD, 2004.

Pelham, Nicolas, "Iraq Ministers Told Only U.S. Can Impose Martial Law," *Financial Times* (London), June 23, 2004.

———, "Rival Former Exile Groups Clash over Security in Iraq," *Financial Times* (London), December 12, 2003.

Perlez, Jane, "Aid Groups Urging Military to Protect Essential Services," *New York Times*, April 12, 2003.

Perry, Tony, "At Least One Iraqi Battalion Is Ready to Help U.S.," *Los Angeles Times*, April 13, 2004.

———, "Authority of Iraqi General Questioned by Myers," *Los Angeles Times*, May 3, 2004.

Peterson, Scott, "In Peacekeeping Mode, U.S. Troops Tested," *Christian Science Monitor*, April 30, 2003.

———, "Iraqi Police Walk Perilous Beat," *Christian Science Monitor*, January 23, 2004.

———, "U.S. Shifting Guard Duty to Former Iraqi Soldiers," *Christian Science Monitor*, July 15, 2003.

Phillips, David, *Losing Iraq*, New York: Westview Press, 2005.

Pickering, Thomas R., and James R. Schlesinger, *Iraq: The Day After*, Washington, D.C.: Council on Foreign Relations, 2003.

Pisik, Betsy, "U.N. Iraq Advisor Issues Vote Warning," *Washington Times*, January 28, 2004.

———, "U.N. Sides with U.S. on Voting in Iraq," *Washington Times*, January 16, 2004.

———, "U.S. Sees Organized Crime in Baghdad," *Washington Times*, May 15, 2003.

Play to Win: Final Report of the Bi-Partisan Commission on Post-Conflict Reconstruction, Washington, D.C.: Center for Strategic and International Studies and the Association of the U.S. Army, January 2003.

Pollack, Kenneth M., "After Saddam: Assessing the Reconstruction of Iraq," The Brookings Institution, January 7, 2004.

"Police Training Fact Sheet," Iraqi Ministry of Interior, April 15, 2003.

"President Names Envoy to Iraq," White House, Office of the Press Secretary, May 6, 2003. As of October 2007:
http://www.whitehouse.gov/news/releases/2003/05/20030506-5.html

Priest, Dana, "Feith's Analysts Given a Clean Bill of Health," *Washington Post*, March 14, 2004.

———, "Iraqi Role Grows in Security Forces," *Washington Post*, September 5, 2003.

———, "Pentagon Shadow Loses Some Mystique," *Washington Post*, Mach 13, 2004.

"Profile: Ahmad Chalabi," *BBC News World Edition*, April 7, 2003.

Quimby, Debbie, "ERDC Contributes to Operation Iraqi Freedom from the Home Front," *ERDC Information Bulletin*, December 5, 2003.

Quinlivan, James T., "Force Requirements in Stability Operations," *Parameters*, Winter 1995–96, pp. 59–69.

Recknagel, Charles, "Iraq: Al-Basrah Riot Underlines Frustration With Energy Shortages," *Radio Free Europe*, August 11, 2003.

Reed, Christina, "Burning Assets: Oil Fires in Iraq," *Geotimes*, May 2003. As of October 2007:
http://www.geotimes.org/may03/geophen.html

Reel, Monte, "Garner Arrives in Iraq to Begin Reconstruction," *Washington Post*, April 22, 2003.

Rhode, David, "America Brings Democracy: Censor Now, Vote Later," *New York Times*, June 22, 2003.

———, "Deadly Unrest Leaves a Town in Northern Iraq Bitter at U.S.," *New York Times*, April 20, 2003.

———, "Iraqis Were Set to Vote, but U.S. Wielded a Veto," *New York Times,* June 19, 2003.

Riccardi, Nicholas, "In Iraq, An Army Day for No Army," *Los Angeles Times*, January 7, 2004.

Richburg, Keith B., "Iraqi Leaders Gather Under U.S. Tent," *Washington Post,* April 16, 2003.

Richey, Warren, "Efforts to Make Progress Tangible to Iraqis," *Christian Science Monitor*, May 12, 2003.

Richter, Paul, "Iraqi Officials Wage Political War in U.S.," *Los Angeles Times*, February 5, 2004.

———, "Security Council Endorses Iraq's New Governing Body," *Los Angeles Times*, August 15, 2003.

Ricks, Thomas E., "Iraq Battalion Refuses to 'Fight Iraqis,'" *Washington Post*, April 11, 2004.

Ried, Robert H., "Bremer to Block Islamic Charter," *Washington Times*, February 17, 2004.

Rieff, David, "Blueprint for a Mess," *New York Times Magazine*, November 2, 2003.

Rikhye, Ravi, "Iraqi Police and Paramilitary Forces," *Orbat.com*, January 25, 2004.

Ripley, Tim, "Mean Streets," *Jane's Defence Weekly*, October 9, 2003.

Risen, James, "Data from Iraqi Exiles Under Scrutiny," *New York Times*, February 12, 2004.

Robson, Seth, "California Police, U.S. Troops Help Equip Iraqi Counterparts," *European Stars and Stripes*, March 2, 2004.

Ronco Corporation, "Decision Brief to Department of Defense Office of Reconstruction and Humanitarian Assistance on the Disarmament, Demobilization, and Reintegration of the Iraqi Armed Forces," March 2003.

Rosenberg, Elizabeth, Adam Horowitz, and Anthony Alessandrini, "Iraq Reconstruction Tracker," *Middle East Report,* No. 227, Summer 2003. As of October 2007: http://www.merip.org/mer/mer227/227_reconstruction.html.

Rosenberg, Matthew, "Iraq Army Is Years Away, General Says," *Philadelphia Inquirer,* January 22, 2004.

"RTI Gets Contract Extension for Iraq Rebuilding," *Associated Press*, April 1, 2004.

Rubin, Alissa J., "Crowds Seize Loot to Return to Owners," *Los Angeles Times*, April 17, 2003.

———, "Select or Elect? Iraqis Split on Constitution Delegates," *Los Angeles Times*, September 29, 2003.

———, "U.S. Struggles in Quicksand of Iraq," *Los Angeles Times*, May 5, 2003.

Rutenberg, Jim, "The Struggle for Iraq: Hearts and Minds," *New York Times*, December 17, 2004.

Sachs, Susan, "An Angry Former Iraqi Officer Says U.S. Is Wasting His Talent," *New York Times*, November 2, 2003.

———, "Iraqi Tribes, Asked to Help G.I.'s, Say They Can't," *New York Times*, November 11, 2003.

Sanders, Edmund, "A Delicate Duet of Policing," *Los Angeles Times*, August 26, 2003.

Sanderson, Ward, "More Iraqi Troops Participating in Raids," *European Stars and Stripes*, January 3, 2004.

Sanger, David E., "Chalabi's Seat of Honor Lost to Open Political Warfare with U.S.," *New York Times*, May 21, 2004.

Santora, Marc, with Patrick E. Tyler, "Pledge Made to Democracy by Exiles, Sheiks and Clerics," *New York Times,* April 16, 2003.

Scarborough, Rowan, "General Optimistic over Iraqi Policing," *Washington Times,* February 3, 2004.

———, *Rumsfeld's War,* Washington, D.C.: Regnery Publishers, 2004.

Scavetta, Rick, "Fair Treatment of Prisoners Crucial to Policing of Baghdad," *European Stars and Stripes,* June 15, 2003.

Schmitt, Eric, "G.I.'s Provide Public Works as Well as Security to Iraqis," *New York Times,* December 30, 2003.

———, "Plans Made for Policing Postwar Iraq," *New York Times,* April 9, 2003.

———, "U.S. Is Creating an Iraqi Militia to Relieve G.I.'s," *New York Times,* July 21, 2003.

———, and Joel Brinkley, "State Dept. Study Foresaw Trouble Now Plaguing Iraq," *New York Times,* October 19, 2003.

———, and Douglas Jehl, "Weapons Searchers May Switch to Security," *New York Times,* October 29, 2003.

———, and Steven R. Weisman, "U.S. to Recruit Iraqi Civilians to Interim Posts," *New York Times,* April 11, 2003.

Schodolski, Vincent J., "In Baghdad, Army MPs Reshaping Wary Police Force," *Chicago Tribune,* June 6, 2003.

Schrader, Esther, "Attack Puts Emphasis on Recruits," *Los Angeles Times,* November 3, 2003.

———, "Fighting Force Is Giving Way to Police Force," *Los Angeles Times,* April 16, 2003.

———, and Paul Richter, "U.S. Delays Pullout in Iraq," *Los Angeles Times,* July 15, 2003.

"Secretary General to Administer Iraq's Oil-For-Food Program," International Information Programs, U.S. Department of State, March 28, 2003. As of October 2007: http://usinfo.state.gov/xarchives/display.html?p=washfile-english&y=2003&m–March&x–20030328183058namfuaks0.3604242

Shadid, Anthony, "Iraqi Parties Move to Fill Security Role," *Washington Post,* November 14, 2003.

———, "Iraqi Security Forces Torn Between Loyalties," *Washington Post,* November 25, 2003.

———, "Troops Test Cooperation with Clerics," *Washington Post,* May 23, 2003.

———, and Sewell Chan, "Iraqi Militia Provokes More Clashes," *Washington Post,* April 6, 2004.

Shanker, Thom, "Allied Forces Now Face Both Combat and Operation Duty," *New York Times,* April 13, 2003.

———, "G.I.'s to Pull Back in Baghdad, Leaving Its Policing to Iraqis," *New York Times,* February 2, 2004.

———, "U.S. Is Speeding Up Plans for Creating a New Iraqi Army," *New York Times,* September 18, 2003.

———, and Eric Schmitt, "Delivery Delays Hurt U.S. Efforts to Equip Iraqis," *New York Times,* March 22, 2004.

———, and ———, "U.S. Considers Recalling Units of Old Iraq Army," *New York Times,* November 2, 2003.

Sheridan, Mary Beth, "For Help in Rebuilding Mosul, U.S. Turns to Its Former Foes," *Washington Post,* April 25, 2003.

"Shiites Report Top Leader One of Bombing Victims," *CNN.com,* August 29, 2003.

Sipress, Alan, "Once-Dominant Minority Forms Council to Counter Shiite and Negotiate Future," *Washington Post,* January 6, 2004.

———, and Carol Morello, "U.S.-Led Gathering to Begin Remaking Iraq," *Washington Post*, April 15, 2003.

Slavin, Barbara, "France Willing to Help Train Police," *USA Today*, September 19, 2003.

———, "Power Could Transfer to an Expanded Council," *USA Today*, February 19, 2004.

———, "Rebuilding Iraq to Start Quickly," *USA Today*, March 20, 2003.

Slevin, Peter, "Baghdad Anarchy Spurs Call for Help," *Washington Post*, May 13, 2003.

———, "Iraqi Opposition Pledges Anti-Hussein Unity," *Washington Post*, August 10, 2002.

———, "A Sense of Limbo in the South," *Washington Post*, May 6, 2003.

———, "U.S. Scrambles to Rebuild Iraqi Army," *Washington Post*, November 20, 2003.

———, "U.S. Wary in Choosing Iraq's New Leadership," *Washington Post*, March 23, 2003.

———, and Bradley Graham, "U.S. Military Spurns Postwar Police Role," *Washington Post*, April 10, 2003.

———, and Dana Priest, "Wolfowitz Concedes Errors on Iraq," *Washington Post*, July 24, 2003.

———, and Robin Wright, "U.N. Backs Plan to End Iraq Occupation," *Washington Post*, June 9, 2004.

Slocombe, Walter B., "To Build an Army," *Washington Post*, November 5, 2003.

Smith, Michael, "British Patrols Vital for Law and Order," *Daily Telegraph* (London), May 26, 2003.

Smyth, Gareth, "Emboldened Opposition Gathers," *Financial Times* (London), February 27, 2003.

———, "Weapons Ultimatum Issued in Northern Iraq," *Financial Times* (London), June 18, 2003.

Soriano, Cesar G., "Iraqi Troops Now Request Postwar Roles," *USA Today,* May 16, 2003.

Spillius, Alex, "U.S. Bolsters Forces to Quell Looting in Northern City," *Daily Telegraph* (London), April 22, 2003.

Spolar, Christine, "Baghdad's Disorder Gets in Way of Reconstruction," *Chicago Tribune*, May 1, 2003.

———, "Iraqi Soldiers Deserting New Army," *Chicago Tribune*, December 9, 2003.

Squitieri, Tom, and Glen C. Carey, "U.S. Pushing to Transfer Security Duties," *USA Today*, September 19, 2003.

Stanley, Bruce, "Iraq to Attend Next Week's OPEC Meeting," *New York Times,* September 17, 2003.

Stecklow, Steve, "Before Iraq War, U.N., U.S. Hatched Plan to Feed Nation," *Wall Street Journal*, September 26, 2003.

Stern, Marcus, "Uncertain Road to Rebuilding," *San Diego Union-Tribune*, April 21, 2003.

Stohl, Rachel, "War Ends, but Iraq Battle over Small Arms Just Begins," *Christian Science Monitor*, May 28, 2003.

Stone, Andrea, "Thousands of Iraqis Enlist in Army," *USA Today,* July 21, 2003.

Strauss, Gary, "Amnesty Begins for Automatic Weapons," *USA Today*, June 2, 2003.

Strobel, Warren P., and Hannah Allam, "U.S. Alters Plans for Iraqi Handover," *Philadelphia Inquirer,* February 18, 2004.

Tavernise, Sabrina, "Kurds Celebrate Elections of Mayor in Kirkuk," *New York Times,* May 29, 2003.

———, "A Northern Iraqi City Chooses an Interim Government," *New York Times*, May 6, 2003.

"Timetable Set for Iraq Transfer," *BBC Online*, November 15, 2003. As of October 2007:
http://news.bbc.co.uk/2/hi/middle_east/3272721.stm

Timmerman, Kenneth R., "Details of the Postwar Master Plan," *Insight on the News,* November 24, 2003.

Torriero, E.A., "U.S. Aims to Tame Gun Riddled Iraq," *Chicago Tribune*, May 31, 2003.

Tran, Tini, "First Iraqi City Handed Over to Civilian Government," *Washington Times,* May 16, 2003.

"Transfer Does Not Affect Troops," *BBC News Online*, November 16, 2003. As of October 2007:
http://news.bbc.co.uk/2/hi/middle_east/3274463.stm

Traynor, Iran, "U.S. Closes Exiles Training Camp After Only 100 Show Up," *The Guardian* (London), April 2, 2003.

Trofimov, Yaroslav, "U.S. Gains Relative Sense of Calm with Quiet Presence in Iraqi Town," *Wall Street Journal*, June 13, 2003.

Tyler, Patrick E., "Iraqi Factions Seek to Take Over Security Duties," *New York Times*, September 19, 2003.

———, "Iraqi Groups Badly Divided over How to Draft a Charter," *New York Times,* September 20, 2003.

———, "Iraqis Set to Form an Interim Council with Wide Power," *New York Times*, July 11, 2003.

———, "New Overseer Arrives in Iraq," *New York Times*, May 13, 2003.

———, "New Policy in Iraq to Authorize G.I.'s to Shoot Looters," *New York Times*, May 14, 2003.

———, "Opposition Groups to Help Create Assembly in Iraq," *New York Times,* May 6, 2003.

———, "Political Leaders Resisting U.S. Plan to Govern Iraq," *New York Times*, June 5, 2003.

———, "U.S.-British Project: To Build a Postwar Iraqi Armed Force of 40,000 Soldiers in 3 Years," *New York Times*, June 24, 2003.

———, "U.S. Steps Up Efforts to Curb Baghdad Crime," *New York Times*, May 16, 2003.

———, and Edmund L. Andrews, "U.S. Overhauls Administration to Govern Iraq," *New York Times,* May 12, 2003.

Tyson, Ann Scott, "Iraq Battles Its Leaking Borders," *Christian Science Monitor*, July 6, 2004.

United Kingdom House of Commons Defence Committee, *Iraq: An Initial Assessment of Post-Conflict Operations, Volume I*, Sixth Report of Session 2004–05, March 16, 2005.

———, *Lessons of Iraq: Government Response to the Committee's Third Report of Session 2003–04*, May 26, 2004.

United Kingdom Ministry of Defence, *Operations in Iraq: Lessons for the Future*, December 2003.

United Nations, *Likely Humanitarian Operations*, December 10, 2002. As of October 2007:
http://www.casi.org.uk/info/undocs/war021210scanned.pdf

"Unstable Iraq Looks to New Security Forces," *Jane's Intelligence Review*, December 2003.

U.S. Agency for International Development, "Democracy in Iraq: Building Democracy from the Ground Up," May 2004. As of February 2008:
http://www.usaid.gov/iraq/pdf/iraq_demgov_0504.pdf

U.S. Agency for International Development, *A Year in Iraq*, May 2004. As of October 2007: http://www.usaid.gov/iraq/pdf/AYearInIraq.pdf

U.S. Department of State, "Opinion Analysis," U.S. Department of State, Office of Research, M-106-04, September 16, 2004.

"USAID Contingency Plans for Humanitarian Assistance to Iraq," U.S. Agency for International Development Fact Sheet, February 24, 2003. As of October 2007: http://www.usaid.gov/press/factsheets/2003/fs030224.html

U.S. General Accounting Office, *Rebuilding Iraq: Fiscal Year 2003 Contract Award Procedures and Management Challenges*, GAO-04-605, Washington, D.C., June 2004.

————, *Rebuilding Iraq: Resource, Security, Governance, Essential Services, and Oversight Issues*, GAO-04-902R, Washington, D.C., June 2004.

"U.S. Humanitarian Planning and Relief Efforts," briefing by Andrew Natsios, Bernd McConnell, et al. Transcript released by the U.S. State Department, February 25, 2003. As of October 2007: http://www.state.gov/p/nea/rls/rm/17963.htm

U.S. Joint Forces Command, "Doctrinal Implications of the Standing Joint Force Headquarters (SJFHQ)," Joint Warfighting Center Joint Doctrine Series, Pamphlet 3, June 16, 2003.

"U.S. Prepares Meeting of Iraqi Exiles and Contenders for Leadership," *Washington Post,* April 9, 2003.

"U.S. to Begin Recruiting for New Army," *USA Today,* July 10, 2003.

Wadhams, Nick, "U.N. Says No to Iraq Elections Until at Least 2005," *Associated Press Worldstream*, February 23, 2004.

Waldman, Amy, "U.S. Struggles to Transform a Tainted Iraqi Police Force," *New York Times*, June 30, 2003.

Weisman, Jonathan, "Iraq Chaos No Surprise, but Too Few Troops to Quell It," *Washington Post*, April 14, 2003.

Weisman, Paul, and Vivienne Walt, "Hostility Towards U.S. Troops Is Running High in Baghdad," *USA Today*, May 7, 2003.

Weisman, Steven R., "U.S. Set to Name Civilian to Oversee Iraq," *New York Times*, May 2, 2003.

————, and John H. Cushman, Jr., "U.S. Joins Iraqis to Seek U.N. Role in Interim Rule," *New York Times*, January 16, 2004.

————, and David E. Sanger, "U.S. Open to a Proposal that Supplants Council in Iraq," *New York Times*, April 16, 2004.

Whitelaw, Kevin, "Law and Disorder," *U.S. News and World Report*, May 26, 2003.

Wilkinson, Tracy, "Fresh Rulers to Emerge, Garner Says," *Los Angeles Times*, April 25, 2003.

Williams, Carol J., "Iraqi Council's Most Pressing Task: Legitimacy," *Los Angeles Times*, August 28, 2003.

————, "U.S. to Expand Iraqi Corps," *Los Angeles Times*, December 8, 2003.

Williams, Daniel, "Few Iraqis Meet the Deadline for Turning in Their Guns," *Washington Post*, June 15, 2003.

————, "Iraqi Exiles Mass for Training," *Washington Post,* February 1, 2003.

————, "Lack of Security Hampers Efforts to Aid Baghdad Police," *Washington Post*, June 9, 2003.

————, "New Ministry to Recruit Paramilitary Force in Iraq," *Washington Post*, September 2, 2003.

————, "Rampant Looting Sweeps Iraq," *Washington Post*, April 12, 2003.

————, "U.S. Army to Train 1,000 Iraqi Exiles," *Washington Post*, December 18, 2002.

————, and Vernon Loeb, "At Qatar Base, U.S. Begins a Test Run for War," *Washington Post*, December 10, 2002.

Wilson, Scott, "Iraqi Council's Leader Is Slain," *Washington Post*, May 18, 2004.

————, "A Mix of 'President . . . and Pope,'" *Washington Post*, May 16, 2003.

————, "U.S. Aids Raid on Home of Chalabi," *Washington Post*, May 21, 2004.

————, "U.S. Delays Timeline for Iraqi Government," *Washington Post*, May 22, 2003.

Wong, Edward, "Iraqi Militias Resisting U.S. Pressure to Disband," *New York Times*, February 9, 2003.

————, "New Iraq Agency to Hunt Rebels," *New York Times*, January 31, 2004.

————, "Once-Ruling Sunnis Unite to Regain a Piece of the Pie," *New York Times*, January 12, 2004.

————, "Policy Barring Ex-Baathists from Key Iraqi Posts Is Eased," *New York Times*, April 23, 2004.

————, "Sunnis in Iraq Form Own Political Council," *New York Times*, December 26, 2003.

Wood, David, "U.S. Forces Scour Iraq in Attempt to Disarm Guerrillas," *Newhouse.com*, June 16, 2003.

Woods, Richard, Tony Allen-Mills, and Nicholas Rufford, "Painful Rebirth of Iraq in Cauldron of Defeat," *The Sunday Times* (London), April 13, 2003.

Woodward, Bob, *Plan of Attack*, New York: Simon and Schuster, 2004.

Wright, Robin, "Iraqis Back New Leaders, Poll Says," *Washington Post*, June 25, 2004.

————, and Thomas E. Ricks, "Bremer Criticizes Troop Levels: Ex-Overseer of Iraq Says U.S. Effort Was Hampered Early On," *Washington Post*, October 5, 2004.

————, and Daniel Williams, "U.S. to Back Re-Formed Iraq Body," *Washington Post*, November 13, 2003.

Yacoub, Sameer N., "Training Begins for Iraq's Militia," *Washington Times*, August 16, 2003.

"A Year After Iraq War: Mistrust of America in Europe Even Higher, Muslim Anger Persists," The Pew Global Attitudes Project, March 16, 2004.

Zarocostas, John, "WTO Expected to Grant Iraq Observer Status Despite Baghdad's Lack of Customs Control," *Washington Times*, February 8, 2004.

Zimmerman, Colonel Douglas K., "Understanding the Standing Joint Force Headquarters," *Military Review*, July–August 2004, pp. 28–32.

Zinni, Anthony C., "Restore Regular Iraqi Army to Assist with Reconstruction," *Atlanta Journal-Constitution*, February 5, 2004.

Zoroya, Gregg, "Danger Puts Distance Between Council, People," *USA Today*, October 21, 2003.

————, "Fallujah Brigade Tries U.S. Patience," *USA Today*, June 14, 2004.

————, "Iraq's Deadliest Month," *USA Today*, April 30, 2004.